The Metallogeny of Lode Gold Deposits

The Metallogeny of Lode Gold Deposits

A Syngenetic Perspective

Ulrich Kretschmar (Late)
Former President, Golden Scarab Corporation, Canada

Derek McBride
Consulting Geologist, Canada

ELSEVIER

AMSTERDAM • BOSTON • HEIDELBERG • LONDON • NEW YORK • OXFORD
PARIS • SAN DIEGO • SAN FRANCISCO • SINGAPORE • SYDNEY • TOKYO

Elsevier
Radarweg 29, PO Box 211, 1000 AE Amsterdam, Netherlands
The Boulevard, Langford Lane, Kidlington, Oxford OX5 1GB, UK
225 Wyman Street, Waltham, MA 02451, USA

ISBN: 978-0-12-803222-0

British Library Cataloguing-in-Publication Data
A catalogue record for this book is available from the British Library

Library of Congress Cataloging-in-Publication Data
A catalog record for this book is available from the Library of Congress

Contents

Additional Online Resources: booksite.elsevier.com/9780128032220

Preface

> Observation is the generative act in scientific discovery. For all its aberrations, the evidence of the senses is essentially to be relied upon—provided we observe nature as a child does, without prejudices and preconceptions, but with that clear and candid vision which adults lose and scientists must strive to regain.
>
> **Peter Medawar, *Pluto's Republic* (1982)**

Dr Ulrich Kretschmar, (Ph.D U of T.1973: Scott) was deeply immersed in the writing of this book when he went for a walk in the woods behind his home on December 16th of 2014 and died suddenly and unexpectedly of a massive heart attack. On the day before his death, Ulrich commented that he knew he was not going to be able to finish the research he had begun with this project in his lifetime. Not because he expected to die any time soon, but because, as he said, "The more I know, the more I realize I don't know." One understanding led to another mystery to be studied and understood, with more and more about earth genesis yet to be researched and revealed in the future.

Dr Kretschmar strove throughout his life for understanding and mastery in all the fields he was passionate about, and he was most passionate about geology. He was determined to write this book, first of all, because he wanted to share the findings he had made through careful observation over 30-plus years, beginning in Meguma in Nova Scotia in the 1970s and continuing to his latest work, which included, an in-depth study of Chester Township in Northern Ontario, north of Sudbury.

During the Chester Township study, he began to formulate the book, plotting out structural maps to illustrate what he had observed, studying the findings of other geologists as well as scientists in related fields. He presented his theories at conferences and seminars, and spent long hours over a 3-year period organizing and writing down his thoughts into chapters and plans for the book.

It was Ulrich's intention to include in this book a strong admonition to future generations of geologists to put away their computer models and return to the observation of the earth, to "walk the ground and see what it is saying."

He also wanted to point out in the book what he saw as a weakness in current practices: that, having assumed the truth of a given model, field work currently comprises looking for ways in which what is being seen on the ground fits the model that is presumed to be correct.

It was Ulrich's strong belief that a geologist must start with an open mind (devoid of preconceptions), study the ground, and base any interpretations on what the ground reveals. That was the scientific method he followed, and that is why he came to believe so strongly in the theory of the syngenetic process in gold deposits.

Laura Carter Kretschmar
March 2015

I obtained my geological education at the Haileybury School of Mines and graduated as a gold medallist in 1965. From there I attended Queen's University at Kingston, Ontario receiving a B.Sc. Eng. degree in Geology in 1968 and an M.Sc. Eng. degree in 1972. My early research was graphitic schists in the Abitibi greenstone belt, which at that time were showing up during regional follow-up to airborne geophysical surveys. The presence of carbon in these ancient rocks suggested some form of Archean life. Regional exploration of porphyry deposits provided my M.Sc. thesis material to apply lithogeochemistry to differentiate deposition and alteration patterns in a Cordilleran porphyry copper–molybdenite deposit. At that time, I became interested in the tectonic setting of these deposits.

From there I moved to eastern Canada to study the structural geology and stratigraphy of the Heath Steele Mines massive sulphide deposit under Dr A. L. McAllister, receiving my doctorate in 1976. As part of this study I published my first paper on mineral deposits and their place in the tectonic evolution of a greenstone belt. While teaching at St Francis Xavier University, my field work on the Meguma of eastern Nova Scotia was carried out.

Back in industry, my interest in field geology as a major exploration tool was complimented by using geochemistry, geophysics, and regional tectonics to define preferred exploration targets, which led to the 1987 paper on gold in the Beardmore-Geraldton greenstone belt. This study produced the exploration model that discovered the Nugget Pond Deposit. Some of my studies were published; however, most of the data has remained in company or government assessment files until now.

I commenced consulting in 1985 and have worked in 17 countries, with most of the work being done on gold deposits. When my friend, Ulrich, passed away, I was approached to complete Ulrich's work. I readily agreed because I knew that together our combined century of exploration experience would produce an exciting new approach to gold exploration.

Derek McBride
October 2015

Acknowledgments

This study is the culmination of almost a century of applied experience by two highly qualified and experienced exploration geologists. Together they covered most aspects of mineral exploration from gold to diamonds. Over their careers they were influenced by many mentors. Ulrich passed away before he could acknowledge his mentors, but would have placed Dr S. D. Scott at the top of the list. Dr Scott, Professor Emeritus at the University of Toronto has been Ulrich's mentor from his time as a graduate student until now, and a strong supporter of this book. Later Dr Scott provided support and advice during the completion of Ulrich's work.

I have been grateful to have studied under many mentors, both in the early years and during the finishing of this book. D. W. Atchison of the Haileybury School of Mines introduced me to the value of field geology and Dr. V. G. Milne was an excellent teacher during my 2 summers as his field assistant. Dr P. Stringer, retired structural geology professor, University of New Brunswick, once told me that if I did not have a scab on my nose, I was not examining the rocks in sufficient detail; and Dr A. L. McAllister, my doctoral thesis advisor, said it was my job to prove that the Heath Steele Mines VMS (volcanogenic massive sulfide) deposit predated the structural history, thus confirming it as a syngenetic deposit. It was my pleasure many years later to show him that my Nugget Pond gold discovery had a geological setting identical to that of the Heath Steele Mines VMS deposit.

During the preparation of this treatise, Drs Stringer and M. Schwerdtner, Professor Emeritus reviewed Chapter 8 on structural geology. Dr Schwerdtner assisted with the documentation of "shear zones" and is recognized as an expert on the subject. Dr S. Scott reviewed Ulrich's progress and made many constructive changes. I am indebted to the three of them, on behalf of Ulrich and myself.

In our quest for geological knowledge, we spent much time in the field, with the continued support from our families. To Laura Carter Kretschmar and Bonnie McBride, and our families, thank you for your understanding and the many years of support to enable us to follow our dreams. Oyuntungalag Demchig prepared many of the figures, putting up with many changes, thank you for your patience. R. Chataway provided the records from his Goldenville, Nova Scotia studies; J. Laidlaw and P. Lebaron assisted with the research, and M. Barton with the editing; we thank you all. A final special thanks to Marisa LaFleur of Elsevier for her help and understanding when I was asked to complete this study.

Derek McBride
29 October, 2015

CHAPTER 1

Historical Perspective on Gold Formation Models

CONTENTS

1.1 MODELS IN GENERAL

It is a daunting task for any geologist to understand gold genesis and to find exploration guidelines in the enormous volume of literature. From the perspective of the exploration geologist, this is worth tackling because there is still a large dichotomy when it comes to their genesis, epigenetic-structural aspects and syngenetic concepts. While the targets will not change with the model (Colvine et al., 1984), a better overall understanding will clearly aid target selection and exploration techniques.

A well-constrained metallogenic model will entail plausible explanations for source and chemical composition of fluid, depth of formation, source of heat, P-T conditions of precipitation and formation, trapping mechanism, structural history, and geologic setting. The historic and tectonic context will be clear. Above all, the processes must conform to known physical laws, the answers must be comprehensible to the average earth scientist, and clear exploration guidelines must be inherent. Applicable elements of current discoveries in ocean and atmospheric chemistry, volcanism, and tectonics should flesh out the model.

Cas (2013) states that deposit models often focus on specific aspects of ore genesis, usually deposition and specific geoscience subdisciplines. Models have been developed in isolation, in academic settings of specialist research, and any of the following subdisciplines such as structural geology, geochemistry, P-T-X conditions, phase equilibria, mineral thermometers, isotope fractionation, chemical process modeling, alteration, metamorphism, fluid tracing (stable and radiogenic isotopes), geochronology, alteration/gangue mineralogy/petrology, or igneous/metamorphic petrology may be used and produce conflicting and often mutually exclusive interpretations.

As well-articulated by John Smoot of 4F GeoViz, LLC (December 12, 2013, LinkedIn Discussion Group: A Model Is a Model), "Most models are based

The Metallogeny of Lode Gold Deposits. http://dx.doi.org/10.1016/B978-0-12-803222-0.00001-0

on a set of assumptions derived from facts, experience, intuition and other sources. Some of these often are simplifying assumptions to allow a reasonable model to be constructed. The degree to which the assumptions are violated by the model is one measure of the usefulness of the model. However, the assumptions can be abused fairly significantly in many cases without destroying the usefulness of the model, but the modeler must be able to understand and articulate the deviations."

Referring specifically to gold deposits, Frimmel (2007), states that no uniform classification scheme exists for gold deposits. This lack reflects uncertainties in our understanding of these deposits and also a bias by different researchers and interest groups. He states that a given deposit may be classified according to host rock (e.g., sediment-hosted), or according to a preferred genetic model (e.g., orogenic), the classification may emphasize a specific metal association (e.g., iron oxide copper gold), or it may be based on a comparison with a large prototype (e.g., Carlin type). The fundamental weakness of the existing classification schemes is that different deposit types do not necessarily exclude each other. Interestingly, Pirajno (2009) considers orogenic lode gold of "amagmatic or of uncertain origin."

Difficulties become evident when we try to define lode gold deposits in the context of the current study. The syngenetic model that is the result of the current work (Chapter 10), is largely based on field observations and new geochemical data, but also substantial historical precedent. The extent to which a primary syngenetic origin can be ascribed to a given deposit or deposit class within currently accepted genetic models is the subject of this book and is discussed in Chapter 11. The Russian perspective brought to classification as expressed by Nekrasov (1995), is that in geology, unlike other natural sciences, we deal with consequences, i.e., the end products of geological processes, and not with the factors responsible for their formation. Therefore, the conjectures of any investigator are largely subjective and hypothetical and cannot serve as a basis for the classification of deposits, since the very concept of "deposit" is tentative.

Pirajno (2009) states that ore deposit models effectively are sets of data that best describe a deposit or family of deposits that share similar features, contain common geological attributes, and are formed in similar tectonic environments. However, the authors think that the literature is replete with confusion and divergent views, some of it due to historical domination of structural epigenetic ideas. One need not look further than Perspectives on Genetic Models for Lode Gold Deposits (Kerrich, 1993), the summary preface in a gold deposit volume in the journal Mineralium Deposita, in which the opening statement is that structurally hosted lode gold vein systems in metamorphic terranes constitute a single, coherent group of epigenetic precious metal deposits. At the expense of starting out with the conclusions of this book, it will become apparent that a syngenetic formation model (Chapter 10) requires neither structure nor metamorphic grade as controlling variable. Therefore, in trying to define "lode gold deposits," it is necessary to cast a wide net that includes a broad class of Archean volcanic-hosted vein gold deposits, Au-rich VMS deposits, turbidite-hosted veins, and felsic-hosted deposits. Poulsen et al. (2000), in an excellent review, simply state that bedrock gold deposits are also commonly referred to as "lode" gold deposits. Nekrasov (1995) states that the classification of an object with a high concentration of economic constituents as a deposit is determined by the economics of raw material in a given period, the geographic status of the region, the extent of technological development, and other factors.

Characteristics of Au-rich VMS deposits are described by Poulsen and Hannington (1996), Mercier-Langevin et al. (2014), Beaudoin et al. (2014), and Dubé et al. (2014). Models for Meguma and other turbidite-hosted deposits are described in Chapter 4, as are felsic-hosted deposits, with specific reference to the Coté Gold deposit.

Features common to the broad class of Archean lode gold deposits, according to Hutchinson (1993), Robert and Poulsen (1997), and Gray and Hutchinson (2001) referring to the ores of the Porcupine camp in Ontario include the following: Archean age of host rocks and mineralization; host greenstone terranes that include ultramafic to felsic volcanic rocks, subvolcanic and/or plutonic equivalents, and clastic and chemical sedimentary rocks; association with major structural zones; great vertical extent (1–2 km) of mineralization, commonly lacking appreciable mineralogical and/or geochemical zoning; associated carbonate alteration at both deposit and district scale; nonspecificity of host rocks within the greenstone terrane; common spatial association with felsic intrusive rocks; common occurrence at or near a volcanic–sedimentary rock interface; and ores of both stratiform and discordant morphology, the latter apparently introduced late in tectonic evolution of the host terrane. Early gold enrichments into supracrustal rocks is another feature common to these deposits according to Moore (1977), Bavinton and Keays (1978), Boyle (1979), Saager et al. (1982), Downes et al. (1984), Gross (1988).

1.2 CLASSIFICATION OF DEPOSITS AND GOLD FORMATION MODELS

Frequent major reviews show that currently accepted models fall into three main categories: (1) those based on fluid composition and source of fluid; (2) those based on structure and deposition site and mechanics of deposition, e.g., brittle-ductile transition or dilational jogs; and (3), those based on grade of metamorphism, e.g., greenschist–amphibolite transition.

Current concepts from major summaries discussed in some detail below are by Hronsky et al. (2012), Robert et al. (2007), Beakhouse (2007), Goldfarb et al. (2005), Bierlein and Crowe (2000), Ridley and Diamond (2000), Kerrich et al. (2000), Poulsen et al. (2000), Groves et al. (2003), Colvine et al. (1988), and Groves (1998).

Excellent and complete historical reviews are by Boyle (1979, 1987). He summarizes the origin of Archean gold deposits by discussing abyssal, ore magma, magmatic hydrothermal, granitization, exhalative and lateral, and metamorphic secretion theories. He points out that in the abyssal theory, the source of constituent elements is the very deep regions (layers or shells) of the earth, a concept already advocated in early Egyptian times. The priest–geologists thought that the gold veins in amphibolites of the Egyptian eastern desert and Nubia had their source in the underworld and were the result of the heat generated by the love of Osiris (god of the underworld) and Isis (goddess of fertility). He also points out that the secretion theory, which suggests that mineral veins were concentrated by meteoric waters that obtained the ore constituents from the host rocks, was proposed by Agricola (1556). Two woodcuts that illustrate this concept, from De Re Metallica, in the translation by Herbert Hoover, are shown in Figure 1.1.

In 1644, René Decartes postulated in Principia Philosophiae that the metals in veins had their source in a deep (molten) shell of metallic matter in the earth

FIGURE 1.1
Woodcut illustrations showing the nature of quartz veining according to Georgius Agricola (1556) De Re Metallica in the translation by Herbert Hoover.

and had been driven outward as exhalations by the earth's heat into fractures in the cooled outer crust, where they were deposited.

Boyle (1987) further describes the evolution of the abyssal theory as developed in papers by J.W. Gregory (1928), Brown (1948), and Quiring (1948). He considers the papers by Cameron (1988) and Colvine et al. (1989) as recent additions to the abyssal theory.

A separate summary of ore deposit models up to 1990 and excellent background insights are presented by Hodgson (1990), on which the following is based.

In the early 1900s, when deposits in Timmins and Kirkland Lake were discovered, a magmatic hydrothermal theory of ore formation was generally accepted. Hydrothermal fluids derived from "Algoman" granitoids moved up along major structures such as the major "breaks" of the Abitibi Belt, and deposited gold and associated minerals in dilatant zones. Porphyries provided competency contrast with enveloping mafic schists and the concept of chemically favorable units was widely accepted, but mechanical properties of rocks were considered more important. For example, gold was thought to have been localized in the iron formations because they behaved as brittle units, relative to the enclosing sediments.

In the period 1960–1980, as models for VMS deposits were developing, and the new plate tectonic theories led to a revolution in basic geological concepts, geologists adapted them to gold deposits. The VMS model provided clear exploration guidelines and defined a role for geophysics and stratigraphy. A magmatic hydrothermal model for gold did not provide a clear explanation, as in the case, for example, of the common association with ultramafic rocks.

Synvolcanic model for gold deposits (Ridler, 1970, 1976; Hutchinson, 1987; Karvinen, 1978; Kerrich and Fryer, 1979; McBride, 1987; Gray and Hutchinson, 2001) explained concentration of gold as primarily the result of sea floor and sub-sea floor hot spring activity. Structurally controlled, epigenetic mineralization was interpreted as the result of remobilization, on the scale of individual deposits, during later deformation and metamorphism. As with VMS deposits, geologically complicated patterns of rocks and mineralization, previously interpreted as the result of complex structures, were re-interpreted as the result of synvolcanic processes: structural truncations became stratigraphic pinch-outs; unconformities became facies changes; porphyries became intrusive–extrusive complexes; shear zones became tuff beds; banded veins and sulphide replacement zones in stratiform shear zones became auriferous exhalites; gold localized in competent conglomerate beds became metamorphically reworked placer gold; and iron formation, previously seen as structurally and chemically favorable hosts for late gold, were widely interpreted as primary auriferous exhalites. The regionally extensive "breaks" of the Abitibi Belt were re-interpreted by some (Ridler, 1970; Karvinen, 1980) as carbonate iron formations, and the influence on exploration of these changes in the gold model was profound. Tracing favorable structures now was redirected into tracing favorable stratigraphic sequences and horizons. Stratiform conductors in favorable parts of the volcanic stratigraphy were considered priority targets, irrespective of the structural environment, as were banded oxide iron formation. Carbonate alteration zones, now seen as favorable stratigraphic zones, were carefully re-examined for facies variations related to mineralization. In the Abitibi Belt, many obvious and long-recognized patterns in the distribution of gold, such as the restriction of all the large deposits to a zone within a few kilometers of the main breaks, were de-emphasized; if the gold was due to volcanism, why should not the area distant from the breaks be as prospective, as close to them, if the volcanic sequence was the same?

Concurrently, a metamorphic hydrothermal theory for the origin of the gold-bearing fluids was developing (Fyfe and Henley, 1973; Kerrich and Fyfe, 1981). Gold-bearing fluids were derived by pro-grade metamorphic dehydration reactions related to the emplacement of subvolcanic granitoids during the waning stages of volcanism.

One consequence of the metamorphic hydrothermal model was the idea that gold deposits do not occur in rocks of high metamorphic grade or in deep level granitoids, because of gold mobilization during pro-grade metamorphism. This, despite the occurrence of many large deposits in granitoids bordering greenstone belts, e.g., the supergiant Kolar deposit of India hosted by amphibolite facies rocks (Hamilton and Hodgson, 1986).

Hodgson (1990) also states that in the period 1985–1990, there was a major swing back to traditional views of gold deposits being epigenetic, structurally controlled, and late in the geological development of the greenstone belts.

However, many of the genetic problems remained unresolved, such as the source of the gold and the fluid from which it is deposited.

An influential paper by Colvine et al. (1984), further developed in Colvine et al. (1988) and Colvine (1989), advanced epigenetic gold genesis concepts. Figure 1.2 is their composite deposit scale schematic plan of ore localization in zones of induced permeability, and Figure 1.3 shows further development of their ideal composite depositional model for Archean lode deposits.

In the following sections, the main models for gold deposit formation are described in greater detail as summarized largely by Goldfarb et al. (2005).

1.2.1 Lateral Secretion and Thermal Aureole Gold (TAG)

The lateral secretion model of Boyle (1955, 1959, and 1966 in Yellowknife, in Meguma, and more universally later (1979, 1987)) postulates a broad source region for ore fluids and metals in deeper parts of the same metamorphosed

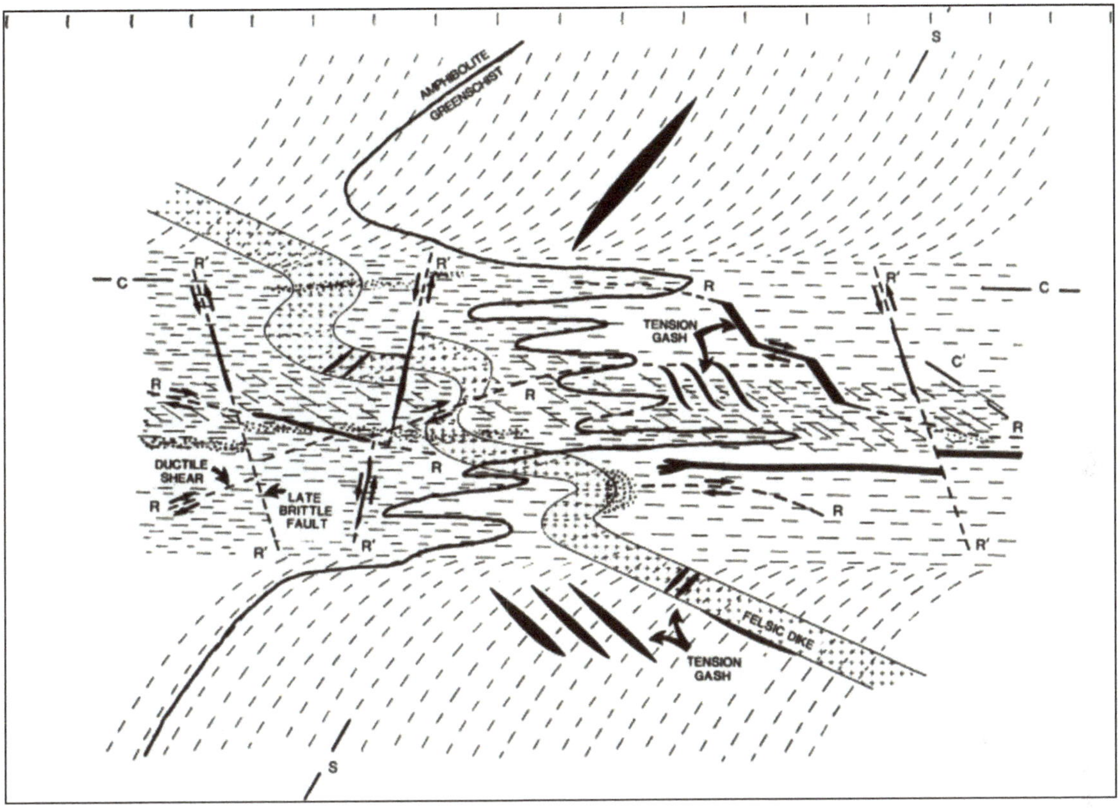

FIGURE 1.2

Composite deposit scale schematic plan of ore localization in zones of induced permeability. *Figure 48 from Colvine et al. (1984).*

rocks. Both metamorphic and meteoric fluids are possible, although Boyle (1979) noted that only metamorphic fluid is feasible.

"Metamorphic secretion" had been suggested for many giant Precambrian orebodies (Anhaeusser et al., 1969, 1976; Vilijoen et al., 1969, 1970; Kirkland Lake: Ridler, 1970; Golden Mile: Travis et al., 1971; Ashanti: Wilson, 1971; and Homestake: Sawkins and Rye, 1974). The gold and metals in the fluid were released from pyrite, magnetite, or mafic silicates in metamorphic rocks, following carbonization of pyrite-bearing mafic rocks and low Fe/(Fe + Mg) argillites. "Depyritization" or devolatilization released sulfur and gold into the hydrothermal fluid and was conducted along flow path conduits (Bőhlke et al., 1992). Boyle (1979) implied that that may have been a direct consequence of pluton emplacement. The same mechanism was entitled thermal aureole gold (TAG) by Wall (1989). According to Bierlein et al. (1998b), the "lateral secretion" model is best viewed as part of the continuum of metamorphic models for ore genesis, since fluid and metal contributions are from the

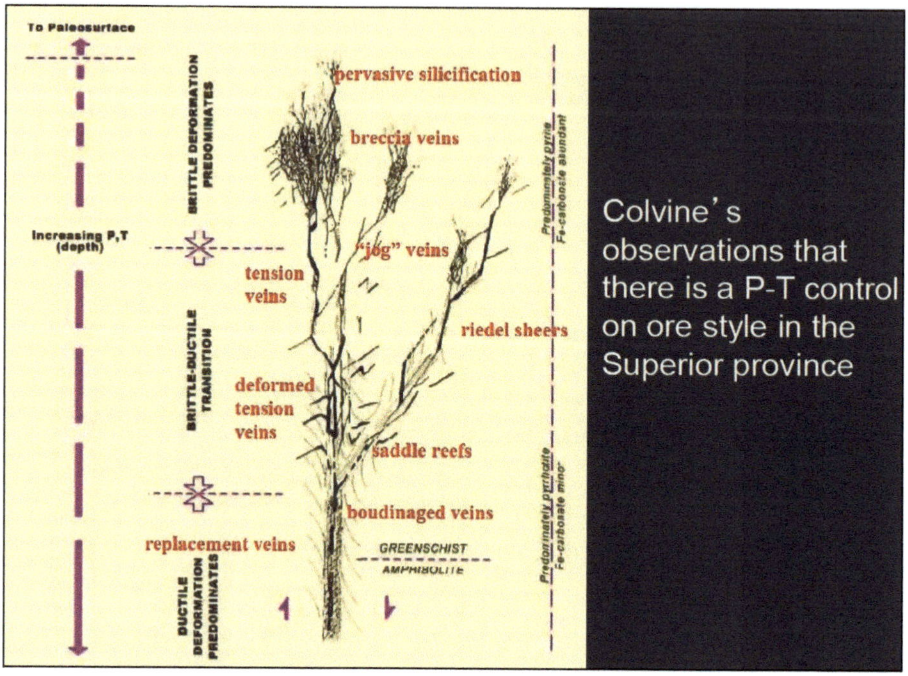

FIGURE 1.3

Ideal composite depositional model for Archean lode deposits from Goldfarb (2013) after Colvine et al. (1988).

metamorphic host terrane, rather than from an underthrust and unrelated lithologic block, but the term lateral secretion is not highly specific regarding the processes of fluid formation and pathways.

In the 1960s and 1970s, Boyle generated pioneering geological and geochemical data for lode gold deposits, principally from the Archean Slave and Superior Provinces in Canada, which were synthesized into the lateral diffusion model. The conjunction of dilatant structures with metamorphism was considered to generate chemical potential gradients that drove ore-forming constituents from contiguous rocks into the dilatant sites (Boyle, 1979). Kerrich (1993) states that the model does not explain the secular distribution of lode gold deposits or metal budget, nor does it stand up to quantitative tests based on the physics of diffusion or isotope disequilibrium between ores and host rocks. However, he states that the observations stand as a landmark contribution.

1.2.2 Deeply-Circulating Meteoric Waters

As summarized by Goldfarb et al. (2005), deep convection of meteoric water to form lode gold deposits in metamorphic rocks was espoused by Agricola (1556), Nesbitt et al. (1986), McBride (1987), and by many workers since then. Nesbitt (1991) pointed out that (1) postmetamorphic peak timing of ores was inconsistent with a metamorphic model, and (2) the lack of a consistent temporal/spatial correlation between gold and granitoids presented problems with a magmatic model. His most significant evidence focuses on the extreme range and very negative values of δ in fluid inclusions in deposits from northern latitudes (Nesbitt et al., 1986; Shelton et al., 1988; de Ronde et al., 2000). Fluid circulation is considered to be driven by high geothermal gradients or plutons; consistently high δ18O values are evidence of low water:rock; and convection was stated to take place under hydrostatic conditions that extended to mid-crustal depths (Nesbitt, 1991). In agreement, Koons and Craw (1991) and Craw and Chamberlain (1996) argue for an influx of meteoric water into a long-term, evolving hydrothermal system as uplift of metamorphic country rocks proceeds during orogenesis, with earlier deposited gold being remobilized by the more oxidized meteoric water incursion.

Bierlein et al. (2006) states that major involvement of meteoric waters in the formation of gold deposits in metamorphic belts is now thought to be unlikely because supporting δ data for mainly secondary fluid inclusions are unrelated to ore-forming events (Pickthorn et al., 1987). Some other workers interpret the data as showing that late, aqueous fluid inclusion trails, commonly of meteoric origin, are the gold-carrying fluids and that earlier, more gas-rich fluids were pre-ore (Cathelineau et al., 1989). Other uncertainties from H—O isotope systematic studies of mineral separates (e.g., Hagemann et al., 1994; Voicu et al., 1999a) are related to temperatures used for calculation of fluid compositions, as well as uncertainties regarding the true limits of metamorphic and magmatic water fields on δ18O versus δ plots.

Other proponents who favor deep circulating waters, e.g., Bouchot et al. (2000), describe an early paragenetic assemblage of quartz, arsenopyrite, pyrite or pyrrhotite and scheelite, in deposits of the French Massif Central view the veins as products of a gas-rich, deep crustal fluid at pressure—temperature conditions anywhere from 450 °C to 5.5 kb (i.e., Salsigne) to 200 °C and 0.4 kb (i.e., Villeranges). In contrast, gold, base metals, bismuth, and latest stibnite were suggested to have been introduced into these same ores by a later meteoric fluid. Vallance et al. (2004) favored the later epizonal stage to have formed orogenic gold deposits when these meteoric waters mixed with a more saline exchanged water of uncertain source.

Goldfarb et al. (2005) suggest that these scenarios, even if correct for the Variscan deposits of southern Europe, are unlikely for these deposits as a whole because it is required that an early gas-rich, W-, and As-bearing fluid event was everywhere succeeded by an ore-forming aqueous fluid flow event within the exact same structural conduits. Also, (1) it is doubtful that all δ18O values would be shifted from the meteoric water line to the same heavy values (i.e., everywhere rock-dominated flow systems), and (2) meteoric water convection under hydrostatic conditions would not achieve the required supralithostatic pressure conditions. Although halogen data have been interpreted as indicative of meteoric water in some younger, small Alpine gold ores of southern Europe (Yardley et al., 1992), He-Ar isotopic data for Alpine deposits argue against this (Pettke et al., 2000).

1.2.3 Metamorphic Models

Lateral secretion-associated genesis of gold ores in metamorphic rocks with metamorphic processes was explored from 1950s to the 1970s. Russian scientists had independently envisaged similar scenarios for Far East Russian gold ores in metamorphic rocks (e.g., Nifontov, 1957; Moiseenko, 1973; Buryak, 1964, 1975; Petrov et al., 1972). The concept was that gold was lost from high-grade rocks and concentrated in low-grade rocks during regional metamorphism. In the mid-1970s (e.g., Fyfe and Henley, 1973; Henley et al., 1976; Norris and Henley, 1976) presented well-reasoned arguments that auriferous fluids were continually released during prograde metamorphic reactions accompanying increases in temperatures of rocks and pore pressures during uplift of metamorphic terranes. The release of fluid during subsequent hydraulic fracturing and a regular pattern of resulting gold districts in a gold province were reflective of focused fluid flow. Common sources of gold were suggested to be (1) rocks with anomalous gold background at depth; (2) seafloor basalts (Keays and Scott, 1976; Henley et al., 1976); (3) pyritic interbedded shales in metavolcanic rocks averaging 80 ppb Au (Glasson and Keays, 1978) suggested sources; or (4) gold in the syngenetic pyrite subsequently mobilized into epigenetic lodes during later metamorphic events.

In the 1980s, metamorphic devolatilization within greenstone belts was stressed by some (e.g., Kerrich and Fryer, 1979; Kerrich and Fyfe, 1981; Phillips and Groves, 1983, 1987) and for turbidite sequences (e.g., Goldfarb et al., 1986) as critical processes for concentration of volatile and metal species within orogenic gold deposits (discussed below). Low-permeability caps (Etheridge et al., 1983, 1984) or fault valve mechanisms (Sibson et al., 1988) were also invoked as critical to the deep crustal hydrogeology.

During fluid production in the middle to lower crust, fluids eventually deposited gold in shallower crustal rocks that were already on a retrograde metamorphic path commonly for the previous 10–50 m.y. (e.g., Goldfarb et al., 1986; Kontak et al., 1990; Robert, 1990; Kerrich and Wyman, 1990; Powell et al., 1991; Stuwe et al., 1993). Goldfarb et al. (2005) suggest that this is consistent with the typical observations that mineralized lodes are discordant with metamorphic foliation, and that alteration assemblages overprint metamorphic phases. This is an argument against the common criticism of a metamorphic model that the disequilibrium between metamorphic and alteration assemblages is incompatible with a fluid produced during metamorphism (e.g., Colvine et al., 1988).

The concept of voluminous fluid release at the greenschist/amphibolite boundary is a critical component of the model (e.g., Fyfe et al., 1978; Powell et al., 1991), which is supported by the chemistry of the hydrothermal fluids. These consist of a mixture of aqueous and nonaqueous volatiles and H_2S. These are considered characteristic of medium-grade metamorphic fluids (e.g., Mullis, 1979). Input of H_2S into such hydrothermal systems is suggested to occur during pyrite-to-pyrrhotite conversion as prograde temperatures rise from 300 °C to 500 °C (Ferry, 1981).

To what extent is an enriched source required? Powell et al. (1991), Phillips and Powell (1992), and Bierlein et al. (1998b) stressed that buried mafic volcanic rocks or interflow sedimentary rocks were significant sources for the gold and CO_2-rich fluid in some deposits of the Victorian goldfields. Others (e.g., Fyfe and Kerrich, 1984; Pettke et al., 2000; Ridley and Diamond, 2000) suggest that medium-grade metamorphic reactions in many rock types, including metasedimentary rocks, can provide the volatiles. Also, background gold, as well as Hg, As, Sb, B, and S concentrations, are stated to show progressive decrease with increasing metamorphic grade in Barrovian metamorphic sequences in metasedimentary terranes (e.g., Moravek et al., 1996 and Pitcairn et al., 2005).

To invoke a metamorphic model, data are consistent with generation of an auriferous ore fluid during upper greenschist and higher metamorphic grade devolatilization reactions in greenstone or slate belts, and deposition of lodes at higher crustal levels in rocks that have long passed peak metamorphic pressures and temperatures.

Geochemical data from situations in which the fluids are derived from rocks that clearly lie below and differ from those in the host terranes, e.g., in the Meguma terrane, prove that much of the ore-forming fluid was exotic to the host sequence (although carbon and sulfur could have been locally added from the Meguma host rocks) but also could not have been derived from spatially associated igneous rocks (Kontak et al., 1990). In the Meguma, the most likely source is considered to be an underthrust terrane, which is now preserved as deep-crustal granulite facies xenoliths in mafic dikes.

McBride (1978) demonstrated that conformable auriferous quartz bands were confined to a limited section within the quartz arenite Goldenvile Formation, which was diagnostic because of its disseminated iron carbonate and up to 2-cm pyrite crystals (See Chapter 8). These rocks had been folded around a series of regional folds.

Similarly, the Renco deposit in Southern Africa occurs in granulite facies rocks, but fluid source areas are thought to be from underthrust lower-grade terranes (Kolb et al., 2000). Many hypotheses suggest that, although there is a crustal source for the ore fluids, as well as likely the gold, asthenospheric upwelling and emplacement of mantle melts into the lower crust may be the ultimate driving force. These concepts combine many features of mantle- and metamorphic-related models.

The possibility also exists that metamorphism of a subducting slab can provide the components that eventually enter the hydrothermal systems. H_2O and CO_2, as well as K_2O, are released during subduction and added back into the crust due to slab devolatilization and melting. Most likely, volatiles are released into the overlying mantle wedge when the slab reaches depths of about 80–120 km, where partial melts are generated and are carried into the lower crust within such arc magmas. Jarard (2003) suggested that from one-third to all chemically bound water may be lost from the metamorphosing slab and added into mantle melts.

Fluid release during slab devolatilization leading to alteration of overlying mafic and ultramafic mantle rocks and serpentinization of forearc mantle rocks is extremely common along deep, steeply-dipping, terrane-bounding faults in Phanerozoic orogens (i.e., California Mother Lode, Chinese Altai Shan). Subsequent devolatilization of the serpentinites, forming olivine, and of associated carbonate minerals can again release H_2O and CO_2 into a fluid phase as geotherms rise. This process leads to a large transitory permeability increase that allows fluid expulsion and eventual focusing into larger structures (Tenthorey and Cox, 2003). Devolatilization of a slab at shallower levels may mobilize fluids directly into the crust, rather than a mantle wedge. Where this occurs, these slab-derived fluids may be the direct source of silica and other components of the auriferous veins; this fluid may leach K and W, and perhaps, Au, as it moves upward along major crustal conduits.

It has been suggested that tectonically-driven changes in far field stresses may be critical for mobilization of fluids released during regional devolatilization events (Kerrich and Wyman, 1990). Such fluids might even have been trapped, in some cases, in pores in low-permeable zones for >100 m.y. (e.g., Thompson and Connolly, 1990), before tectonically-induced changes, such as those causing widespread uplift or major changes in fault kinematics, trigger fluid migration.

A change in plate motions from reverse to strike-slip along terrane-bounding faults has been suggested to lead to formation of the Eocene gold deposits in the Juneau gold belt Goldfarb et al. (1991) and Middle Archean gold deposits of the Zimbabwe craton (de Ronde and de Wit, 1994).

1.2.4 Orogenic Gold

The following description of "orogenic" gold deposits relies heavily on Goldfarb et al. (2001). "Orogenic" gold deposits also known as "mesothermal" are a distinctive class of mineral deposit (e.g., Böhlke and Irwin, 1992; Groves et al., 1998) that have been the source for much of world gold production. The ores are widely recognized in both Phanerozoic mobile belts and older cratonic blocks. Consistent geological characteristics considered by these authors are that (1) they include deformed and variably metamorphosed host rocks; (2) they are sulfide poor; (3) alteration assemblages consist of carbonate sulfide "sericite"-chlorite in greenschist-facies host rocks; (4) fluid inclusions show low salinity, CO_2-rich ore fluids with d18O values of 25–10%, and formation from fluids at supralithostatic pressures; (5) they show a spatial association with large-scale compressional to transpressional structures (e.g., Colvine et al., 1984; Hodgson, 1989; Robert et al., 1994); (6) they consist of abundant quartz-carbonate veins and the mineralized lodes formed over a uniquely broad range of upper to mid-crustal pressures and temperatures, between about 200 and 650 °C and 1–5 kbar (Groves, 1993); and (7) the hydrothermal systems are typically widespread throughout an orogen and represent a regional fluid type inherent to tectonism along convergent margins (Groves et al., 1998).

There are other auriferous deposit types, including epithermal precious metal–bearing veins and gold-bearing porphyries in orogens that are closely associated with magmatic–meteoric hydrothermal systems of more local extent and differ from orogenic gold deposits. For instance, according to Goldfarb et al. (2001), the distinction between orogenic gold deposits and some Carlin-like and intrusion-related gold deposits remains problematic. They also state that there appears to be little point in further subdividing orogenic gold deposits into categories such as "Korean" intrusion-related, "Motherlode," "volcanic-hosted," "Grass Valley," "plutonic-hosted," "Bendigo" turbidite-hosted, and "Homestake" iron formation–hosted as proposed by Poulsen (1996). Goldfarb suggests that divisions based on host rock and/or formation depth provide little additional advantage, because even a single hydrothermal system can evolve over a broad depth range or interact with a variety of country rock types.

Another perspective on the term "orogenic" deposits is provided by Tomkins (2013), who states that these deposits likely form in accretionary and collisional orogens as originally stated by Groves et al. (1998). Transfer of weakly oxidized, low-salinity fluids to the sites of gold deposition is controlled by earthquake events (Cox, 2005; Sibson et al., 1988), allowing fluids to rapidly traverse large thicknesses of crust. This rapid rise takes the fluids out of equilibrium with their surroundings, promoting destabilization of the gold-carrying hydrosulfide complexes [$Au(HS)_2$ and AuHS].

Robert et al. (2005), on the other hand, state that the term orogenic was originally introduced by Groves et al. (1998) because quartz-carbonate vein gold deposits in greenstone and slate belts, including those in BIF, have similar characteristics and are therefore formed by similar processes. Originally, the orogenic model applied strictly to syntectonic vein–type deposits formed at mid-crustal levels in compressional or transpressional settings. The term has progressively broadened to include deposits that are postorogenic relative to processes at their crustal depth of formation. Robert et al. (2005) state that this has led to significant ambiguity in the definition of the boundary between orogenic and reduced intrusion-related deposit models, with many type examples being ascribed to one model or the other by various authors (Thompson and Newberry, 2000; Goldfarb et al., 2001). They further show, as illustrated in Figure 1.4, that the orogenic clan is defined to include only syn-tectonic quartz-carbonate vein-type deposits and their equivalents, formed at mid-crustal levels.

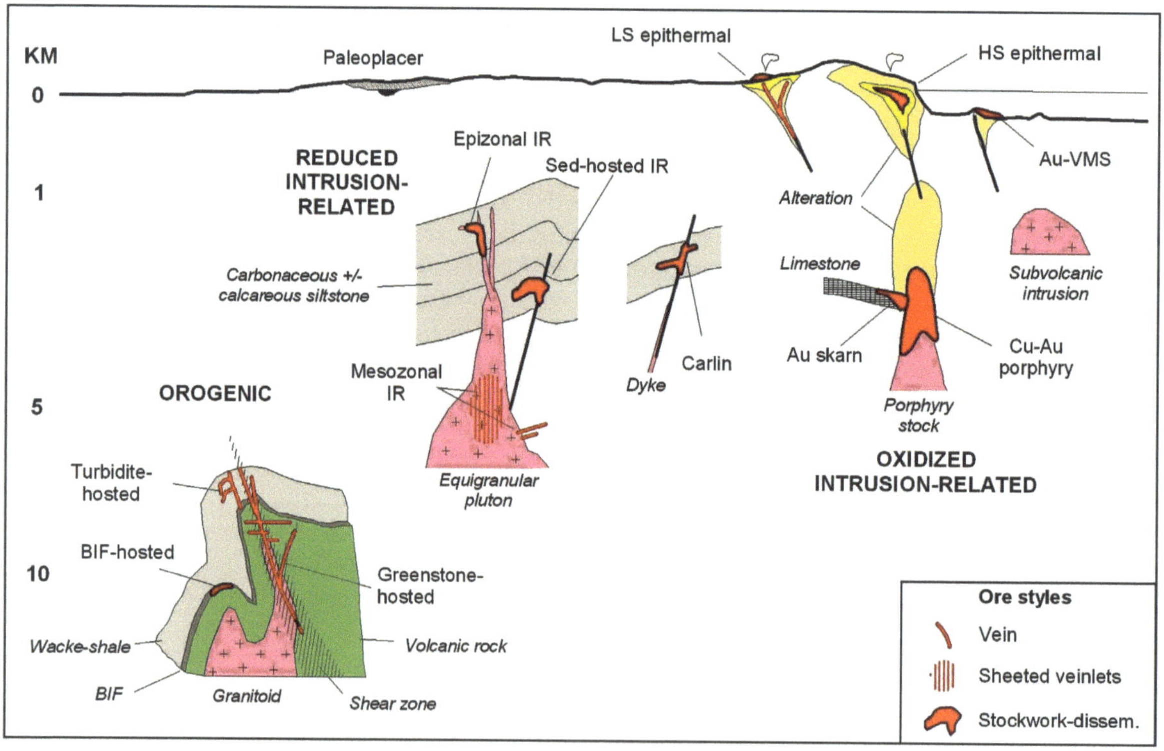

FIGURE 1.4

Schematic cross-section showing the key geologic elements of the main gold systems and their crustal depths of emplacement. Note the logarithmic depth scale. *From Poulsen et al. (2000).*

Specific deposit types in this clan include the turbidite-hosted and greenstone-hosted vein deposits, as well as the BIF-hosted veins and sulphidic replacement deposits. A confusing issue is that greenstone belts also contain gold deposit types that do not fit the orogenic model as defined (Groves et al., 2003; Robert et al., 2005), and also that there is no consensus on the origin of these atypical deposits.

A short video: Orogenic Gold Deposits by Clive Willman and Richard Goldfarb, may be seen at: (https://www.youtube.com/watch?v=RBGuaCDMPzM).

1.2.5 Magmatic Models

Goldfarb et al. (2005) summarize magmatic concepts by stating that classic work by Hutton (1795), Lindgren (1933), Emmons (1937), and Bateman (1950) focused on a magmatic origin, based on the common spatial and temporal association between gold deposits and some type of igneous rock. More recent models suggest that exsolution of ore fluid occurs late in the magmatic process from shallow intrusions (e.g., MacDonald and Hodgson, 1986). Spooner (1993), however, suggests that the more CO_2-rich nature of the ore fluids in gold deposits in metamorphic belts is more consistent with an early fluid release event. Some studies suggest a close relationship between gold and TTGs, particularly in the Abitibi belt (e.g., Burrows and Spooner, 1989; Burrows and Jemielita, 1989; Beakhouse, 2007a,b). δ13C data are used to support such a model (Burrows et al., 1986), although Goldfarb et al. (2005) consider that these data are equivocal. Cameron and Hattori (1987) show gold-bearing fluids related to oxidized felsic magmas, spatial association with I-type stocks, interpretation of carbon and sulfur isotope data, and rare oxide phases in the hydrothermal assemblages. In contrast, other workers have argued that it is reduced, I-type granitoids that are important to the ore-forming process (e.g., Rowins et al., 1997). Goldfarb et al. (2005) also point out that in many metamorphic belts, it is flysch-melt, S-type granitoids that are associated with gold ores (e.g., Chugach terrane: Goldfarb et al., 1997).

The specific geochemical or mineralogical signatures of gold deposits indicative of magmatic exsolution include the presence of scheelite, galena, tellurides, tourmaline, and molybdenite (e.g., Burrows and Spooner, 1989; Hodgson and Troop, 1988). Similarly, geochemical anomalies for Bi, Mo, Te, and W are stated to support fluid released from a magma. Carbon (Burrows et al., 1986) or sulfur (Walshe et al., 2003) isotope data may be used, but Goldfarb et al. (2005) state that these data are equivocal. Sulfur and carbon isotope data may overlap the magmatic field, but this does not preclude other sources. They also suggest that recent precise geochronology, including that of the TTGs in the Late Archean of Canada (Kerrich, 1989), indicate that gold ores are younger than their spatially-associated intrusions and that the association of the two is solely a structural phenomenon.

In detail, Goldfarb et al. (2005) point out that gold ores may be associated with various plutons, including I- and S-types, reduced and oxidized bodies, and shallowly or deeply emplaced magmas. Within a single large and complex orogen, gold ores may be associated with a variety of granitoids (e.g., Victorian goldfields: Zhou et al., 2003). Another reservation, based on geochemical data that shows that Au was depleted in residual magmas in calc-alkaline differentiation sequences (Tilling et al., 1973), presents serious problems for a magmatic model calling upon late exsolution of ore fluids. Similarly, elevated K, Rb, and Ba, and related lithophile ratios, in many alteration assemblages surrounding Archean gold deposits, are inconsistent with any type of magmatic source area (Kerrich, 1989). On the other hand, W, Bi, and Te are mobile in many types of crustal fluids and need not be indicative of a proximal magmatic fluid source (e.g., Goldfarb et al., 2000).

Boyle (1979) points to other observations that must be taken into account before a magmatic exsolution model can be fully evaluated:

1. How does sulfur, which should be enriched in early minerals, make it to final exsolution stages, if a late magmatic scenario is selected?
2. Gold ores are rarely sited in the centers or within the cupolas (but see Fort Knox below) of igneous masses, but almost everywhere along the margins.
3. Hydrothermal carbonates and sulfides are typically only adjacent to veins occurring along these margins, and not consistently along all margins that have shown significant amounts of fluid exsolution.

In cases where the gold ores are associated with crustal melt granites, sulfur and metals are expected to be driven out early in the thermal event (i.e., during metamorphism), and these species should be depleted in the melts.

Where giant deposits such as Muruntau and the Golden Mile have been often suggested to be of magmatic origin, models are required to better address the size of the igneous body for release of such a great volume of fluid and gold; most likely, the required magma mass would be enormous.

Finally, there are clearly a few excellent examples where no spatial association exists between gold and granitoids, such as Macraes in New Zealand.

Despite these problems regarding the universal application of a magmatic model to gold deposits in metamorphic belts, there are some low-grade deposits and prospects in such environments that seem to have a definite affinity to magmatic processes: these are the IRGD discussed above. Deposits such as Fort Knox and Timbarra occur not along the sheared margins of intrusions, but within the apex of the magmatic bodies. How these few examples that occur along the margins of cratons and inboard of accreted terranes, and which more clearly have a magmatic genesis, relate to the majority of gold-forming systems in metamorphic belt (e.g., the orogenic gold deposits) is still not clear. They do, however, indicate that magmatic fluids can, at least in one case (Fort Knox), form important gold deposits at mesozonal depths in metamorphic belts.

An alternative magmatic model is one in which the heat from a rising pluton is critical in lode gold formation. Fluids could obviously be magmatic, metamorphic, or meteoric in such a model that stresses a local, rather than a regional, heat source. Such a model, termed the thermal aureole gold (TAG) model (Wall, 1989), has been applied to Archean (Yilgarn: Witt, 1991), Paleoproterozoic (Pine Creek: Matthai, 1995), and Phanerozoic (Muruntau: Wall et al., 2004) orogenic gold systems. In some cases, such as Pine Creek, new geochronological data indicate that gold ores postdate intrusion by about 50–100 m.y. (Sener et al., 2003). Nevertheless, such a model might still be possible in environments such as Pine Creek, if long-term radioactive decay from U-, Th-, and K-enriched, highly-fractionated granites results in anomalous geotherms that are capable of generating hydrothermal flow for hundreds of millions of years (e.g., Klominsky et al., 1996).

1.2.6 Mantle-Related Models

There are a variety of other scenarios that have been proposed for gold ore genesis, which are related to magmatic processes but do not relate ore genesis to direct fluid exsolution from granitoids fractionating in the crust. During the late 1980s, particularly reflective of the consistent spatial association, lamprophyres were suggested as important components to the ore-forming process (e.g., Rock and Groves, 1988; Rock et al., 1987, 1989). The model stresses that the lamprophyres are inherently rich in CO_2, S, and Au. It was suggested that the lamprophyres brought these components from the deep mantle into the lower crust, and resulting interaction with crustal rocks forms metamorphic fluids and/or felsic melts that then further transported the gold from the mantle. In addition, gold shows a consistent association with many alkalic magmatic systems (e.g., Jensen and Barton, 2000). Arguments against this model have been presented by Wyman and Kerrich (1988, 1989) and Kerrich and Wyman (1994). These include the following: (1) there is a lack of consistent gold enrichment in lamprophyres; (2) no enrichment of many of the other elements that are consistently found in gold deposits occurs in lamprophyres; (3) lamprophyres occur over a much broader range of tectonic settings than most gold deposits in metamorphic rocks; (4) C and Sr isotopes of ores and lamprophyres are very different; and (5) gold would be lost quickly during interaction of lamprophyres with continental crust, and would unlikely be present during eventual lamprophyre crystallization and fluid exsolution. Because of these concerns, most workers now agree that the spatial association between gold deposits and lamprophyres solely reflects the fact that they shared a connection with the same crustal conduits. Dubé et al. (in press) indicate that the lamprophyre dikes at the Red Lake deposit are spatially associated with the high-grade gold mineralization, but are at least 10 m.y. younger, based on cross-cutting relationships and high-precision U/Pb zircon geochronology.

Deep-seated tectonothermal events, with some type of mantle heat source and mantle-lower crust devolatilization process, have also been widely proposed (e.g., Kerrich, 1986; Perring et al., 1989). Most commonly, advection of mantle CO_2, causing granulite facies metamorphism of the lower crust and loss of K, Rb, Au, etc. to a deep fluid phase, has been suggested (e.g., Fyon et al., 1984; Cameron, 1988; Colvine et al., 1988; Colvine, 1989). The model stresses that δ13C data are consistent with a mantle CO_2 migration during asthenospheric upwelling. Noble gas data along presently active, major crustal fault zones indicate a connection of these transcrustal shears with mantle devolatilization (e.g., San Andreas fault zone: Kharaka et al., 1999). Other workers suggest that rather than CO_2 streaming from the mantle, simply mantle underplating or input of mantle melts at the base of the crust initiates crustal devolatilization (e.g., Powell et al., 1991). Alternatively, passage of a crustal plate over a mantle plume (e.g., Murphy et al., 1999) or subduction of a crustal slab window (e.g., Haeussler et al., 1995), have been suggested as the ultimate causes for generation of gold-forming fluids. Whereas many of the above models suggest a mantle trigger to produce a high metamorphic grade auriferous fluid, some models indicate that gold and other components lost during granulization of the lower crust are mobilized within lower crustal melts. For example, Bouchot et al. (2000) suggest that the spatial association of gold ores with Li-, F-, and tourmaline-rich leucogranites throughout the Variscan gold districts of southern Europe support transport of some W, As, and Au within the ascending lower-crustal melts.

1.2.7 Crustal Continuum (Continued Direct Quote of Goldfarb (2005))

The continuum model for gold deposits proposed by Groves et al. (1993, 2003) and Groves (1993) has as a basis that lode gold deposits in subgreenschist, amphibolite, and granulite facies rocks, share numerous characteristics, such as (1) structural hosting, (2) metal inventory, (3) element association of Au, Ag, W, As, Sb, Se, Te, Bi, and (4) ore fluid source (Figure 1.5).

The term lode gold is used for this single class of structurally hosted Au, Ag vein deposits, which in the view of Groves et al. (1993) and Groves (1993) can be

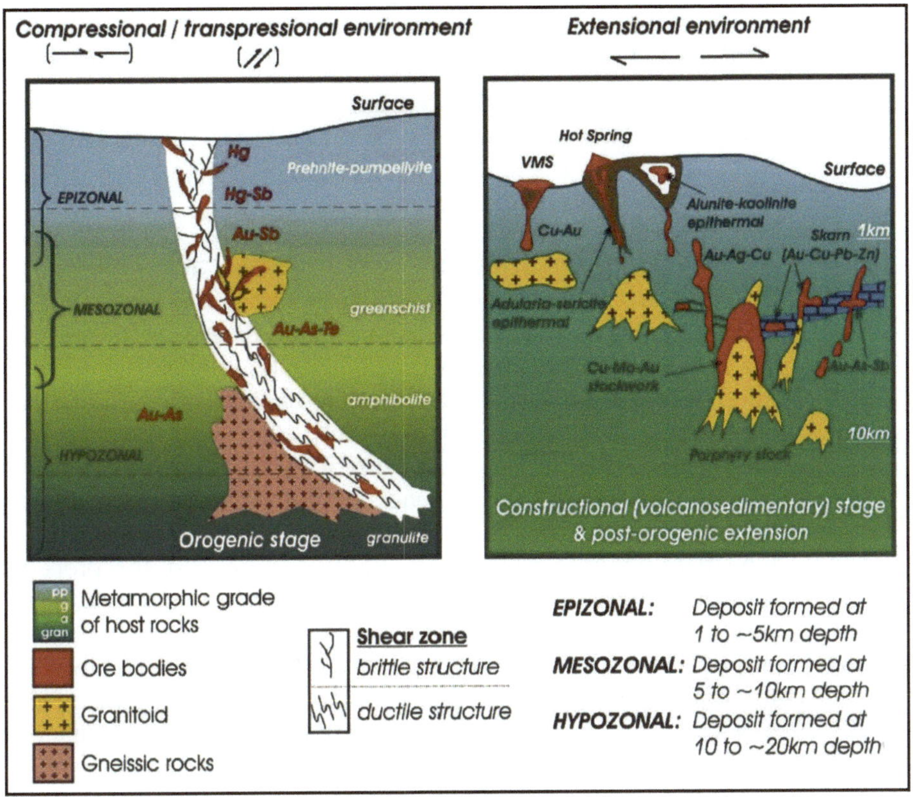

FIGURE 1.5

Main elements of the Crustal Continuum Model of Groves (1993). *From Goldfarb (2013).*

considered as forming over a crustal depth range, or crustal continuum, extending from granulite to sub-greenschist facies environments—hence the term crustal continuum model.

Phillips and Powell (2009) and Tomkins and Grundy (2009) question the continuum model. The problems lie with applying the continuum model to rocks metamorphosed to upper amphibolite and granulite facies (for example the Griffin's Find in Australia). From petrographic observations and advanced thermodynamic modeling of mineral assemblages they state that gold mineralization occurred prior to peak metamorphism. Tomkins and Grundy (2009) point to the difficulty of transmitting hydrothermal fluids through rocks above 600°–650° without causing partial melting and consider that this temperature range should be regarded as an upper limit for the formation of orogenic gold deposits. This has implications for targeting gold systems as it implies targeting for gold deposits within upper amphibolite-facies or higher metamorphic grade belts must take into account the overprinting of the primary mineral assemblages, and also structures associated with the gold mineralization.

The authors think these issues are less of a problem if the gold deposits formed at the same time as the enclosing rocks and were subsequently deformed and metamorphosed. The mode of mineralization is then independent of the metamorphic grade that is superimposed on it.

1.2.8 Tectonic Diffusion of Gold Abstract

The tectonic-diffusion approach, which forms the basis for an estimate of orogenic gold resources, is based on the fact that the vertical tectonic displacement of rock bodies (including ore deposits) relative to the Earth's surface is effectively random at global and geological scales of consideration (Wilkinson and Kesler, 2007; Kesler and Wilkinson, 2010). Once a deposit forms at any crustal level, tectonic processes cause it to move upward by exhumation and erosion, downward by subsidence and burial, or to stay at the same level in the crust as time passes. If we consider a large number of deposits of any specific type that form at some characteristic level in the crust, this process is similar to diffusion and can be simulated numerically by continuously emplacing the deposits at a specific depth and then allowing them to undergo random vertical displacement. This approach replicates the random walk of individual deposits through crustal time-depth space and produces a theoretical age-frequency distribution of deposits at the computational 'Earth surface,' which is then compared to the age-frequency distribution for known deposits. The comparison is done iteratively until a best fit is obtained between the calculated and the real-Earth age-frequency distributions. Since the model calculation keeps track of the movement of all deposits through the crust, the number of deposits at any depth can then be specified.

1.3 "SHEAR ZONES" AND THE ROLE OF STRUCTURE IN LODE VEIN MODELS

Other factors are considered imperative, namely, structural preparation of the host beds, especially shearing, fracturing, faulting, and drag folding, all of which produce dilatant zones that concentrate the gold and its associated gangue elements. The present understanding of these concepts is summarized in Figure 1.6 and shows schematic development of veins during folding and regional metamorphism.

The purported role of structure, dilational jogs, transcurrent faults, splay faults, shear zones, intersecting shear zones, or flexures within shear zones that may have concentrated the mineralization, brittle–ductile transition textures greenschist-amphibolite facies transition is discussed in (Econ Geol. 49, Figure 5)

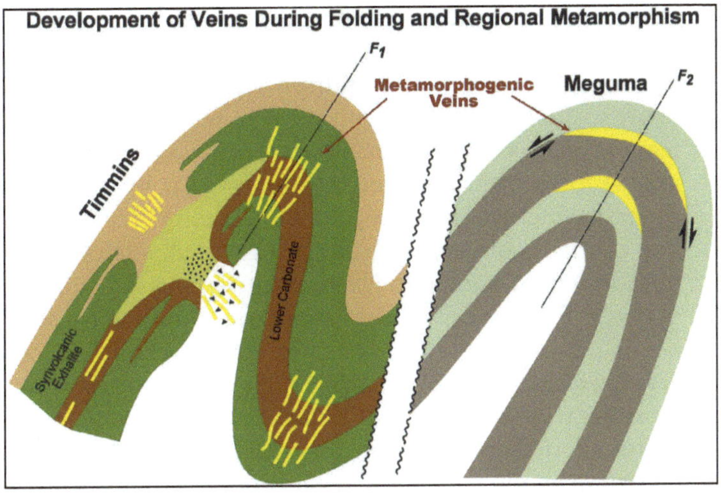

FIGURE 1.6
Schematic summary of the effect of shearing, folding, and production of metamorphogenic veins in gold rich VMS (Karvinen, 1980) and Meguma turbidite-hosted veins (Proudlove et al., 1989).

One concept of the role of "shear zones" is presented by Cameron (1993), who states that deep shear zones were a source for gold at Kirkland Lake, where long-term displacement along a major strike-slip fault resulted in localized extension and thinning of the crust, which produced extensional basins at the surface, a broad deformation zone at depth, and upwelling of mantle. Partial melting and decarbonation of the upwelling mantle are shown at Kirkland Lake by alkaline magmatism and a long period of carbonatization. When extension ceased, there was uplift, during which overpressured, CO_2-bearing fluids, which carried gold from depth, were focused into the narrow, upper part of the shear zone. The Bamble shear belt in Norway is a lower crustal analogue of this extensional terrane, characterized by a counterclockwise P–T–t metamorphic path. During extension, mafic magma carried CO_2 into the shear belt; the CO_2 may have been a cause for oxidation of its rocks during prograde metamorphism. There was a long period of isobaric cooling, when fluid movement was inhibited, because inclusions formed during peak metamorphism were underpressured. During this period, internal buffering by the mineral assemblage caused oxidation of pyrrhotite to pyrite. Pyrrhotite is a principal host of gold in mafic lower crust; its transformation to pyrite made the gold accessible to later fluids. Isobaric cooling was followed by rapid, isothermal uplift. Fluid inclusions containing CO_2 now became overpressured, causing their release. It was during this period that gold was likely extracted from the deep shear belt, the rocks of which now contain only 0.06 of the crustal abundance gold.

Another, more structurally and theoretically oriented perspective on shear zones, is presented by Hodgson (1989) who states that the structure of vein-type gold deposits is defined by the shapes and geometrical relationships of mineralized bodies, the form of the mineralization making up these bodies, and the sequence of vein-forming events. Most mineralized zones occur within or are spatially associated with shear zones, especially shear zones in larger systems of intersecting shear zone sets. They range in shape from tabular to linear, and in form from disseminated, to breccia, to stockwork or sheeted veinlet zones, to single veins. There typically is a complex history of mineral deposition that overlaps, and is genetically related to, the deformation that generated the host structural zone. Hodgson discusses the structural features of vein-type deposits in relation to the geometrical properties and evolution of vein-hosting fractures, particularly shear zones. Veins are localized by tectonically-generated dilatancy in an environment of low mean stress caused by high fluid pressure. It is proposed that a major cause of tectonic dilation of shear zones is the interference between intersecting shears during bulk, inhomogeneous flattening by movement on systems of intersecting shear zone sets.

Hodgson also addresses the formation of typical laddered ribbon veins and chevron folding (Figure 1.7), both of which are of interest because a syngenetic origin explains many of the same features.

Regarding the importance of structure in the Meguma, Henderson et al. (1989) state that Mawer's conclusion (Mawer, 1986), which says that gold-bearing bedding-concordant quartz veins are late-tectonic saddle-reefs restricted to

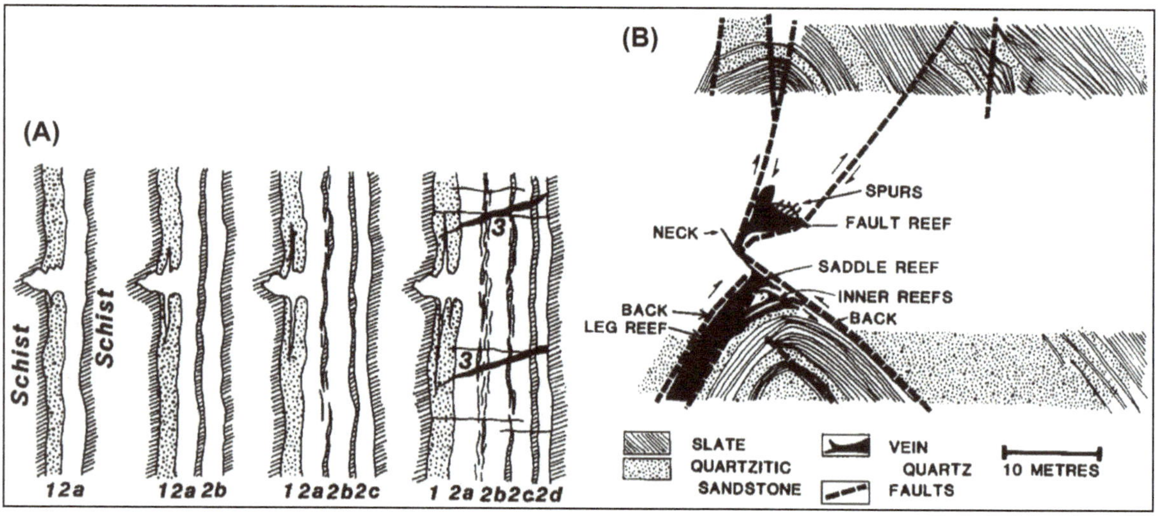

FIGURE 1.7

Sequence of events in the development of typical laddered ribbon veins in the Hollinger mine, Timmins, Ontario. (A): Stippled (1) early, tourmaline-bearing vein; clear (2a, 2b, etc.) main, multi-stage, crack-seal veins; black (3): late ladder veins (Hurst, 1935). (B): Structural features and terminology of mineralized zones associated with chevron folds, Bendigo area, Victoria, Australia. *After Hodgson, 1989, Figures 26 and 27.*

the hinge areas of domical anticlines, is incorrect, and is not a good exploration strategy as stated because it ignores early-tectonic gold mineralization, and limits exploration to anticlinal hinge areas that have received considerable attention for over a century (e.g., Faribault, 1889).

Another example of the extent to which shear zones are considered important is reflected in the contribution by Chattopadhyay (2010) in the book Gold Metallogeny, India and Beyond (Deb and Goldfarb, 2010). He summarizes deposit characteristics and then finds that many of the Indian deposits do not fit the criteria. His generalized description adequately summarizes current thinking that orogenic gold deposits show strong structural control and typically occur in the brittle–ductile transitional rheological domain of the upper crust. Shear zones, faults, and fracture networks, produced in a brittle–ductile regime, host the majority of such deposits. They exhibit complicated structural patterns arising out of repeated shearing, reactivation of pre-existing fractures, fluid migration, and associated chemical alteration of country rocks, opening and resealing of precipitated veins, etc. Structurally complex Archean greenstone belts in different parts of the world host such orogenic gold deposits mainly in moderately metamorphosed rock units. From extensive studies of such deposits, it is now well understood that they form in compressional/transpressional domains along the margins of continental magmatic arcs and/or collisional suture zones in adjacent terranes. Low-salinity fluids, created at depth through dehydration of subducted hydrous sediments, carry the gold and deposit it at shallower levels adjacent to major fault conduits. A cyclic fault-valve mechanism, reflecting the episodic build-up of high fluid pressure within a sealed deep-rooted fault zone, followed by seismic failure along the fault plane, rapid fluid escape, sharp pressure drop, and resulting gold deposition, has been suggested for emplacement of gold-bearing quartz lodes. Laminated quartz veins with multiple crack-seal events, horizontal extensional "flat" veins, and wall-parallel (in shear fractures) and wall-normal (in extensional fractures) mineral fibers serve as evidence of such fluid pressure fluctuations associated with ore formation.

Most of the present and past major producing gold mines of India (e.g., Kolar, Hutti, Ramagiri, Gadag) are orogenic gold deposits. Chattopadhyay (2010) concludes that despite numerous studies of these deposits, structural interpretations of the mineralized zones and related controls on gold mineralization have remained controversial.

1.3.1 Unanswered Questions and Poorly Constrained Aspects of Current Models

It is apparent that timing of veining, source of fluids and source of gold, while considerably better understood and delineated, are still issues central to a comprehensive understanding of lode vein formation.

1.4 SYNGENETIC GOLD FORMATION CONCEPTS

A significant role for syngenetic processes in the formation of volcanic-hosted gold deposits was widely discussed in the 1980s, when the synvolcanic nature of VMS deposit formation became apparent. Among the most important references is Knight (1957), who suggested a favourable source bed concept for gold and other metals, which initiated discussion of syngenetic ore-forming processes. Others are as follows: Ridler (1970), Akande (1982), Hutchinson and Burlington (1984), Ridler (1975, 1976), McBride (1978), Karvinen (1978, 1980, 1981, 1982), Kerrich (1983), McBride (1987), Hodgson (1990), Hutchinson (1987, 1993), and McBride (2009). In Chapters 2 and 4, the authors show that concepts of a syngenetic origin to gold ores in metamorphic belts date back to T. Sterry Hunt (1868) for concordant auriferous veins in the Meguma terrane. However syngenesis is not mentioned in current mainstream gold deposit literature (e.g., Goldfarb et al., 2005; Hronsky et al., 2012).

Kretschmar first presented a syngenetic lode vein formation model—developed after detailed mapping of gold-bearing veins in Nova Scotia—in an unpublished Geological Survey of Canada report (Kretschmar, 1983). An oral presentation of the findings was made at the NE Section of the Geological Society of America (Kretschmar, 1984), and a summary was presented to the Toronto Geological Discussion Group in 1984. His more recent treatment includes Kretschmar (2011, 2012).

In a book dealing with turbidite-hosted gold deposits (Keppie et al., 1986), Simon Haynes published a synsedimentary hydrothermal sea-floor model in which exhalative silica gel is released from seafloor hot springs (Haynes, 1986) His schematic representation of his model is shown in Figure 1.8.

The authors think that syngenesis fell out of favor or failed to be integrated into the mainstream literature due to an onslaught from articulate proponents of deformation and remobilization, as well as concerns such as vein age dates younger than the host rocks, and very high apparent fluid inclusion formation pressures, which are not readily resolved. As seen in the discussion of models, these include Colvine et al. (1984) for the abyssal theory, Goldfarb et al. (2005) for metamorphogenesis, Beakhouse (2007a,b) for magmatism, and recently Groves et al. (1998) and his students for orogenesis.

A common spatial association of vein-type gold deposits with base-metal–rich VMS deposits and with oceanic volcanic rocks, within a given gold province, led to the idea that the gold was highly concentrated into ores during seafloor exhalative events. For gold-rich VMS deposits, a solely syngenetic origin was proposed by Marquis et al. (1990) but argued otherwise by Sillitoe et al. (1996). Current characteristics of Au-rich VMS deposits are described by Poulsen and

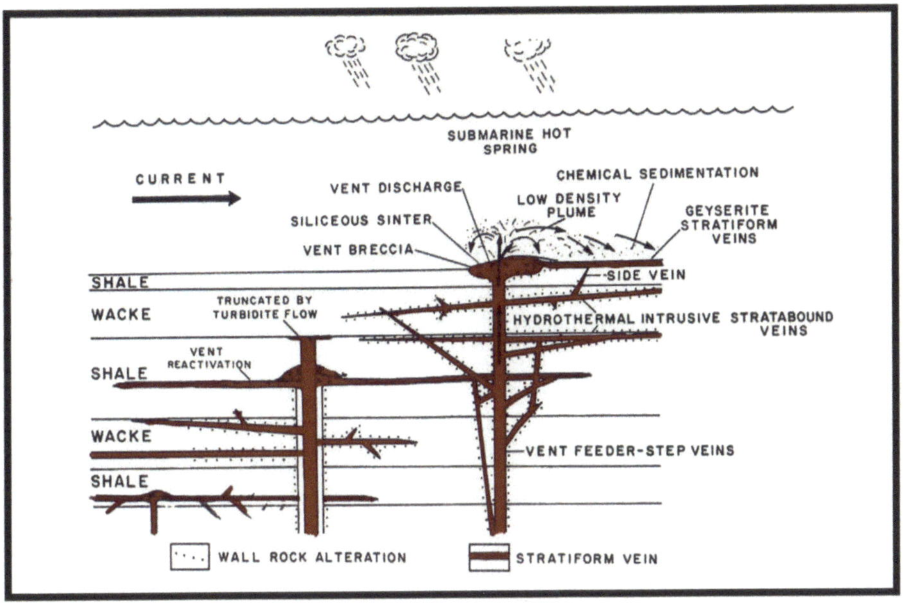

FIGURE 1.8
Submarine hydrothermal vent model of gold-quartz vein formation. *From Haynes (1987).*

Hannington (1996), Dube and Gosselin (2007a), Mercier-Langevin et al. (2014), Beaudoin et al. (2014), and Dubé et al. (2014).

For BIF-hosted gold deposits, a similar dichotomy is apparent, e.g., Lupin (Kerswill, 1993, 1996 vs Poulsen et al., 2000) and Morro Velho (Ladeira, 1991 vs Lobato et al., 2001). Goldfarb et al. (2005) state that according to Chamberlain (2003), the Bulyanhulu deposit in Tanzania presents features in which some components of the ore system could potentially be derived from a spatially associated VMS body. A geochemical argument for the association with VMS is presented by Goldfarb et al. (1988), who noted at the southern end of the Juneau gold belt, in an area with an abundance of small VMS deposits, exceptionally high levels of H_2S (2–3 mol%) in vein-forming fluids of some of the younger epigenetic lode-gold deposits. These anomalous sulfur concentrations were interpreted to indicate significant contributions of volatile species from regional metamorphism of nearby VMS lenses within metasedimentary rock sequences.

Thus, although commonly ignored in arguments against the model, it should be recognized that many workers who favored such a syngenetic model, actually describe and support epigenetic features in younger lode-gold ores (e.g., Gray and Hutchinson, 2001).

As we have seen above, many metamorphic, magmatic, meteoric, or mantle fluid source models suggest that gold was leached from syngenetic and diagenetic sulfide minerals in the metavolcanic and/or metasedimentary pile. Therefore the concept that gold in the hydrothermal systems first entered oceanic crustal strata through sea floor processes before later concentration into epigenetic lodes is widely held.

1.4.1 Why Lode Veins Cannot Be Syngenetic

Specific objections are rarely articulated, and syngenesis is either completely dismissed or blindly accepted according to Emsbo (2005). Reasoned responses given by Bierlein and Crowe (2000, p. 136) as to why lode veins cannot be syngenetic are provided to questions from the audience of how viable the syngenetic models of the 1970 and 1980s are in view of current geologic knowledge. Bierlein and Crowe (2000) cite the following as evidence that lode gold deposits cannot be of syngenetic origin:

1. High-precision geochronological data place upper, post-peak metamorphic constraints on the timing of mineralization
2. Hydrothermal alteration invariably overprints metamorphic assemblages and fabrics
3. Wall-rock alteration mineralogy, geochemistry, and isotopic halos all indicate that the hydrothermal assemblages are not in equilibrium with the surrounding host rocks
4. Regarding the significance of the association between graded-bed sedimentary host rocks and lode gold deposits, they answer that lode gold deposits are epigenetic in association and largely structurally controlled. The nature of the host rock cannot play a vital role in controlling ore formation, but it can have an effect on the structural framework that develops, as well as modifying the fluid composition and alteration assemblages. The association between Phanerozoic lode gold deposits and turbiditic meta-sedimentary rocks is controlled by the geotectonic setting.

Objections based on the sedimentation environment (Hayk Khanzatyan, personal communication, 2013) were these:

1. It is not possible to get it in a turbidite environment where intensive inflow and deposition of terrugenous material is going on the sea floor due to frequent underwater storms
2. In case of syngenetic hydrothermal activity in that environment, the silica gel will get mixed with the terrigenous material on the sea floor, and in case of recrystallization of the deposits, we can have silica-clay-carbonate mixed matrix (cement) in the rock varying from mudstone, calcareous siltstone, to sandstone

3. Well-bedded cherty layers of hydrothermal origin may be present also, but not bedded quartz layers
4. Quartz veins may form in the fractures of the host rock just after hydrothermal activity when the hydrothermal fluids pass through the feeder zone, keeping the temperature and the pressure constant
5. When the pressure and the temperature slowly get decreased, the highly concentrated silica-gel infill re-crystallizes to quartz of mesothermal or epithermal origin depending on original high or low temperature and the rate at which it cools down.

According to Kontak and Horne (2010), the syngenetic exhalative model of Haynes is incompatible with vein mineralogy and textures as discussed by Mawer (1986, 1987). Mawer (1987) studied textures and states that in thin section, the quartz truncates slate foliation, wall rock fragments are included in the vein, and that Henderson (1983) has established that veins postdate cleavage formation. On the other hand, Henderson et al. (1989) state that a late-tectonic origin for bedding-concordant veins is inconsistent with their observations that many bedding-concordant veins formed early in the tectonic history of the Meguma zone.

According to a summary by Robb (2004), syngenetic versus epigenetic-diagenetic origins have also been debated for stratiform sediment-hosted copper deposits (e.g., Sweeney et al., 1991 for the Zambian copper belt). He states that the syngenetic view, in which sulfides are attributed to precipitation directly from anoxic sea water, has been largely superseded by models that view ore genesis in terms of either early or late diagenesis. A diagenetic origin is substantiated by many prominent features of stratiform sediment-hosted copper deposits, especially the broadly transgressive nature of ore zones relative to lithologies, the clear relationship of metal deposition to a redox front, and replacement textures in the sulfide mineral paragenetic sequence.

Exploring the extent to which these and similar criteria apply to lode veins is the subject of this book. Textures are addressed in Chapter 2, and Meguma veins (Section 4.1) and a general syngenetic formation model is proposed in Chapter 10. The question of whether lode veins are syngenetic or epigenetic should really be rephrased: In the Meguma terrane, our eyes are telling us that the predominant bedding parallel veins formed during ongoing turbidite sedimentation. Among the objections to this concept are that fluid inclusion pressures in vein quartz indicate 5–10 km formation depths (Mona Sirbescu, Dan Kontak, and Kingsley Burlinson, personal information), and that both K/Ar and Re/Os formation ages from vein arsenopyrite give dates that are significantly later than Meguma sedimentation (Morelli and Creaser, 2005; Creaser, personal information). Although the authors have found no entirely satisfactory explanation for these discrepancies, caveats to geological interpretation of fluid inclusion pressures and age-dating techniques are presented in Chapter 7. The present work examines lode deposit genesis in a wider and more holistic geological context.

1.4.2 The Two-Stage Model: Syngenetic Gold Upgraded during Metamorphism

A prominent feature of lode gold deposits is their high degree of variability in ore form and host lithology, with virtually every rock type present in Archean greenstone terranes. This variability is considered compelling evidence by Colvine et al. (1984) for the widely held belief that Archean gold deposits are the product of at least multiple stages of concentration and probably also of more than one basic genetic process.

In this regard, the literature presents a united front: some form of shearing, metamorphism, deformation, and structural upgrading is required to form gold-bearing lode veins. Even the rara avis who acknowledge a syngenetic precursor (e.g., Ridler, 1976; Karvinen, 1978; Valliant and Hutchinson, 1982; Hutchinson and Burlington, 1984; Wood and Large, 2007; Large et al., 2007; McBride, 2009) require a final stage of remobilization and upgrading from a lower-grade source. Indeed, extensional veins in conglomerate in the Hollinger mine (e.g., Figure 34 in Poulsen et al., 2000) appear to attest to this process.

Bierlein and Crowe (2000) states that gold was then remobilized from the massive sulfide ores, or "enriched proto-ore exhalite," into veins and other epigenetic ore styles during subsequent deformation and/or metamorphism (see also Akande, 1982; Hutchinson and Burlington, 1984; Proudlove et al., 1989; Hodgson, 1990). Obvious questions therefore are as follows: What is the scale of remobilization?; To what extent does secondary enrichment contribute to making a deposit economic, and what is the timing of this remobilization?

1.5 SO WHY BOTHER?

It is clear from these widely divergent models that one cannot readily lump together all gold deposits. It should also be apparent that a satisfactory model must go beyond description of structure or metamorphism, because we can always ask what is the proof that the ore deposit was not there before it was deformed or metamorphosed. No single avenue of investigation has so far yielded a totally satisfactory comprehensive model, which must integrate heat source, sources for gold, favorable fluid characteristics, precipitation mechanisms, as well as final ore deposition and preservation. It is equally important to incorporate detailed structural analysis of the deposit and its host environment to differentiate the influence of regional deformation and shearing on the timing of and modification of mineralization. Contrary to past interpretations it will be illustrated in Chapter 8 that the structural component can be deciphered.

An essentially syngenetic environment inherently may satisfy most criteria, and the place to look for evidence is as always on the outcrop, where we can start to examine vein and rock textures. This is the subject of Chapter 2.

CHAPTER 2

Interpreting Textures in Outcrop

CONTENTS

2.1 READING THE ROCKS

Within the current context of the metallogeny of lode veins, evidence for syngenesis, and exploration applications, these are key questions: (1) How should quartz veining at the outcrop scale be interpreted? (2) What do the textures and internal stratigraphy of lode vein indicate about formation processes? (3) To what extent have subsequent deformation and metamorphism affected the veins and host rocks (i.e., what are recognition criteria)? (4) To what extent do subsequent events obscure primary processes?

In this chapter, the authors present some general observations and show photographs from excellent seashore exposures and stripped outcrops in Cambro-Ordovician Meguma and Archean metavolcanics. These aspects are also discussed later from different perspectives: age dating and fluid inclusion composition in Chapter 7, and specific deposits including those hosted in turbidite sequences, in Chapter 4. Alternative interpretations have been presented by Haynes (1983), Mawer (1986, 1987), Henderson and Henderson (1986), Horne and Culshaw (2001), Kontak et al. (2001), Kontak and Horne (2010), and Kontak et al. (2011). There is no comparable detailed geological description for the Hercules and Jacobus vein systems in the Beardmore greenstone belt, which will be described in greater detail in Chapter 4.5.

2.2 IS THE TEXTURE PRIMARY OR SECONDARY?

Nowhere is it truer that we see what we have been trained to see than in the interpretation of textures. Regardless of our background, we have a tendency to try to force field observations into our models, when ideally descriptions should be our starting point, and presumptions and existing models should be secondary (Haakon Fossen: www.geosociety.wordpress.com/2013/11/28/let-the-rocks/talk).

The Meguma lode veins have been the subject of extensive study and a significant aspect of the debate on genesis centers around the timing and nature of

The Metallogeny of Lode Gold Deposits. http://dx.doi.org/10.1016/B978-0-12-803222-0.00002-2

vein formation. As early as 1868, Sterry Hunt (1868) stated that he could not see how anyone could doubt that the interbedded quartz was deposited at the same time as the argillite. The authors concur with this premise and present evidence in its favor. McBride (1978) presented evidence that the auriferous veins were bedding parallel and were confined to a section of the Goldenville stratigraphy that also contained disseminated iron carbonate and pyrite crystals up to 2 cm. The stratigraphy had been subsequently folded into tight anticlines and open synclines. The bedding-parallel veins were folded as part of this stratigrphy. In particular, the occurrence of veins in Cambro-Ordovician turbidites on Nova Scotia seashores and in stripped trenches in Archean mafic volcanics in the Beardmore-Geraldton greenstone belt show that these veins formed at the same time as the enclosing rocks and that deformation and metamorphism was a subsequent event. Discussions of vein textures and classification are extensive (e.g., Haynes, 1983, 1987; Mawer, 1986) and excellent recent photos of vein types are shown by Kontak and Horne (2010) and Kontak, Horne, and Kyser (2011).

2.2.1 Bedding-Parallel or Bedding-Concordant Veins

There is some agreement that bedding-parallel or bedding-concordant veins, which host 90% of Meguma gold deposits, are synsedimentary (McBride, 1978; Haynes, 1987), and a synvolcanic model is presented by McBride (1987) for lode vein mineralization at the Northern Empire mine in Beardmore. Key evidence that Meguma and Archean lode veins are formed by a similar seafloor hydrothermal process is that they share many common features, mainly that they occur primarily within the E division argillite of Bouma turbidites or at the contact with overlying greywackes or volcanics. This is clearly shown in Figures 2.1 and 2.2, which show barren quartz veins at Markie Point, and Figure 2.3 the "Bishop Lead," a gold-bearing lode vein sequence in the Harrigan Cove gold deposit in Nova Scotia.

Internal structure of veins is illustrated in Figure 2.3, and a polished slab of vein stratigraphy from the Beaver Dam Gold Deposit (location on Figure 4.1) showing four or more domains. Kretschmar interpreted these as representing different episodes of deposition and secular evolution of the hydrothermal fluid. The bottom 7-cm bed is massive white Q with a few apparent shale wisps. Overlying are thin shale partings deposited on an indented surface; hollows are filled with 3- or 4-mm thick accumulations of black shale, chlorite, or pelitic material. Q_2 is 1–2 cm of darker recrystallized quartz. They are separated by a thin, poorly-developed bedding surface with small amounts of clay-like material. Separating the lower three quartz "domains" from the overlying fourth is a 3-mm-thick pyrite–galena cap. The upper quartz "domain" consists of 7.5 cm of coarsely crystalline quartz with irregularly shaped conical inclusions, generally elongated perpendicular to bedding. The fragments are delicately

FIGURE 2.1
Markie Point, South Shore, Nova Scotia. Typical shoreline exposure of Meguma turbidites with synsedimentary quartz veins (tops are to the left of photo). Turbidites are numbered and subdivided sequentially from bottom T_1 to overlying T_3. *From Kretschmar (2011).*

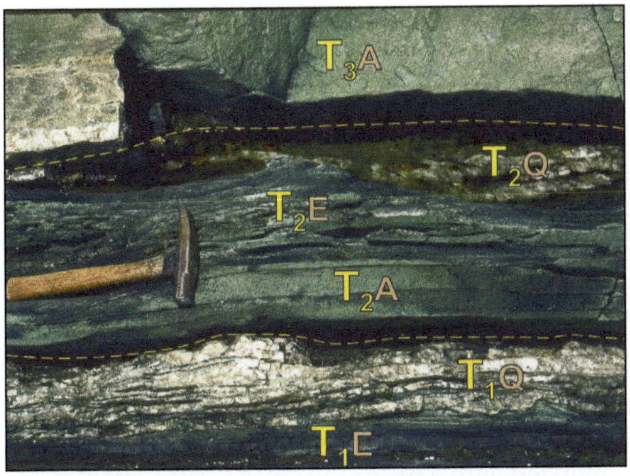

FIGURE 2.2
Upper Ec division of Bouma turbidite (GC for gold-bearing veins) showing quartz vein with complex internal structure, overlain by 30-cm-thick turbidite with graded bedding. Massive base of next overlying turbidite (T_3A) is at top of photo. Individual units are numbered from base upward. Outcrop location: NAD 83. UTM 20T, 0543478/496511 (Figure 4.1). *From Kretschmar (2011).*

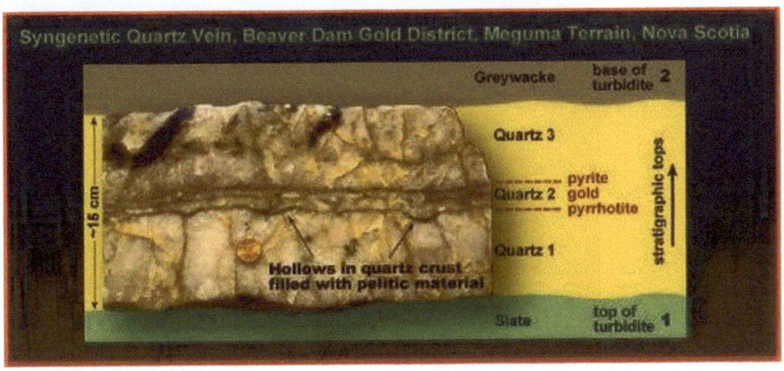

FIGURE 2.3
Polished slab of quartz vein, Beaver Dam Gold Deposit. The vein is a crust on pelitic substrate (upper E division of a turbidite). Monomineralic sulfide layers and probable abiogenictic carbon require multiple pulses of fluid deposition (see Section 7.6.1 on Strelly Pool cherts).

FIGURE 2.4
Kodiak Golden Mile vein and Gold Cycle (Chapter 3) exposed in a stripped trench in the Elmhirst "intrusion." Complex laminated textures that resemble fine bedding parallel to that in the host rocks. View to SW Q is 1.5-m thick and E is 3 m thick. Facing direction to SW; dips steeply to SW. *From Kretschmar (2011).*

laminated; individual laminae consist of graphite, sericite, or chlorite. The quartz filling interstices between these structures is whiter and less crystalline. The significance of the observed features is further discussed in Chapter 7.

Archean lode veins from Beardmore show very similar features (Figures 2.4 and 2.5). The Golden Mile vein exposed in trenches is 1–3 m wide and is at the top of a 2–3 m thick Gold Cycle "E" unit (defined in Chapter 3). The quartz shows striking and systematic differences in textures along its >1000-m length. At the center of the vein, there is a massive high-grade quartz knob, whereas the peripheral extensions are clearly bedded (laminated) and lower grade (Figure 2.6). Gold Cycle top determination (Chapter 3) shows facing directions to the SW for the Golden Mile and Sage Gold Jacobus Extension veins (Section 4.5).

The cyclical, asymmetric nature of the quartz and the underlying pelagic-pelitic unit provides empirical field-based evidence for synsedimentary and synvolcanic formation, which is systematized with a proposed Gold Cycle (Chapter 3).

FIGURE 2.5
Steeply dipping Vein 04, Jacobus Golden Extension of Sage Gold Inc. View along strike of vein to NW. Gold Cycle units are mafic-chloritic tuff "E," laminated Q, and overlying gabbroic flow (A). Gold Cycle top determination criteria shows tops to NE. Geological context in Section 4.5.

FIGURE 2.6

Markie Point, Nova Scotia. Two quartz beds exposed in a glaciated seashore exposure. Host rock consists of massive, vaguely bedded greywacke. Quartz occurs as an upper 10-mm-thin continuous unit and a lower boudinaged mass, which is probably a parasitic fold. Asymmetric bleaching is shown by a thin, lighter-colored margin underlying upper quartz unit. Note the isoclinal parasitic fold in the lower greywacke beds.

2.2.2 Angulars, Boudins, Cross-Cutting Breccia Veins, and Alteration

It has been stated that cross-cutting veins are one of the reasons why lode veins cannot be syngenetic. In response, the authors think that one must distinguish between different types of veins and the extent to which they transgress stratigraphy. Above all, however, the scale of observation is important. An excellent review of Meguma veins types and a historical overview is provided by Kontak and Horne (2010). Haynes (1987) has classified veins into 11 subtypes.

"Angulars," the historical term in the Meguma gold camp, are clearly cross-cutting veins. They may emanate from either the top or bottom of a bedding-concordant stratiform vein. A small-scale, gold-bearing angular from Beaver Dam is shown in Figure 2.7. It consists of sugary quartz and native gold in a 2.5-cm-thick vein that cuts 10 cm of arsenopyrite-bearing siltstone.

Boudins are common in the more competent quartz bands (Figure 2.8) and clearly demonstrate that the quartz bands were present prior to the regional deformation (see Chapter 8).

Cross-cutting, later veins have a different texture. In the Kenty Mine, several breccia-textured veins that are possible feeders transect pillow basalts and terminate at the mine horizon. The Kenty Mine was one of the vein deposits in the Swayze Greenstone Belt (Rickaby, 1934). A sketch is shown in Figure 2.9.

The main Kenty vein consists of a complex quartz-carbonate horizon at the contact between pillow basalts and tuffaceous sediments (Figure 2.10), a setting reminiscent of the Stawell Mine (https://www.youtube.com/watch?v=gropjZwl3_k). The breccia vein textures are reminiscent of the Morning Star

(a) (b)

FIGURE 2.7

Left: Beaver Dam Gold Deposit, Nova Scotia. Small-scale gold-bearing "angular," consisting of sugary quartz and native gold in a 2.5-cm-thick vein that cuts 10 cm of arsenopyrite-bearing siltstone. Right: Markie Point, Nova Scotia. Barren angular vein originating and emanating upward from quartz vein.

(a) (b)

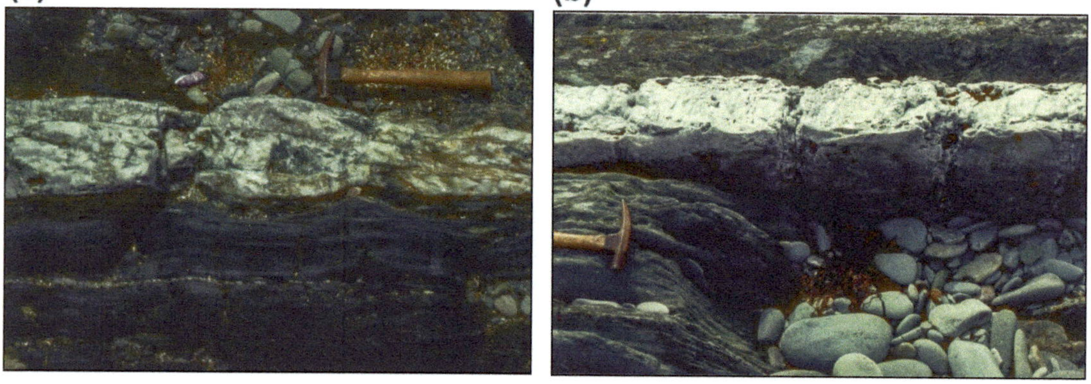

FIGURE 2.8

Nova Scotia Seashore exposures of boudins of quartz mats and vein textures. Left: Mink Island. Right: Softwood Island. Quartz appears to contour onto ripple marks or sedimentary structure in siltstone. Overlying coarse greywacke shows minimal contouring. Metamorphic grade is lower greenschist facies and features can be interpreted as probably due to contemporaneous sedimentation of gel-like silica and siltstone.

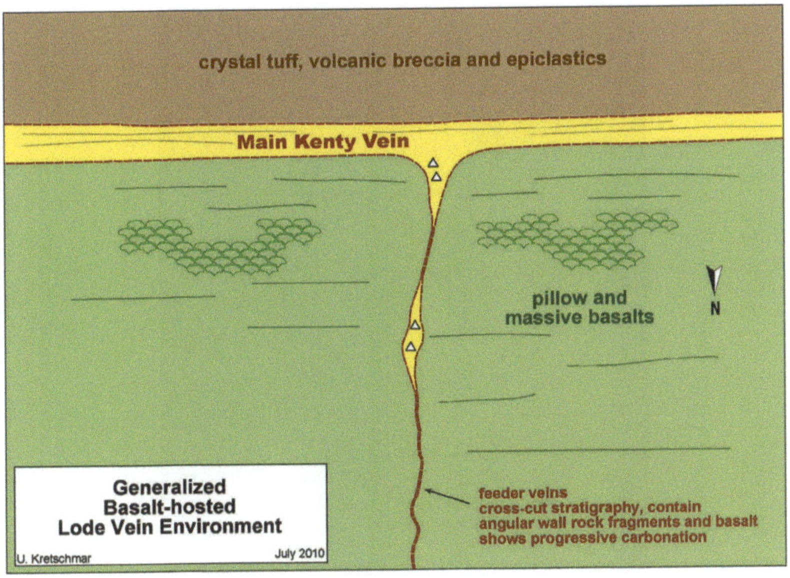

FIGURE 2.9

Kenty Mine, Swayze Greenstone Belt, Ontario. Sketch shows formation model for the Main Kenty quartz-carbonate-gold ore horizon. Fragmentation or breccia veins transect underlying massive pillowed basalt sequence and were probable feeder veins. Unpublished sketch from personal archives of U. Kretschmar.

Mine at Woods Point (https://www.youtube.com/watch?v=4OcwyR1j4fo). The interpretations presented in these two videos are by well-known authorities.

In the Meguma, cross-cutting veins that are possible feeders are massive, unstructured, or vaguely structured, with an appearance substantially different from the laminated bedding-parallel veins. Representative textures from Goose and Softwood Islands are shown in Figure 2.11.

2.2.3 Harrigan Cove Gold

Figures 2.12 and 2.13 are trench maps from the Harrigan Cove Gold District—the sequence youngs to the south. Figure 2.13 is detail of part of Figure 2.12.

In the main area of trenches at Harrigan Cove (Figure 2.12), there is excellent albeit discontinuous exposure of a 50-m stratigraphic section with numerous quartz veins, some of which are known to be gold-bearing. These relationships can be summarized as follows. Quartz vein occur in one of three ways:

1. entirely within E division slates
2. between the top of E division slate and the base of the overlying A division greywacke or
3. between the top of foliated greywacke and the overlying A.

FIGURE 2.10
Gold-bearing iron carbonate horizon, Kenty Mine.

FIGURE 2.11
Goose Island, Nova Scotia. Cross-cutting breccia vein textures displaying felted mass of euhedral and subhedral quartz crystals. Vuggy character indicates that these veins postdate the deformation.

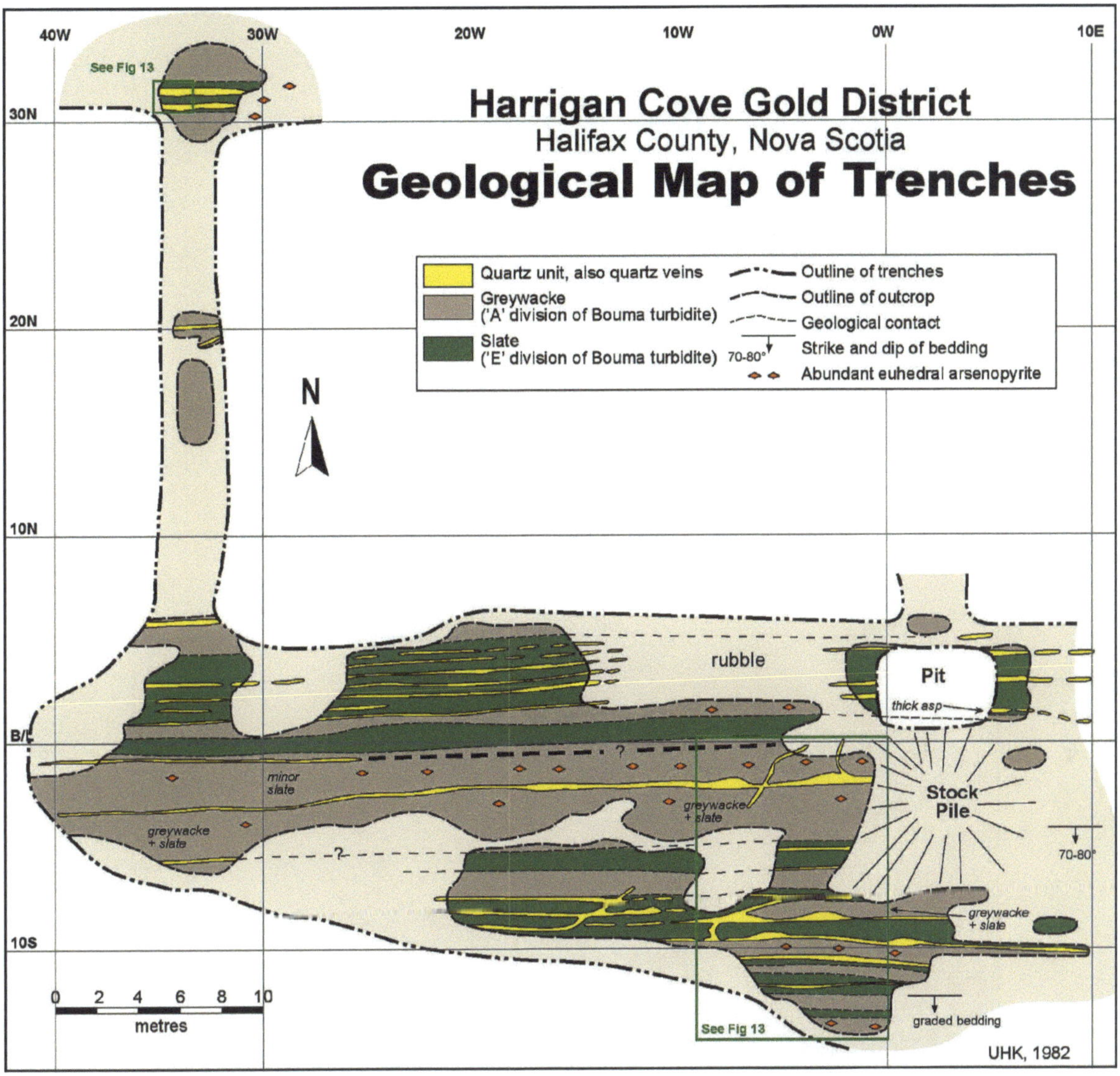

FIGURE 2.12

Harrigan Cove Gold District, Nova Scotia. Lode gold vein sequence is exposed in a trench. Map is oriented with north at top of page.

The thickness of individual quartz veins does not appear to be obviously related to the thickness of the containing turbidite or to the thickness of any division within the turbidite. However—and this may be important from a mining point of view—in a general way, the amount of quartz is the greatest in the thickest slate.

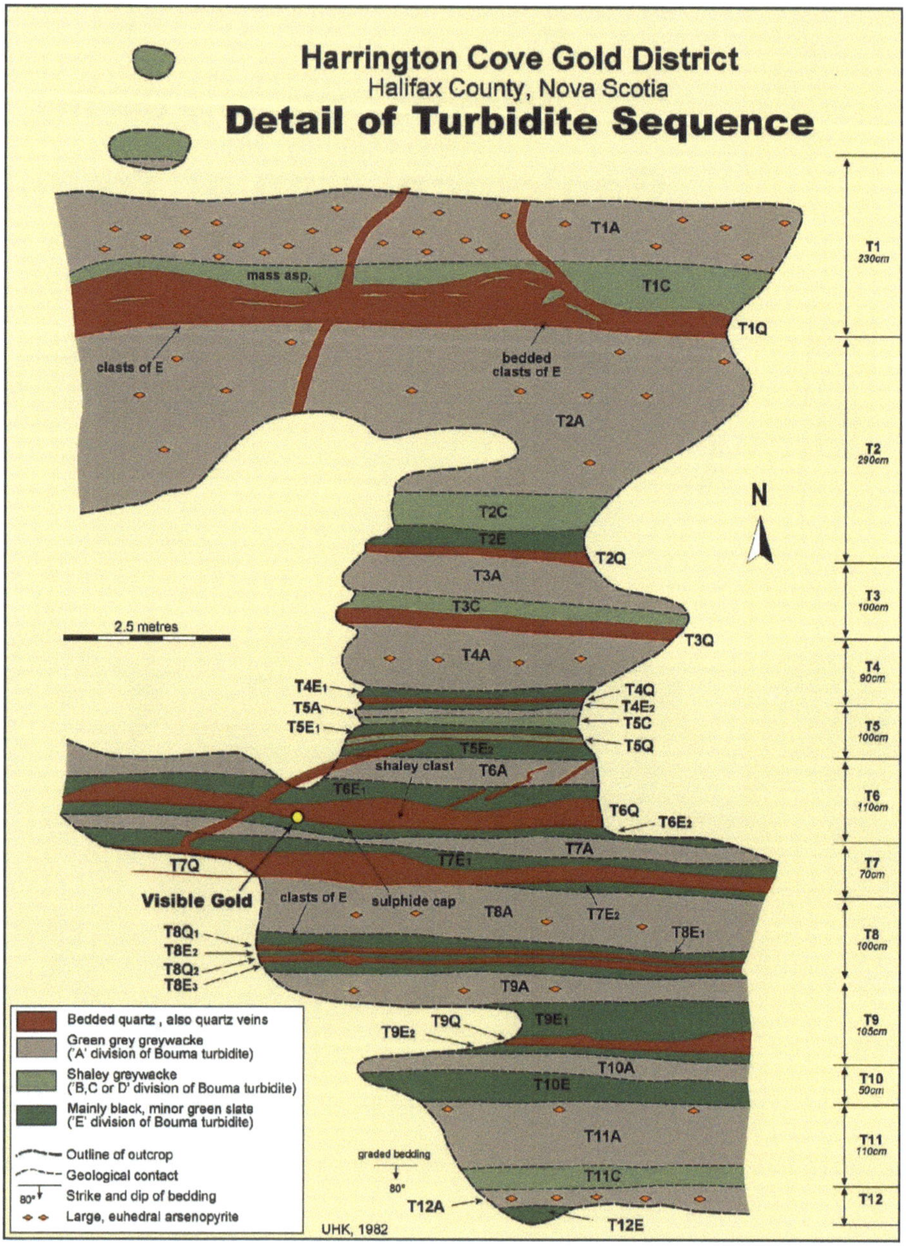

FIGURE 2.13

Detail of part of Figure 2.12. Map isoriented with north at top of page Source: Fig. 2 of Crocket et al. (1986), after Kretschmar, (1983).

The basal A division greywackes are green, chloritic, and massive. E division slates are generally greenish-gray, but locally they are dark green to black. Foliated greywacke occurs at the top of A division and may represent original C or D division. Asymmetric bulbous features are developed in several quartz veins. Cross-cutting relationships are rare but there are several "angulars" and/or quartz veins (Figure 2.14).

Arsenopyrite is common to very abundant throughout the section and occurs in all host rocks to a greater or lesser amount, but not in all quartz veins. Sulfides are rare and generally occur as thin beds capping quartz veins or in thin beds in the slate.

A detailed description of 12 turbidites shown in Figure 2.13 is presented in Table 2.1, and they are also shown in Figure 2.14.

The most notable features displayed are:

1. clear stratigraphic control to the occurrence of quartz veins ($T_1 - T_{12}$),
2. asymmetric development of textures such as:
   a. bulbous "troughs" that appear to be filling indentations in the underlying slate or foliated greywacke (T_1Q, T_6Q, T_7Q, T_9Q)
   b. a thin sulfide "cap" to T_6Q
   c. accumulation of slate clasts in the base of T_2Q
   d. internal laminations (bedding planes) separating massive, milky-white quartz from greenish, finely-laminated quartz
3. textures relating to primary deposition environment such as the partial erosion of T_7E_2 by T_8A
4. the complete absence of greywacke fragments in the quartz veins regardless of the host rocks containing them
5. the "draping" and slight rotation of thin slate fragments in the asymmetric bulbous depressions in T_1Q, and
6. the ubiquitous presence of arsenopyrite in the turbidites, but variable and apparently stratigraphically controlled concentrations in different slates and greywackes (e.g., T_1A and T_3A).

The most productive lead at Harrigan Cove was the "Bishop Lead" (Malcolm, 1912). It occurs about 30 m stratigraphically below the section illustrated in Figures 2.14 and 2.15. The "South Whidden" Lead occurs another 10 m lower in the section (Figure 2.12). Both of these quartz veins show a variety of textural relations that are described below.

A 2-m square outcrop of the "Bishop Lead" exposed in a trench is shown in Figure 2.15, as are the locations of arsenopyrites sampled.

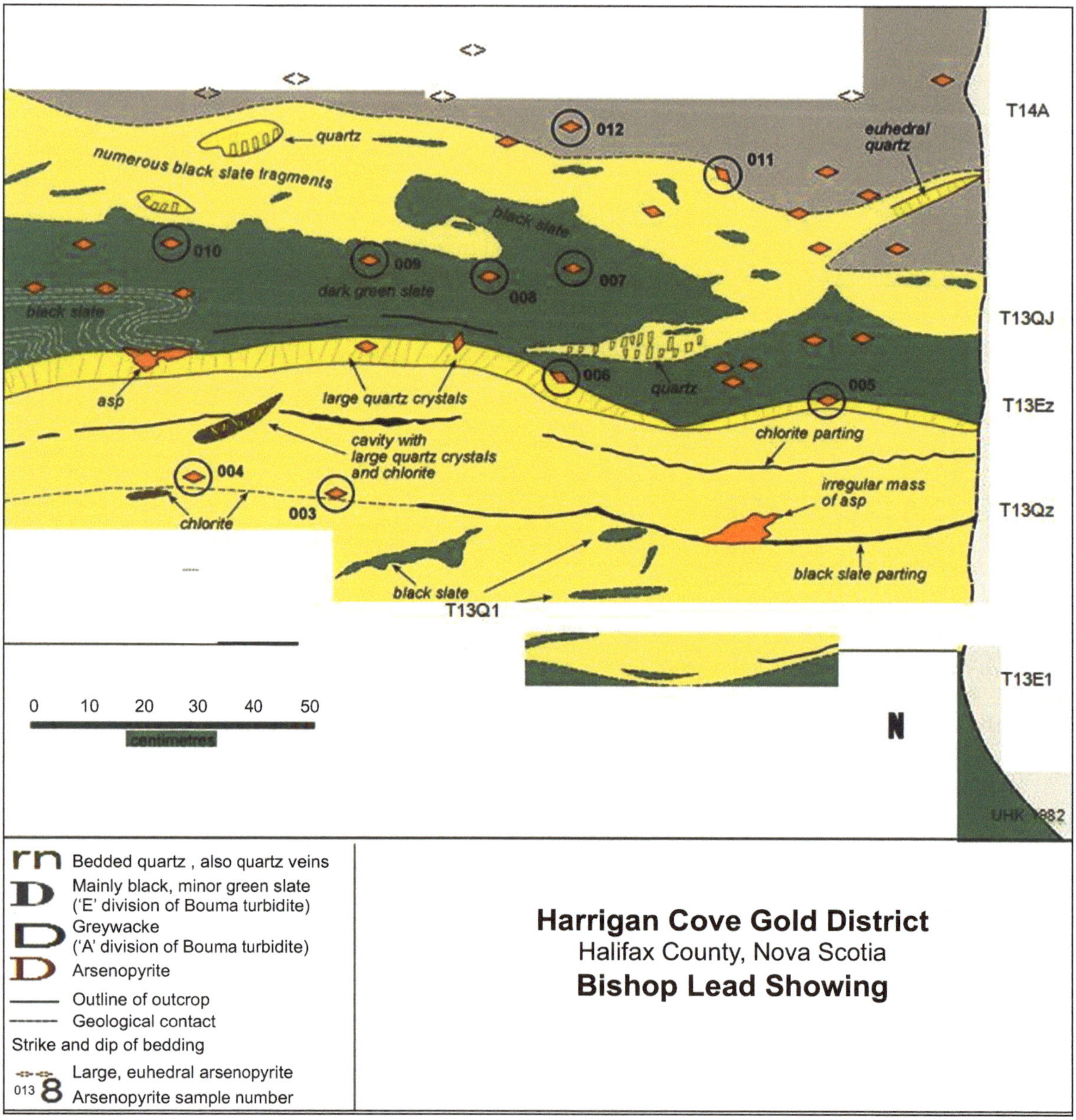

FIGURE 2.14

Harrigan Cove Gold District. Detailed geological plan of "Bishop Lead" Showing.

Table 2.1 Description of Stratigraphic Section Including 12 Turbidites, Harrigan Cove District (see Figure 2.13)

Unit	Thickness (cm)	Description
$T_{12}E$	30	Grayish-green slate, no asp apparent
$T_{12}A$	25	Coarse, euhedral asp in greenish greywacke
$T_{11}C$	25	Gritty slate
$T_{11}A$	85	Graded greywacke
$T_{10}E$	35	Greenish-gray well bedded massive slate
$T_{10}A$	25	Massive greywacke
T_9Q	15	10- to 20-cm thick, massive quartz, minor slate inclusions bulb development to east of section in Figure HC–7
T_9E	70	Grayish-green slate; contains T_9Q
T_8Q_2		1–3 cm finely laminated clear quartz and asp. Asp occurs as nodules 1–3 cm thick
T_8Q_1		1–3 cm white quartz with irregular black quartz throughout
T_8E	40	Gray slate, no obvious asp or sulfides; contains T_8Q_1 and Q_2
T_8A	70	Massive greywacke, green color
T_7Q	22	20–24 cm thick white quartz, locally red stain, vugs and open cavities common; v.f.g. py and trace cp.
T_7E	15	Gray slate; contains T_7Q
T_7B	25	Gritty greywackes, massive asp (2–3%) up to 5 mm
T_6Q	15	v.f.g. visible gold in vug with chlorite thin py, cp, asp cap at top of T_6Q red stain (hematite?) on some quartz, locally bulbous, thickness 10–20 cm. Quartz mainly white, locally laminated, some vugs with euhedral quartz
T_6E_2	4	Thin black slate, 4 cm
T_6E_1	30	Black argillaceous slate
T_6C	2	Gritty greywacke, greenish-gray, no obvious sulfides
T_6A	35	Gray greywacke: f.g. asp, about 1 mm; some euhedral asp; white chalky mineral (feldspar) brown weathered surface
T_5E_2	20	20 cm Black slate above T_5Q, slightly weathered
T_5E_1	4	4–5 cm Black slate below T_5Q, no obvious sulfides
T_5Q	3	2–5 cm Laminations parallel to bedding, thin asp bed on top of quartz. One well developed chlorite bed with trace of pyrite in it.

There are three quartz veins each with distinctly different features. They occur in black and gray slate and are overlaid by massive greywacke. In summary, $T_{13}Q_1$ and $T_{13}Q_3$ are characterized by irregular, very ragged fragments of black slate, whereas $T_{13}Q_2$ contains no black slate fragments. $T_{13}Q_2$ is massive white quartz

FIGURE 2.15
Harrigan Cove Gold Deposit. Stripped trench, exposing "Bishop Lead," which was the most productive "lead" in the old days, and displaying complex bedding-parallel quartz, interbedded with black slate and greywacke. See subdivision of stratigraphy into $T_{13}Q$, $T_{13}E_1$, and $T_{13}Q_2$ in Table 2.2 and on geological map Figure 2.14. Hammerhead tip points to stratigraphic top.

with several 1-mm internal laminations or chlorite partings. There are several cavities with euhedral quartz and some light green chlorite in them. Its upper surface is marked by coarsely crystalline, subhedral quartz. $T_{13}Q$ has very irregular contacts and shows "angular" projections into both over- and underlying units. Quartz in the "angulars" is more coarsely crystalline than the quartz veins from which they emanate.

The two most important features of the host slates are:

1. that there is a 5-cm bed of black slate underneath both $T_{13}Q_1$ and $T_{13}Q_3$, but the rest is gray (more silty) as well as black slate. This reflects a change in sedimentation prior to deposition of the quartz veins.
2. that there are disrupted beds and small-scale internal folds that seem to be due to soft sediment deformation.

From 1% to 2% euhedral arsenopyrite (asp) occurs throughout except apparently in $T_{13}Q_1$. Crystals up to 2 cm long are uniformly distributed in $T_{13}E$. In $T_{13}Q_2$, asp occurs largely as euhedral crystals or as irregular masses on the internal slate parting or on its upper surface where crystals grew with their long axes both parallel and perpendicular to the bedding. The amount of euhedral, 4–5 cm asp crystals decreases from about 7% at the $T_{14}A - T_{13}Q_3$ contact to none visible, about 150 cm stratigraphically higher.

Two segments of the "S. Whidden Lead" are illustrated in Figure 2.16 and the stratigraphy, which is a 9-m section below and about 30 m along the strike to the west of the "Bishop lead" exposure, is described in Table 2.3 and illustrated in Figure 2.15.

The quartz vein-bearing black slate ($T_{22}E$) is immediately underlaid by a gray foliated greywacke. $T_{22}Q_1$ is composed of 3- to 5-cm-wide quartz laminae

Table 2.2 Bishop Lead Stratigraphy (see Figures 2.14 and 2.15)

Unit	Thickness (cm)	Description
$T_{14}A$	200	Massive greenish-gray greywacke, 7% asp in lower part to 0% at 150 cm above base
$T_{13}Q_3$	15–25	Massive white milky quartz, 40–50% very irregular (ragged) black slate fragments throughout; irregular contacts, projections, and "angulars" emanating both above and below
$T_{13}E_2$	20–30	Slate; at base slate is sandier than at top; asp, 1–2%, up to 2 cm long disrupted bedding and soft sediment structures, black for 5 cm underneath $T_{13}Q_3$
$T_{13}Q_2$	8–17	Purer quartz than $T_{13}Q_1$, no irregular slate fragments internal bedding planes defined by chlorite-filled laminae and 1–2 cm vugs lined with subhedral quartz; upper part locally coarsely crystalline subhedral quartz
	1–3	Black slate bed dividing Q_1 and Q_2; changes along strike (to E) to green chlorite; 3- to 5-cm mass of asp crystals and several separate crystals project upward into $T_{13}Q_2$
$T_{13}Q_1$	20–25	Milky white massive quartz filled with 30–40% irregularly shaped black slate fragments; rare 1- to 2-cm light green chlorite blebs; no asp
$T_{13}E$	125	Gray (silty) and black slate; soft sediment deformation structures, 1–2% asp; black slate only for 5 cm underneath $T_{13}Q_1$ and alternating 1- to 2-mm thick gray (sandy) and black slate below that

interbedded with thin slate and arsenopyrite beds. In area (a), about 20% of arsenopyrite occurs as crystals and masses symmetrically distributed on stratigraphic foot wall and hanging wall. In area (b), arsenopyrite constitutes about 50% of $T_{22}Q_1$ (Figure 2.15).

$T_{22}Q_2$ is 3–4 cm of white quartz with 30% irregular masses of arsenopyrite throughout. Four meters to the east in area (b), $T_{22}Q_2$ is a massive arsenopyrite bed that thins and disappears in black slate.

$T_{22}Q_3$ is 20 cm thick. The lower 10 cm is massive milky white quartz with internal laminae and 1–2% arsenopyrite irregularly scattered throughout. The middle marker bed has euhedral arsenopyrite crystals, which have their long axes oriented parallel to the bedding. The upper half of this unit has 40–50% of very irregular black slate fragments in massive milky white quartz. These appear to have been "dropped" into a semisoft silica medium that was then squeezed, compacted, and crystallized. The upper surface of $T_{22}Q_3$ is highly irregular and convoluted, and there are finger-like projections and "angulars" of quartz reaching upward into the overlying black slate and $T_{23}B$

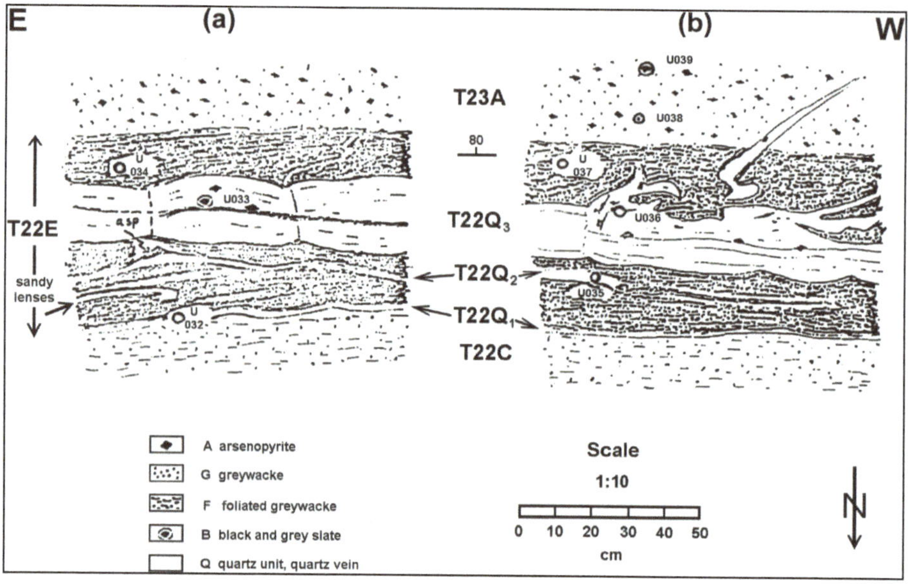

FIGURE 2.16

Detailed geology of two segments (a and b), part of "S. Whidden Lead," Harrigan Cove gold district.

Table 2.3 Stratigraphy of the "South Whidden Lead" (see Figures 2.16 and 2.17)

Unit	Thickness (cm)	Description
$T_{24}Q$	0.4	Two quartz veins with 30–40% irregular black slate fragments probably $T_{13}Q$ ("Bishop Lead")
$T_{24}E$	0.5	Black slate with soft sediment deformation textures, asp 7–8% up to 2 cm diameter
$T_{24}A$	3.5	Grayish-green massive greywacke; asp throughout, 1% 1 mm at base to 5% up to 1 cm near top; fragments of black slate up to 2 cm thick, 10 cm long at top
$T_{23}E$	0.25	Gray slate, well bedded, alternating 1–3 mm light and dark beds, asp 2–3% up to 2 cm
$T_{23}A$	0.25	Gray greywacke, asp 7–8% up to 2 cm diameter
$T_{22}E$	0.8	Black slate with 3 quartz veins; Q_1 at base; Q_2 13 cm above Q, and Q_3 3–5 cm above Q_2 (see text)
$T_{22}C$	1.0	Gray foliated greywacke 5% asp up to 1.5 cm in diameter
$T_{22}A$	2.0	Gray greywacke, s1 foliated near $T_{22}C$

FIGURE 2.17
Harrigan Cove Gold District. "South Whidden Lead." Details of soft sediment deformation textures and the nature of arsenopyrite occurrence. Photo HC-36. See also Figure 2.16.

(Figure 2.16(b)). In area A, a 1-cm folded massive quartz "angular" originates at a "pinched" point in $T_{22}Q_3$.

The black slate host, $T_{22}E$, shows moderately well- to poorly developed bedding consisting of alternating gray (more sandy) beds and black slate. The black color is attributed to elemental carbon. In area a), irregular sandy lenses that show soft sediment deformation features occur beneath $T_{22}Q_2$. $T_{22}E$ is black slate for only 5–6 cm above Q_3 and well-bedded in gray slate for 13 cm above the black slate. This is an important observation, as it demonstrates a stratigraphic control to the quartz vein that is specific not only to the slate unit but, in this case (also the "Bishop lead"), to the black, more organic and reducing beds in the slate division (Figure 2.17).

Figure 2.14 illustrates the quartz textures and the nature of diagenetic and hydrothermal arsenopyrite. Host rock is green and black slate (pelagic-pelitic or E division of Bouma turbidite). Large diagenetic euhedral arsenopyrite crystals grew within grey and black shale, occasionally cross-cutting quartz-shale contacts. This diagenetic arsenopyrite is strongly zoned from sulfur-rich centres to arsenic-rich rims. Irregular masses of arsenopyrite grow on thin black shale partings internal to quartz. Arsenopyrite temperature and Re-Os age determinations must take into account zoning, a mixed sedimentary-diagenetic-hydrothermal and general disequilibrium formation conditions.

2.3 THE EFFECT OF DEFORMATION

Meguma vein formation and, indeed, all vein formation cannot be readily separated from mechanics and structure, because some form of deformation and remobilization is required by all other formation models. A thesis of this book and a key concept underpinning syngenetic lode vein formation is that it is possible to see through metamorphism, deformation, and structure, and that frequently geologists can distinguish between what is primary and what is secondary. The authors think that unless there is clear evidence to the contrary, textures should be considered primary. This position is diametrically opposite to that in much of the literature on gold deposits, in which it is assumed that all quartz is secondary. It is made more difficult to defend by the work of Ramsay (1980b) on chevron folding and Ramsay (1980) on the crack–seal mechanism, which provides a theoretical basis for their formation and shows quartz vein textures that resemble those that the authors consider primary and syngenetic, but on a very small scale. The classic study of chevron folding by Ramsay (1974) also demonstrates that bulbous hinge zones or layer boudinage may reflect local modifications of fold style.

As noted by Henderson et al. (1989) in response to a discussion by Mawer (1989), a prevailing formation theory considers gold-bearing quartz veins in the Meguma as weakly deformed late-tectonic anticlinal saddle reefs, resulting from a hydraulic-fracturing mechanism. This was proposed originally by Faribault (1899) and elaborated on by Graves and Zentilli (1982). Henderson et al. (1989) state that a late-tectonic origin for bedding-concordant veins is inconsistent with their observations that many bedding-concordant veins formed early in the tectonic history of the Meguma zone (Henderson, 1983; Henderson et al., 1986; Henderson and Henderson, 1986), and that their data clearly show that most of these veins predate cleavage and folding in the host strata, and are not saddle reefs. The veins show more than 60% layer parallel shortening, which is not confined to major fold hinges. The veins also are folded on the major fold limbs and do not taper downdip. They suggest that bedding-concordant veins formed in the Meguma strata before cleavage and folding, and underwent essentially the total tectonic strain recorded by other strain gauges. This is illustrated by deformation textures in a thin section of greywacke from the Goldenville Gold Deposit (Figure 2.14) and a weathered surface of quartz veins in a bedded greywacke from Softwood Island (Section 4.2).

An interesting theory of vein formation is presented by Mawer (1986, 1987), which has as its basis intermittent hydraulic fracturing and healing ("crack–seal"). Phyllosilicate inclusion bands, "columnar" quartz grains, and included sheets of host rock are considered as precipitation features. Mawer states that deformation structures that overprint these textures include pinch-and-swell (boudins), tectonic stylolites, cross-cutting extension and hybrid fractures. The problem that he faces is putting solutions into the unconsolidated muds and sands; they do not have the competence to remain intact during hydraulic fracturing.

Examples of what the authors consider syndepositional slumping are shown in Figures 2.18 and 2.19.

FIGURE 2.18

Harrigan Cove Gold District, Nova Scotia. Syndeformational slumping in the "South Whidden Lead" (in centimeters, detail of Figures 2.18 and 2.19).

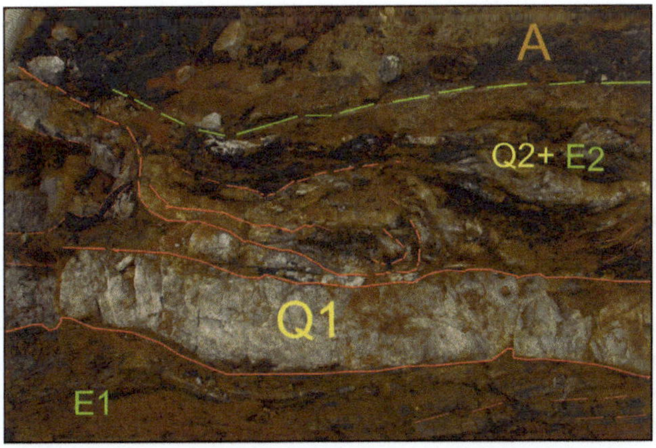

FIGURE 2.19

Beardmore area, Sage Gold Vein 03 exposed in trench. Gold-bearing Q_1 overlying E_1 chloritic tuff is 30 cm thick. Quartz Q_2 and chloritic tuff E_2 overlay Q_1 and overlaying gabbroic flow A.

2.4 TEXTURES FROM VOLCANIC AND FELSIC-HOSTED LODE VEINS

Vein textures from lode veins in the Maskwa batholith in Manitoba, the Hercules vein systems in the Elmhirst "Intrusion," and the Côté Gold deposit in Chester Twp are presented in Chapters 4.3 and 4.4.

2.5 DISCUSSION

In Chapter 2, we have seen veins and textures from the Meguma turbidite terrain and equivalent volcanic terrain in the Beardmore area. Kretschmar presents these examples as evidence that quartz and textures, show the following: (1) some quartz was deposited syngenetically; (2) it displays widely variable texture, to which equally wide interpretation on its origin can be brought; (3) clearly there is later, cross-cutting quartz that can readily be interpreted as feeder veins; (4) the scale of observation is important; (5) a common process of quartz generation is indicated in the Meguma, because quartz veins are ubiquitous and the textures are the same, whether the veins do or do not have gold; (6) some quartz was originally gel-like; and (7) the effects of metamorphism and deformation were minor and can probably be distinguished from primary features. Contrary interpretations are presented, however, by Kontak and Horne (2010).

Outcrops, textures, and, in particular, the timing of veining generates more geological debate than most other fields of investigation. Did that vein form at the same time as the host, or later? Is that black fragment an inclusion, a xenolith, a mafic enclave, or a pyroclastic breccia fragment? Subjective experience and background will determine the answer. As was apparent, diametrically opposed views are expressed with equal elegance by Hunt (1868) and Mawer (1986) for the Meguma veins.

It is instructive to briefly examine sulfide mineral and ore textures that have been widely studied and textures of quartz veins in low grade greenschist facies rocks. Ore microscopy involves identification of individual mineral grains and interpretation of ore mineral textures. Some major references are Edwards (1947), Bastin (1959), Ramdohr (1969), Stanton (1972), and Barton (1991), as listed by Craig and Vaughn (1994), and a more recent treatment by Taylor (2009). Barton (1991) noted that interpretation of textures is simultaneously one of the most difficult and important aspects of the study of rocks and ores, and this is certainly true in the case of lode vein gold deposits. Interpretation of textures is notoriously subjective, and sulfide textures are commonly unreliable for determining timing relations. Whereas chalcopyrite and pyrrhotite are highly mobile, arsenopyrite, pyrite, and, to a lesser extent, sphalerite and galena, are competent and rarely recrystallize or change composition except under granulite facies metamorphism. Whitehead (1973) demonstrated, in a lithogeochemical study of the Heath Steele Mine B-1 zone host rocks, that remobilization of metals in the ore zone through dewatering and compaction under burial and subsequent deformation was only on the geochemical anomalous levels to distances of less than 100 m. McBride (1976b, see Chapter 8) demonstrated that in the B-1 massive sulphide deposit, remobilization did not take place over the sample interval of 1.5 m under the upper greenschist grade metamorphism coupled with deformation, that produced the isoclinal folding. Thus, zoning in sulphide mineral deposits can be considered evidence that metamorphism and deformation have none or very little effect on depositional textures. The lessons of this basic geology must be kept in mind at all times.

The nature of quartz formation in general, P-T limits on deposition, and "veins" are discussed in Chapter 7 and Section 7.3. To separate, clearly synsedimentary quartz from cross-cutting "veins" may prove useful, when the synsedimentary nature of some quartz is better accepted and the concept of "Quartz Unit" (Kretchmar, 1983) can be applied. The entire concept of "veins" and "lode" needs to be re-examined, as was pointed out by Hutchinson (1993, p. 360), referring to the diversity of veins described by Hutchinson and Burlington (1984) and others adhering to syngenetic concepts, e.g., Karvinen (1978). The nature of quartz deposition is also further described in the context of individual deposits, Meguma (Section 4.1), Côté Gold Deposit (Section 4.2), and the Maskwa high-grade gold veins in Section 4.3. The scale of cross-cutting features, possible telescoping, and superposition of effects will also become apparent (Figure 2.20).

FIGURE 2.20

Sober Island, south end, Nova Scotia. Upper surface of folded synsedimentary quartz vein overlying fine muddy pelitic upper portion of E-unit of turbidite. This outcrop is an "M" parasitic fold of thin sandy beds in the nose of a larger anticline. Hammerhead points to stratigraphic tops.

Geologists must learn to recognize critical field relations. An example is provided in a Chester Complex field mapping report (Siragusa, 1993, p. 30), in which the author informs the reader that, due to the widespread occurrence of moderate to severe shearing and grain-sized gradations, textures suggestive of cross-bedding noted in a few outcrops were not regarded as reliable primary features. As will become apparent, this is actually key evidence that the Chester Complex is not an intrusive.

Oscillatory zoning in minerals is a common variations in fluid temperature, pressure, or phenomenon (Shore and Fowler, 1996), but the chemical reactions do not intrinsically require large composition (Gray and Scott, 1990).

Extremely fine, laterally continuous oscillatory zoning is present in sphalerite in many veins that show textural evidence of open-space growth with only a limited number of discrete events (e.g., brecciation or change in mineral assemblage.

Kretschmar has investigated submicrometer zoning of Ag, Sn, and Sb in decimeter-thick sphalerite veins from British Columbia (studied by Beaudoin et al. (2014)). Attributing each of these tens or hundreds of thousands of layers to an event of episodic flow, such as a crack—seal cycle, may not be necessary, if one considers that similar zoning can be produced in minerals grown in isothermal gels or unstirred solutions (Reeder et al., 1990; Putnis et al., 1993).

The authors' experience has shown that complex mineral zoning is common in many syngenetic-bedded deposits (McBride, 1976b, 2014). The depositional environment for many of these deposits is well documented in the literature on volcanogenic massive sulphide and sedex deposits, to the point that it is considered as evidence for a syngenetic origin. It is only a small step to apply the same logic to other bedded mineral deposits. We shall see in Chapter 8 that these textures are preserved throughout the deformational history.

CHAPTER 3

Introduction to Gold Cycles

CONTENTS

Exploration geologists in Canada have unparalleled access to glaciated outcrops, drill core in mining camps, excellent research, and leading-edge genetic models. In the Meguma of Nova Scotia, gold mine stratigraphy and nonmineralized turbidite-hosted quartz veins trend onto glaciated seashores. In Ontario, veins are exposed for kilometers from millions of dollars worth of mechanical stripping of key outcrops and many kilometers of core. Exploration and mining company reports and websites often provide excellent maps and cross-sections. From this, during many years of fieldwork, it has become apparent that one must distinguish clearly cross-cutting veins from bedded quartz, which was designated (Kretschmar, 1983) as quartz units (QU). Geometry and host rock relations can be further systematized into a "gold cycle" (GC), loosely described by turbidite terminology and sketched in Figure 3.1.

The key observation is that, common to Meguma veins in turbidites, and to Archean equivalents in crystal tuffs, epiclastics, and flows, veins occur within deposition cycles, commonly displaying graded bedding. The thickness of a cycle can vary from centimeters to several meters to 200 m, as in the case of the Kerr Addison (Chesterville) Mine in Virginiatown, Ontario (Buffam and Allen, 1948).

3.1 PROVISIONAL DEFINITIONS: QUARTZ UNITS AND THE GOLD CYCLE

In Bouma turbidites, lode gold veins have common characteristics that Kretschmar has systematized in the concept of a "gold cycle" (GC) Figure 3.1, in Kretschmar (2011), a provisional expedient in the absence of detailed studies. Its basis lies in the subaqueous nature of these rocks and the observation that >90% of bedding parallel quartz in Meguma turbidite-hosted gold deposits occurs in the E division of them, as was apparent in Chapter 2, Kretschmar (1983). In this unpublished work, bedding parallel veins were called quartz units (QU) to distinguish them from clearly cross-cutting veins.

The Metallogeny of Lode Gold Deposits. http://dx.doi.org/10.1016/B978-0-12-803222-0.00003-4

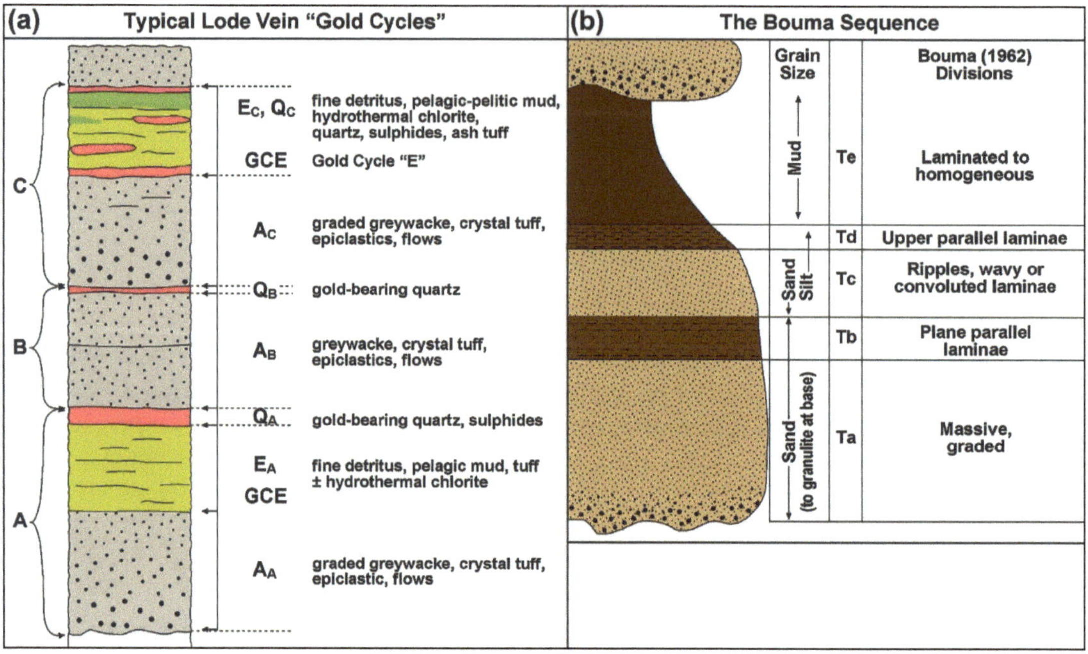

FIGURE 3.1

A: The "gold cycle." Schematic representation of three typical lode gold vein settings, which, by analogy with Bouma cycles Bouma (1962) and Shanmugam (2006), are provisionally called "quartz–gold cycles" or gold cycles. Gold cycle "E" units are called GCE in the current work. In example A, EA may consist of pelagic–pelitic mud, ash or crystal tuff, carbonate, sulfides, and hydrothermal chlorite in various proportions, and EA is capped by massive or laminated quartz–gold–sulfide ("vein" in the literature). Constituents of EC are similar, with more complex textures, which may include laminations, irregular beds, short cross-cutting angulars, and ladder veins. B: Comparison of gold cycle to turbidite subdivisions, originally described by Bouma (1962), redrawn from Figure 8.1 of Shanmugam (2006).

This term is useful where the syngenetic bedded origin of quartz can be established. It clearly addresses the genesis of the quartz and, as such, is a powerful descriptor. A more detailed description of physical properties of this syngenetic quartz is presented in Chapter 7.

In a GC, the lower part consists of generally coarser lithologies, which may fine upward (e.g., greywackes, volcaniclastics, or flows). Overlying is chloritic (ash) tuff and/or pelagic/pelitic sediment or abyssal mud. Gold-bearing quartz (the lode vein) occurs within or caps this generally dark green or black material, which is frequently and mistakenly identified as either lamprophyre or "shear zone." In turn, this may be overlain by the basal unit of the next cycle, as shown in Figure 3.1.

Cycle A: Aa consists of graded greywacke, crystal tuff, epiclastics, of felsic flows and fines upward into Ea which is a mixture of fine detritus, ash tuff, and pelagic mud. Qa may be massive quartz, laminated (ribbon quartz), or sulfide-rich (e.g., Doyon-type veins), or may display these and other features in numerous combinations. It is a syngenetic quartz crust or QU.

Cycle B: Ab displays amalgamated (i.e., massive, ungraded) bedding. Unit Eb is thin or absent. Qb is generally massive with little chlorite, sericite, or sulfides on internal partings.

Cycle C: Ac is similar to Aa. Ec is a complex volcanic sediment. Quartz may be laterally continuous or occur as silica drops or short, discontinuous disc-shaped units.

Quartz in gold cycles is, by definition, largely syngenetic. In a proximal vent environment, the genesis of quartz increases in complexity, and geological, chemical, or isotopic criteria will be needed to distinguish various types of quartz.

In the Meguma, there is a continuum of quartz from microscopic droplets to units several meters thick (Kretschmar, 1983). The amount of quartz in QC reflects the interplay of silica–gold (plus sulfide–carbonate–chlorite) deposition from hydrothermal solution and pelagic/pelitic sedimentation. Ec is a combination of hydrothermal chlorite, ash, or crystal tuff (in volcanic terrain) and pelagic/pelitic mud. The thickness of Ec is a function of the pelagic/pelitic sedimentation rate and time available for its accumulation, and may vary from centimeters up to 200 m (e.g., Kerr Addison Mine, Larder Lake, Ontario, 1300-ft (400-m) level, as shown in Figure 3.2 and Figure 8 of Hinse et al., 1986). In the literature, Ec or Ea units of the gold cycle are frequently termed "shear zones" (or, in some instances, are auriferous "exhalites" as in Hutchinson, 1993) and, as will become apparent (Chapter 5), are commonly misidentified as "lamprophyre." Characterization of interflow sediments has been carried out by Crocket et al. (1980), Kerrich (1983), and Bierlein et al. (1998b). In the present work, E units in gold camps are referred to as gold cycle "E" or "GCE."

3.2 EXAMPLES FROM THE LITERATURE

A substantial component of the present thesis in this book consists of the characterization of gold cycles. The entire Chapter 5 explores the link with lamprophyres. Only a brief introduction is presented here. Their nature varies substantially in thickness and complexity, as shown by the following examples. One of the earliest description of the connection between lamprophyres and lode veins is shown as Figure 3.3. McLennan (1915) considers that the lamprophyre controlled the localization of the lode quartz veins in some of the California Motherlode occurrences.

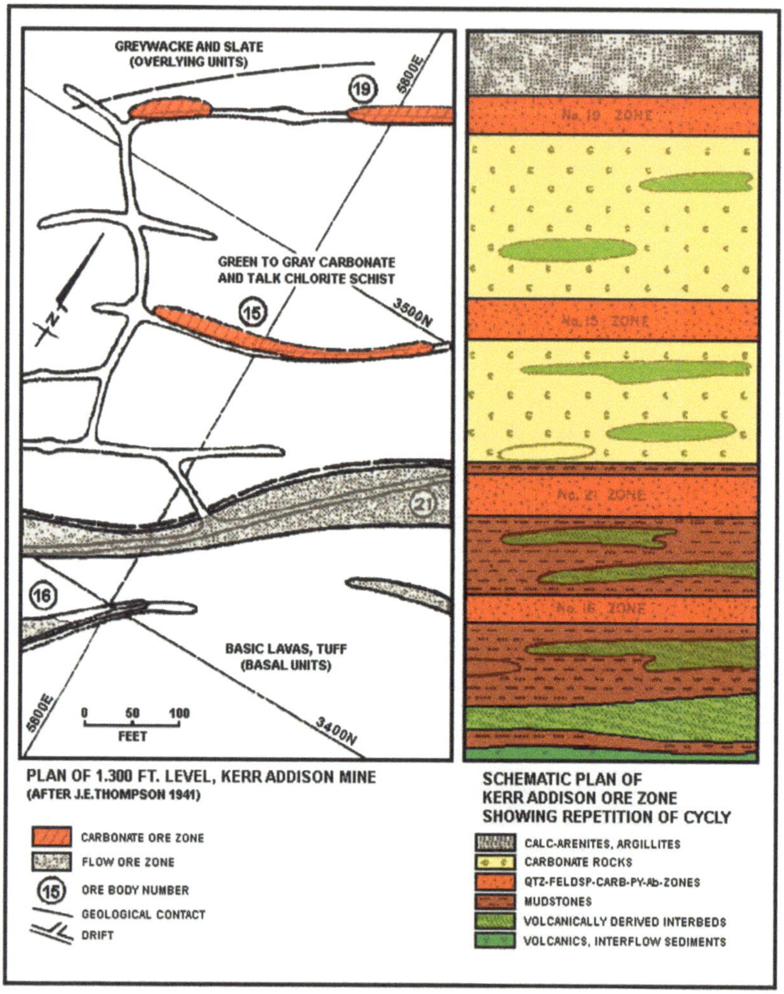

FIGURE 3.2

Kerr Addison, Larder Lake stratigraphy showing gold cycle stratigraphy that is 200 m thick. Several separate mineralized horizons indicate continuing gold–silica–carbonate hydrothermal activity, interspersed with mafic and felsic volcanism. *Redrawn from Hinse et al. (1986), after Thompson (1941).*

Other examples from the volcanic environment in the Timmins and Larder Lake gold camps are shown in Figures 3.3–3.6. Gold cycles in the felsic-hosted Côté Gold deposit and the Maskwa batholith (Figure 4.18) are described in Chapters 4 and 6.

On the larger scale, economic concentrations of gold mineralization have been shown to preferentially occur near or at the top of mafic volcanic sequences in tuffaceous rocks that are commonly interbedded with iron carbonate exhalite, or in the adjacent clastic sedimentary rocks (McBride 1987, Karvinen 1980).

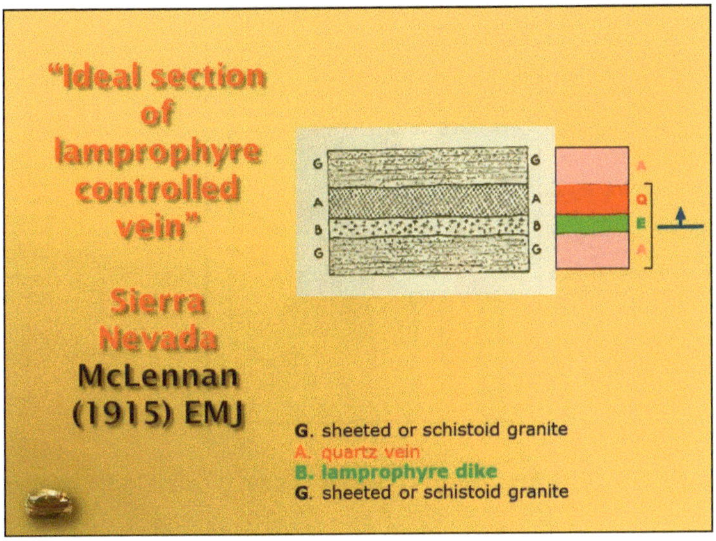

FIGURE 3.3

Ideal generalized section of a minette lamprophyre controlled vein from the California Motherlode gold fields. Figure 1 of McLennan (1915). Both quartz vein and lamprophyre have variable thickness. These lamprophyres are probably pelagic-pelitic interflow sediments or tuffs (Gold Cycle E or GCE), and their role, in the genesis of lode and veins forms the subject of Chapter 5.

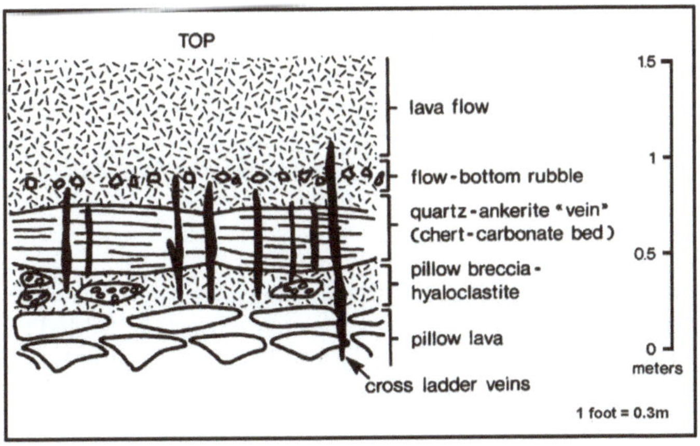

FIGURE 3.4

Aunor Mine, 1000 ft level. Timmins. Gold cycle generalized stratigraphic section, showing synvolcanic quartz—ankerite vein, boudinaged, and metamorphogenic cross-ladder veins. *Figure 3.5 of Karvinen (1980).*

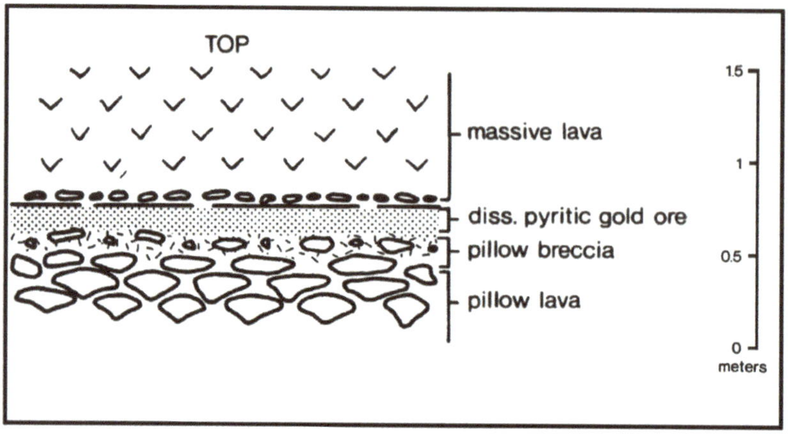

FIGURE 3.5
Moneta Mine, McIntyre dacite ore, Timmins. Gold cycle stratiform pyritic gold ore in massive carbonaceous and/or carbonate-rich interflow sediments. *Figure 3.6 of Karvinen (1981).*

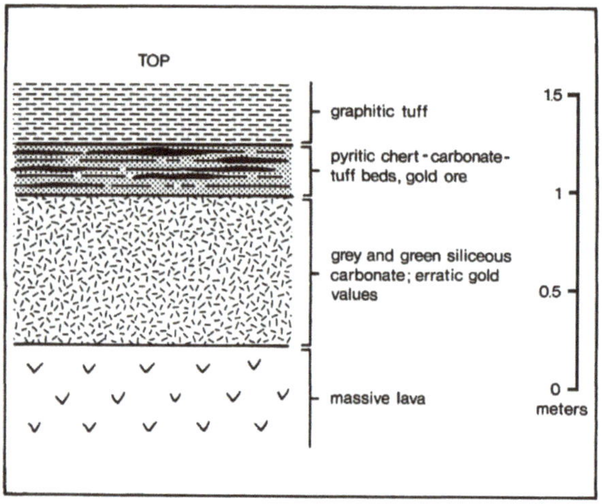

FIGURE 3.6
Porcupine Peninsular Mine, 80–88 zone, Timmins. Gold cycle stratigraphy showing synvolcanic stratiform pyritic gold in well-bedded chert–carbonate tuff beds. *Figure 3.9 of Karvinen (1982).*

The variations in the auriferous quartz band in the Northern Empire Mine were controlled by paleochannels in the footwall rocks. The band was later folded to a vertical position that resulted in boudinaging of the bedding parallel band and minor folding of the feeder veinlets to it (see Chapter 9).

CHAPTER 4

Field Examinations in a Variety of Gold Settings in Canada: The Meguma, Nova Scotia; Chester Twp and Beardmore, Ontario; Maskwa, Manitoba, and Nugget Pond, Newfoundland

CONTENTS

Both authors have carried out extensive field work in the study areas in the following discussion. They have never worked together on any exploration project or been active in any camp at the same time. Each has reached conclusions independently, based on his experience. Both have published numerous papers and reports presenting evidence on the generation of gold and base metal deposits and are firm in their conclusion that gold deposits are formed syngenetically. The foundation for the syngenetic perspective lies in careful and detailed geological mapping and logging of core. Geological mapping should look at the mineralization, the host rocks, regional setting, and tectonic history. Such studies are rarely done. Much of the authors' work has included considerable time mapping the study areas as well as performing detailed deposit analysis. Five geographically and geologically diverse areas with exploration advanced to different degrees are described in this chapter.

4.1 MEGUMA AND TURBIDITE-HOSTED GOLD

Chapter 2 shows bedding-parallel or conformable vein textures from the Meguma and Archean equivalents and what the authors consider compelling evidence for a syngenetic origin. Such veins constitute more than 90% of lode vein deposits, and a geological summary for Meguma veins will provide background for alternative interpretations of their origin. Deposits and outcrop areas described in the text are shown in Figure 4.1. McBride's (1978) map area is shown by the rectangle.

The Metallogeny of Lode Gold Deposits. http://dx.doi.org/10.1016/B978-0-12-803222-0.00004-6

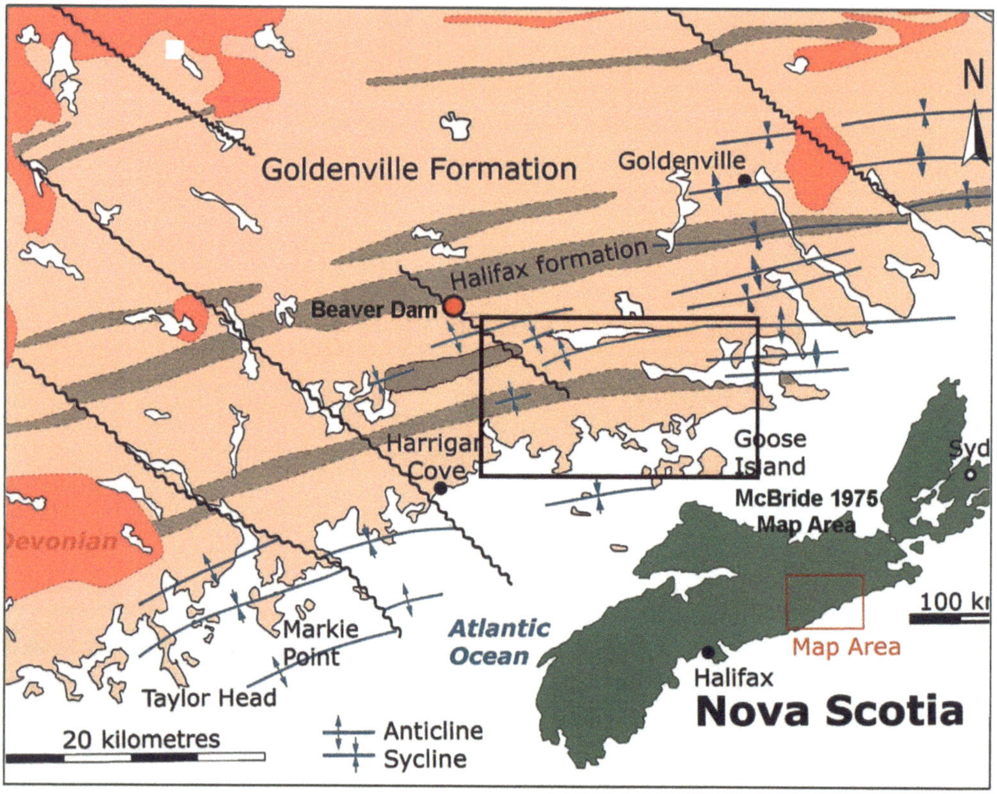

FIGURE 4.1

Locations referred to in the text, of seashore exposures of the Meguma, Kretchmar and McBride study areas, Southeast Coast, Nova Scotia.

4.1.1 Deposit Summary

The deposits are widely researched, and important descriptions of Meguma turbidite-hosted deposits are by Boyle (1986), Kontak et al. (1990), Kontak and Smith (1993), Sangster and Smith (2007), and Kontak et al. (2001). Dubé and Gosselin (2007) include the Meguma deposits in the "greenstone clan" formed at depth from solutions and remobilized during deformation. The formation model and deposit descriptions have progressively evolved, with Kontak and Horne (2010) providing a recent and most complete epigenetic–structural view. To understand these deposits, key elements are regional geology, mine cross-sections, vein characteristics, stratigraphic relations, and structural overprinting. Although modern age dating of veins and intrusive granites as well as fluid inclusion studies provide constraints on formation character for regional and local geology, the descriptive work of Faribault (1899), Malcolm (1912), and McBride (1978), in particular, is very relevant; however, it is often is ignored!

The host Meguma Supergroup, discussed by Waldron et al. (2009), is described as a thick (>10-km-thick) Cambrian to Early Ordovician turbiditic clastic succession, with a lower, coarser-grained Goldenville Formation containing numerous quartz and lode gold veins and an upper, predominantly fine-grained Halifax Formation.

A stratigraphic section Figure 4.2 (Sangster and Smith, 2007, Figure 6) shows the position of major gold districts. There is a concentration of deposits in a 500-m-thick interval at 1500–2000 m, but they are widely separated and cannot be demonstrated to be in the same stratigraphic interval. In the Ecum Secum area, McBride (1978) defined a 200-m-thick section of host stratigraphy with its base 1300 m below the Halifax contact.

A plan map of the Goldenville Gold District is shown in Figure 4.3 (Sangster and Smith, 2007, Figure 7), which, although larger than most other gold districts,

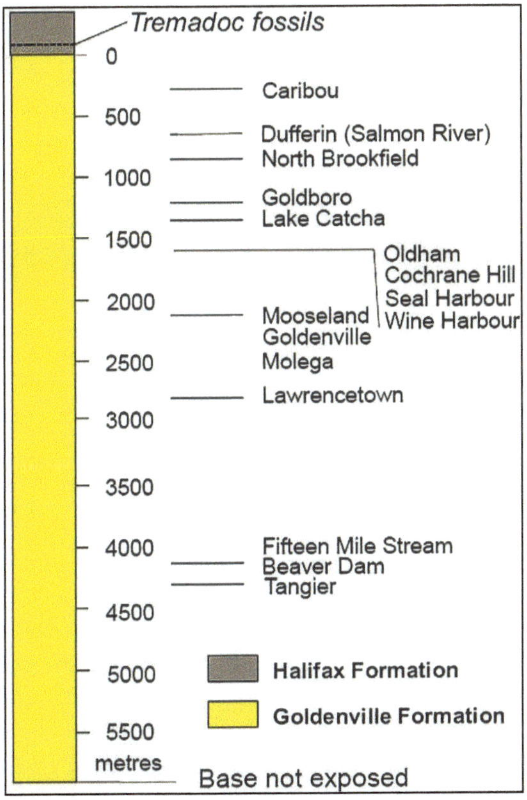

FIGURE 4.2

Meguma stratigraphy. *From Sangster and Smith (2007).* The top of the section is defined by Ordovician Tremadoc fossils in overlying Halifax Formation, which span an age of 485.4 ± 1.9 and 443.4 ± 1.5 Ma (*Wikipedia*). The lower part of the Goldenville may extend into the Neoproterozoic (1000-541 Ma) (*White and Barr, 2004*). The base of the Meguma is not exposed. Shown on the stratigraphic column are the locations of selected gold deposits referred to in the text.

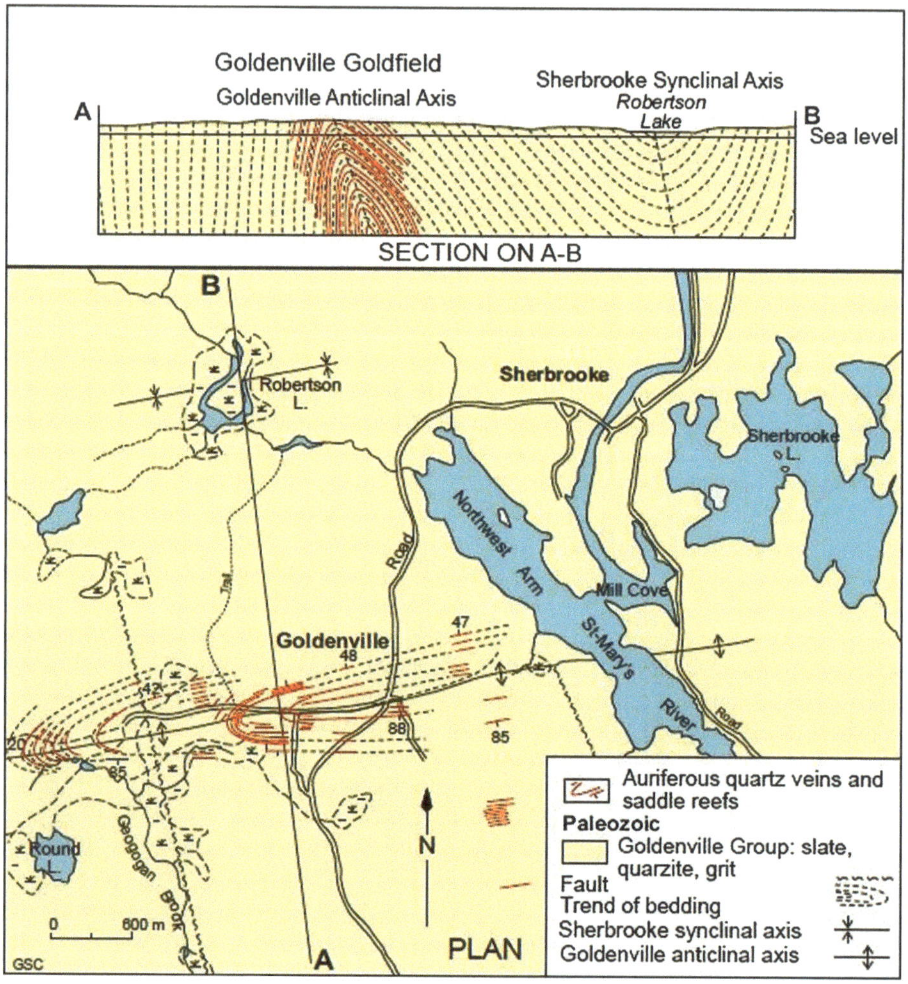

FIGURE 4.3

Geological plan of Goldenville Gold District. *From Sangster and Smith (2007).*

shows features that are typical of Meguma gold occurrences. Bedding is outlined with the dotted lines and the auriferous quartz veins by red lines. The interbedding and folded nature of this sequence on the regional scale is obvious. It stands out as the most important controlling guide to gold vein exploration.

The detailed geology and observations from mining of this deposit are discussed in chapter 9. It is apparent from the historical record that all the mining was from sections of bedding-parallel quartz bands and that "angulars" were feeder veins that only contained gold at their intersections with the bedding-parallel bands (Hedley, 1941).

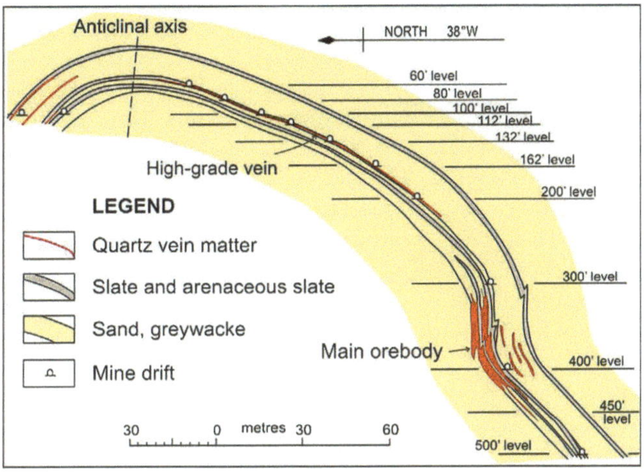

FIGURE 4.4

Geological section of the Caribou Mine. Bedding-parallel veins occur laterally from the main ore body, which is reported to be quartz-healed breccia ore. *Sangster and Smith, 2007, Figure 11.*

The Caribou Mine shows how stratabound bedding-parallel veins (QU in Chapter 2) occur laterally from the main ore body. The ore horizon is folded into an anticline, but there is no thickening in the hinge area. The main zone appears to be a lens on the limb of the anticline, and there is a suggestion that it has a minor fold of the "S" sense. Both mineralized horizons are stratabound, and the main zone lies well off the "saddle reef" portion of the anticline (Figure 4.4).

Slight apparent thickening of stacked bedding–parallel veins (QU) in the hinge area of the Dufferin anticline is indicated (Malcolm, 1912, Figure 9). This figure shows the multiple bedding parallel veins folded over the anticline and well down the limbs. Note that only some of the veins are auriferous. Records indicate that production predated 1909, indicating that the mine section in Malcolm includes most of the active mining to that time. The break in the main ore zone on the left limb probably represents a barren section of the vein. The quartz stringers reported from the fold nose or in some of the slates were reported to have no gold values of interest, which indicates that the gold in the adjacent bedding parallel is not remobilized with the quartz (Figure 4.5).

A detailed 20-m geological map of the underground workings of the Beaver Dam gold deposit (east wall 1100 cross-cut) at the Papke, Millet Seed, and Austen Zones, shows (1) the nature of turbidite sedimentation, (2) the stratabound character of arsenopyrite, (3) both synsedimentary quartz (QU) and locally cross-cutting quartz, and (4) later offsetting faults. Unfortunately it does not differentiate between gold-bearing and barren quartz veins; it does show that the arsenopyrite is bedding controlled (Figure 4.6).

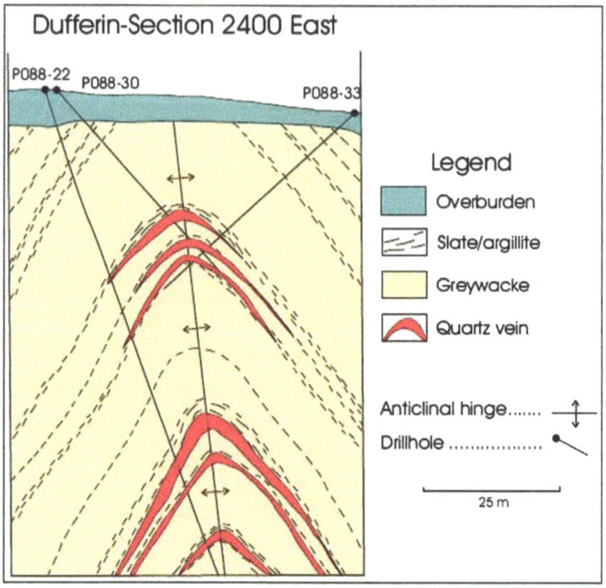

FIGURE 4.5

Stacked reefs at the Dufferin Mine, Salmon River gold district. Composite section through vertical shaft. Section shows exploration mine workings and the result of some of this exploration. A mine tunnel does not necessarily indicate gold ore. Remember exploration drilling was unknown or rare prior to the early 20th Century (Malcom, 1912).

Notable features that have a bearing on genetic models are (1) the consistent lateral continuity of bedding-parallel veins, which ranges from hundreds of meters to >1 km, and (2) the stated preference for hinge regions of regional folds. These are discussed below and the structural implications are analyzed in Chapter 8.

4.1.2 Vein Textures and Meguma Vein Origins Based on Textures

Vein textures are complex, and all formation models use as a basis each author's interpretation of textures and timing of vein emplacement relative to folding. Kontak and Horne (2010) and Kontak and Smith (1993) present a historical overview. Vein types are described by Haynes (1983, 1984, 1986, 1987).

A hydraulic pressure process is invoked for the buckled or parasitic nature of veins, and, with buckling, reflects layer-parallel shortening prior to regional folding (Graves and Zentilli, 1982; Henderson et al., 1986, Sangster, 1990). On the other hand, Henderson et al. (1990) interpret laminated veins as pre-folding fossil water sills. Haynes (1986, 1987) interprets laminated texture and dark columns of quartz in a matrix of white crystalline quartz as being

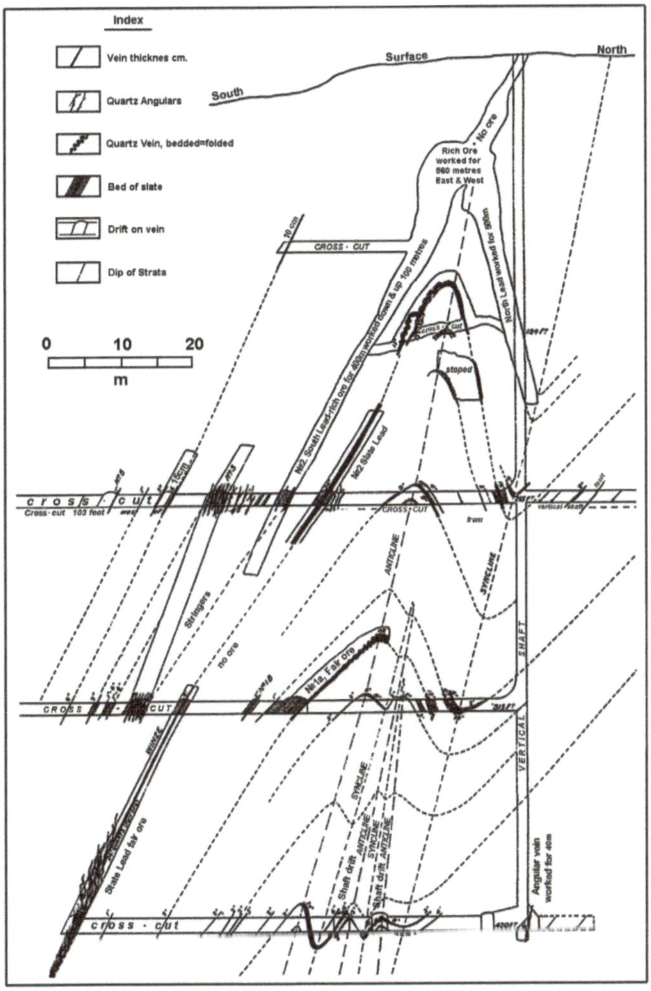

FIGURE 4.6
Beaver Dam deposit, east wall of 1100 cross cut, showing both stratiform and cross-cutting nature of quartz (Malcolm, 1912).

of organic origin due to sub–sea-floor hydrothermal hot springs (for Kreschmar's interpretation, see discussion of the "Rosetta Vein," Section 7.9, Kretschmar, 1984). Early prefolding vein formation and stratigraphic controls also underlie lateral secretion models of Graves and Zentilli (1988) and Ryan and Smith (1998). They and McBride (1978) argued for a stratigraphic control on mineralization. McBride did not see any evidence that deformation had any effect on the mineralized quartz bands other than that expected from the deformational history of the rocks throughout the Ecum Secum Map Area (McBride, 1987). Probably the best evidence for the gold being deposited

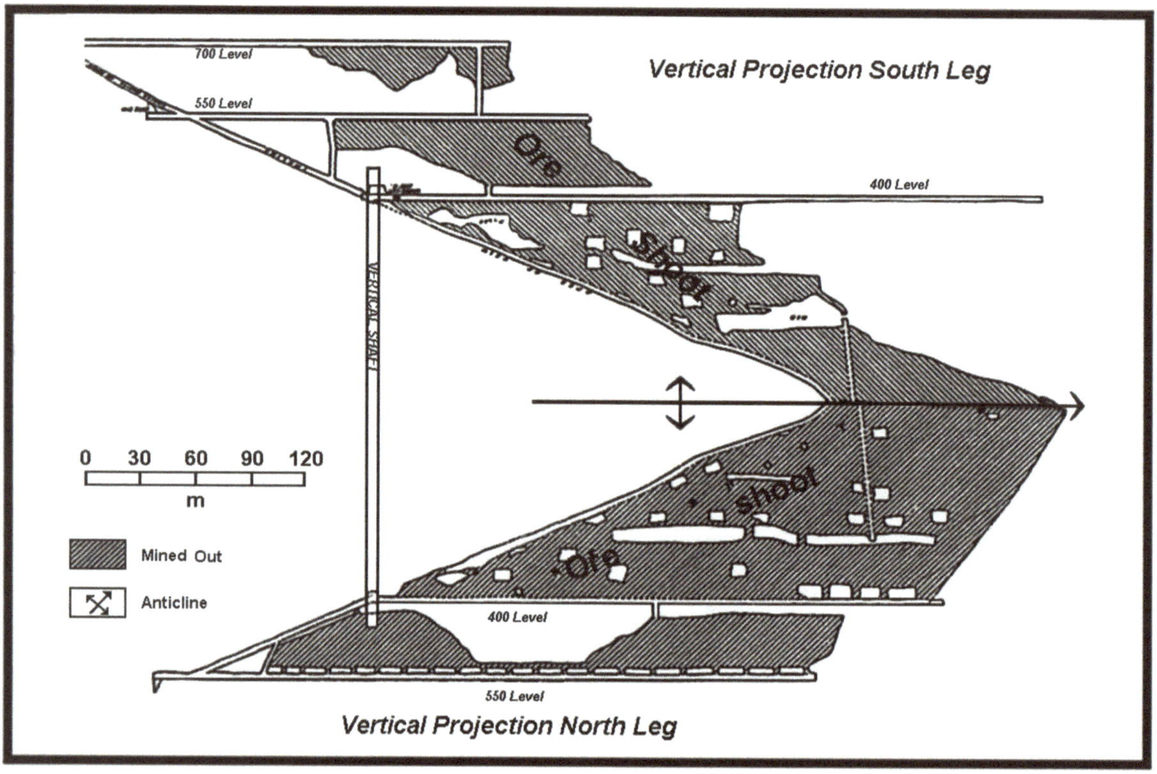

FIGURE 4.7

Richardson Vein, Upper Seal Harbour, N. S. showing the difference in rake of the conformable gold horizon across the hinge of the anticline. (Figure 14, Malcolm, 1912, p 246.)

syngenetically and the band being preserved during deformation is the Richardson Mine (Figure 4.7, Figure 14 from Malcolm, 1912, page 246). On the north limb, the mineralization rakes east at a steep angle of about 60°; whereas on the south limb, its rake is a much shallower at about 30°. In the hinge of the fold below the 400-foot (130-m) level, there is no economic gold mineralization in the hinge to fold, but there is down the limbs. There are two possible reasons for this distribution: the first is that the mineralized band is folded obliquely across the anticline as a ribbon on a surface; the other is that there are two remobilized bands coming from different sources and fortuitously meeting on the crest of the anticline. The lack of ore at depth on the fold hinge precludes the second possibility. Later, McBride presented evidence that the deformation of rocks during regional metamorphism did not remobilize either metals or minerals out of their depositional environment (McBride, 1991).

Synfolding models for vein formation were based on the assumption that vein fluids were of metamorphic origin, that veins cut cleavage, entrained cleaved wall rock fragments, were confined to hinge zones of regional folds, and were deformed (Taylor and Schiller, 1966; Keppie, 1976; Mawer, 1986, 1987). A composite origin with some veins forming early in the deformation history, but the bulk of the bedding concordant and saddle reefs late with respect to folding history, was proposed by Williams and Hy (1990). Kontak et al. (1990a) also propose a late-stage origin, because the veins were confined to the steep, highly-strained parts of the folds. Horne and Culshaw (2001) noted that veins were emplaced synchronously with late fold tightening, during periods of high fluid pressure in concert with flexural-slip along steep bedding planes. The major problem with the above models is that they ignore the structural history. All units in the Meguma have been folded by a ductile deformation that has produced tight anticlines and more open synclines. Faribault's years of mapping in the Nova Scotia gold fields clearly shows this. McBride's (1978) mapping in the Ecum Secum Map Area demonstrates the pattern and shows that the mineralized stratigraphy is very much a part of these events. Malcolm's 1912 summary shows many photographs of these structures where they include mineralized and unmineralized conformable quartz bands (Chapter 8). The pattern of these parasitic and major folds is a classic fold system, and the slatey cleavage is axial planar to all the folds, major and parasitic; in laminated veins, this slatey cleavage can be seen in the internal layers of slate and slate inclusions. All contacts are sharp. Considering that these rocks were unconsolidated sands, muds, and sand and quartz layers, it is unlikely that deformation did anything but deform what was already present. Any mass movement of fluids and elements would have destroyed the primary textures.

A magmatic origin for the veins was proposed by Newhouse (1936). Age dating of granites and some veins established similar age of 380 Ma (MacDonald and O'Reilly, 1989). This was considered evidence for a magmatic origin of some veins by Kontak et al. (2004). Subsequent Re-Os dating on arsenopyrite and 40Ar/39Ar from several deposits (e.g., Beaver Dam: Kontak and Smith, 1993) indicates at least one instance of a 408-Ma vein age. These bimodal ages coincide with two major tectonothermal events in the Meguma terrane. McBride, on a field trip along the south side of Chetabucto Bay, observed a granite just west Halfway Cove that had a pronounced foliation parallel to that in the enclosing Meguma rocks. This observation suggests that some of the granites predate the Devonian deformation.

4.1.3 Alteration, Mineral Chemistry, Stable Isotopes, and Age Dating

Key data on age dating, mineral chemistry, and stable isotopes for the Meguma deposits with particular reference to the Beaver Dam deposit is summarized in Kontak and Smith (1993) and Kontak and Horne (2010). Aspects relevant to discussion of syngenesis versus epigenesis are discussed in the following sections.

Kontak and Horne (2010, p. 75) cite evidence for vein formation late in folding history that includes the following: (1) veins are confined to domal structures; (2) vein development was controlled by a flexural-slip folding mechanism, which is favored when and where fold limbs are steep; (3) discordant veins cut folds; (4) fragments of cleaved wallrock are found within the veins; and (5) at Beaver Dam and Mooseland, veins cut hornfelsed granite aureole. They suggest the absolute timing is confined by model and isochron Re-Os ages on arsenopyrites, giving 408 Ma and 380 Ma for the Ovens and Dufferin deposits, respectively. This and other dating of veins demonstrate to them that there were two distinct ages of vein formation, one at 408 Ma, which is coincident with regional metamorphism, and the second at 380 Ma, which overlaps emplacement of granites within the Meguma terrain.

There is some apparent inconsistency regarding the veins at Beaver Dam (BD). Kontak and Horne (2010) state that there is no evidence of heating to higher temperatures at Beaver Dam after vein formation, and that similar ages of 370 Ma were obtained for amphibole, muscovite, and biotite. The large differences in blocking temperature between these minerals are consistent with a single thermal event of short duration (Kontak and Horne, 2010, p. 510) for Beaver Dam. It is also stated that at Beaver Dam, which is within 1 km of an exposed granite intrusion, quartz veins contain more complex silicates and sulfides, for which these researchers point out a similarity to intrusion-related gold deposits.

Bedding-concordant veins are considered to be composite in origin and to have formed in two distinct episodes that overlap in time. The crack–seal or laminated quartz is early and reflects extreme silicification of adjacent wallrock. Advanced stages of crack–seal texture and superposition of stylolitic textures reflect the dynamic environment is which these veins formed (Kontak and Smith, 1993, Figure 5); or, rather, does it simply reflect features of the overprinting deformation?

Vein mineralogy consists predominantly of quartz, carbonate, chlorite, rare tourmaline, and Cu-Fe-As sulfides. Associated wall rock alteration varies from absent to cryptic to intense. Kontak and Horne (2010) describe the following alteration phases adjacent to veins in order of decreasing abundance: arsenopyrite, pyrite, pyrrhotite, carbonate, silica, muscovite, biotite, tourmaline, and rare garnet and apatite. They state that there is a strong correlation among gold, arsenic, and sulfur in both vein and wall rock samples. It is difficult to see this as alteration, because the coarse arsenopyrite crystals are disseminated throughout certain slate beds without any evidence of solution ways.

Fluid inclusion studies show that a dominant fluid type in the auriferous quartz veins is a mixed $H_2O-CO_2 \pm CH_4$ fluid with X $(CO_2) = 0.05-0.15$ and with salinities of 5 wt% NaCl.

The stable isotopes are as follows: δS34 variation intradeposit is minimal δS34 < 1−2 0/00, which indicates that the fluid was buffered, whereas interdeposit variation is large, from δS34 + 9 to δS34 + 25, with a consistent upward increase through the stratigraphy that mimics the δS34 signature of the wall rock (op. cit. Figure 4.8). δD values are typical of metamorphic fluids, but with some overlap with a magmatic reservoir. For Sr, there is a large spread for many deposits that suggest that there has been wall rock influence and that the fluids must have originated from an underlying reservoir. The lead

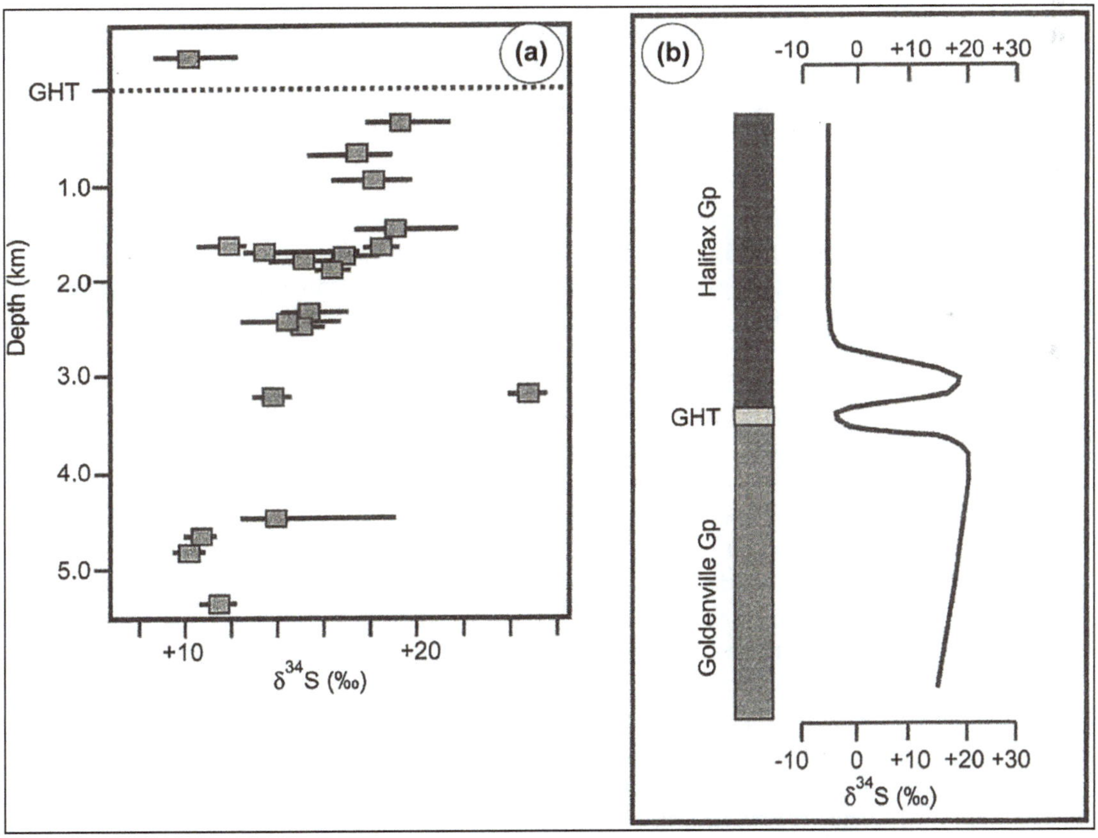

FIGURE 4.8

Illustrates the pattern described in the Goldenville Formation, which appears to mirror sedimentation parameters. A similar systematic increase in δS34 shown by arsenopyrite is attributed to ocean anoxia. δO18 and δC13 variation in carbonates are considered an artifact and not a primary signal. Both carbon and sulfur are considered to have been derived from Meguma metasediments (Kontak and Horne, 2010).

isotopic signature of the vein fluids indicate mixed reservoir for Pb, with end members equating to those of reservoirs that correspond to the Liscomb Gneiss and Meguma Supergroup. Combined Pb and Sr isotopic data reflect extraction of Pb from crustal reservoirs by a long-lived hydrothermal system, which accounts for the homogeneity of the data.

At the Beaver Dam deposit, there is a wide range of Sr and Y contents and a large positive Eu anomaly from carbonate and plagioclase separates (Kontak and Smith, 1993, Figure 10). Arsenopyrite shows a limited composition range that they interpret as mineral equilibrium between pyrrhotite–pyrite and arsenopyrite. Calcite does not fractionate the REE and chondrite-normalized patterns for this mineral, but must reflect changes in REE content of the vein-forming fluid. Minerals in equilibrium with respect to REE chemistry and local wallrock with a pattern typical of shale are not the source of REE. For Beaver Dam, Kontak and Smith also state that mineral paragenesis and chemistry suggest there was a single hydrothermal event that resulted in the formation of two vein types that represent a continuum, and that probably reflect the normal progression of a hydrothermal system, which includes thermal and chemical equilibrium with the surrounding Meguma metaturbidites.

Constraints on vein formation models have been suggested by Kontak and Smith (1993, p. 515) for the following reasons: (1) the syngenetic exhalative model of Haynes is incompatible with vein mineralogy and textures as discussed by Mawer (1986, 1987); (2) multiple episodes of vein formation as suggested by Kerswill (1988) are unlikely, because Kontak et al. (1993) suggests a single vein-forming event; (3) the suggestion of Sangster (1990) that the source of vein constituents is the Goldenville Formation is untenable given the REE data, which does not show a similarity to the North American Shale Composite (NASC) of Gromet et al. (1984); (4) the Goldenville–Halifax transition is not considered to be an important source of metals, because there are no Mn-rich minerals in the veins; (5) the isotopic data do support a source other than the Meguma and fluid, which was subsequently modified during ascent and reaction with wallrock; and (6) the presence of delicate zoning in many of the silicate minerals is consistent with vein formation during the waning stages of the Acadian Orogeny, rather than a pre- or synmetamorphic origin (Kontak and Smith, 1993, p. 516).

4.1.4 Discussion

The interpretation of timing and textures is central to lode vein formation models and it is fortunate that there are excellent glaciated seashore exposures where we can debate the fine points of quartz and turbidite deposition. These textures are compellingly in favor of formation at the same time as the enclosing rocks. Although this may be dismissed as a subjective opinion, the authors think, in part, that the "enigmatic" nature (Sangster and Smith, 2007) of the Meguma deposits is the result of K/Ar and Re/Os dates of veins that give ages that are younger than those of the Meguma. If the quartz veins formed during sedimentation, how is this possible?

In part, the answer lies in the details of the dating technique and sampling, since Poole (1971) reports K-Ar of 475–500 Ma on detrital muscovite from the Goldenville Formation, and Farquhar and Haynes (1986) report model ages of 550 ± 50 Ma from Pb isotope dating of galena from the earliest stratiform veins, which is remarkably close to the deposition age of the Goldenville.

Also, fluid inclusion formation pressures appear to indicate mesothermal formation depths (Mona Sirbescu, Dan Kontak, Kingsley Burlinson, personal communication). What one may conclude from fluid inclusion data and quartz formation is discussed in Chapter 7.5. Similarly, some perils in interpreting age dates are discussed in Chapter 7.

Virtually all data can be interpreted from whatever perspective one brings to it. With a syngenetic perspective as a premise—and without either negating or ignoring the widely accepted structural and clearly articulated epigenetic views on their origin discussed above—.the authors want to further explore the implications of syngenesis (and a nongenetic role for deformation and metamorphism). A central question revolves around the P-T-x formation parameters of hydrothermal quartz and what formation mechanisms are suggested by textures. Equally important, is there anything in recent careful work on mineralogy, age dating, and stable isotope analyses that unequivocally mitigates against a syngenetic origin for the veins? To bring clarity to the nomenclature involving quartz, Kretchmar introduced the term quartz unit (QU) in Chapter 3 for clearly stratabound quartz (Kretschmar, 1983). This designation is retained throughout the present work.

The lateral and vertical continuity of veins reflects a constant, similar formation process, evidence for which are quartz textures from drop-shaped to millimeter-sized veins, to veins several meters thick. The synchronicity of vein formation indicated by cross-cutting relationship of all vein types is consistent with a singular formation model (Kontak and Smith, 1993). Preferential quartz vein formation in hinges of folds is only apparent, since that is what we see in outcrop. McBride's (1978) mapping demonstrated that the favorable section of the Meguma stratigraphy continues around the Ruth Falls syncline and that the nearest gold deposits are exposed on the Goldenville anticline some 10 km to the north on the other side of a major syncline (Figure 4.1). This area was not investigated for historical pits or trenches.

The entire Meguma is deformed, and if there is a preferential location for veins and gold districts, the loci of folding may reflect synsedimentary faults, seeps, depressions, or alterations that have weakened the strata. The resulting laminated veins (QU) that occur as saddle veins at the crests of folds that are no different from any other laminated veins on the limbs. Breccia-textured veins may represent cross-cutting feeders or veins that have slumped downslope. Lateral variation in lode veins is typical and reflects emanation of fluid from a point source, seep, or fault, and secular variation in P-T-X parameters (Chapter 7). Changes in vein textures along strike do not necessarily reflect different degrees of dynamic recrystallization as suggested by Kontak and Smith (1993, p. 475), but are highly likely to reflect primary temperature and compositional variation (See discussion of ore deposition, Chapter 9). Vein textural complexity most likely reflects proximity to source vents or seeps. A most important characteristic of stratiform laminated veins (QU) is their asymmetry.

In Goldenville and most districts, there is no thickening of veins in the hinges. If thickening in fold hinges is due to deformation and lateral secretion, then all veins should be thickened in all the fold hinges. The Dufferin deposit (Figure 4.5) seems to be an exception, but vein thickening in a hinge zone may reflect original vein thickness and would not appear to be generally applicable to turbidite-hosted vein deposits.

The deposits occur essentially throughout the Goldenville Formation in slate, greenschist, and amphibolite facies metamorphic rocks. On a macro-scale, gold deposit location shows no apparent preference (Kontak and Horne, 2010). δO18 of quartz veins from gold districts, plotted as a function of stratigraphic position, shows an essentially systematic and gradual increase upward.

Figure 4.8 illustrates this pattern, which appears to mirror sedimentation parameters. A similar systematic increase in δS34, shown by arsenopyrite, is attributed to ocean anoxia. δO18 and δC13 variation in carbonates are considered an artifact and not a primary signal. Both carbon and sulfur are considered to have been derived from Meguma metasediments (Kontak and Horne, 2010).

Stratigraphic control in Goldenville is indicated by mineral variation between veins (Goldenville, Kontak and Horne, Figure 4.8), with each vein apparently characterized by a specific mineral assemblage. There is a concentration of occurrences at >1500 m stratigraphic and >2000 m below the upper Goldenville Formation contact, but these occurrences are not on the same stratigraphic horizon (Sangster and Smith, 2007, Figure 3). McBride (1978) reported a 200-m-thick pyrite-iron carbonate-rich sequence 1300 m below the Goldenville–Halifax contact, with several small gold occurrences for which he proposed a syngenetic origin. This pyrite-iron carbonate sequence can be traced for more than 80 km across four anticlines and synclines (Chapter 8, Figures 8.22–8.24).

FIGURE 4.9
Part of anticline Coxheath Mine showing parasitic folding with an axial plane slaty cleavage, McBride Photo, 1989.

If the Meguma Supergroup formed in a rifted basin (Waldron et al., 2009), perhaps the rifts are the locus of hydrothermal vents and seeps, and reflect subtle anisotropy in rock strength. How long did it take for the >10-km-thick Meguma to accumulate? If vein ages reflect time of original sedimentation, ages for Moose River should be older than for West Gore (neither one is currently dated), and Dufferin (380 Ma) should be older than Ovens (400–405), which is not the case (Kontak and Horne, 2010, Figure 4.9 and Sangster and Smith, 2007, Figure 3, shown as Figure 4.2). Although this discrepancy provides a valid argument for a younger episode of vein formation, the possibility that vein ages have been reset during granite intrusion and metamorphism is very real. A careful study of vein ages as a function of stratigraphic height and proximity to granite contacts and metamorphic grade might provide further insights.

4.1.5 A Syngenetic Model for Meguma Lode Veins

Integral to any genetic gold model is evidence from the Meguma, which, in a simplified form, the authors see as follows: Heat is provided by the subsiding basin and elevated geothermal gradient attending an extensional tectonic regime that existed along the NW margin of Meguma Terrane in SW Nova Scotia over a period of about 150 million years (White and Barr, 2004).

Silica-rich fluids are generated from dewatering, lithification, and heating of the turbidite sequence. If the underlying source region lithologies (either Meguma or the Liscomb gneiss) are pyritic, reducing, and gold-enriched, these fluids will be gold rich. Fluid generation is likely more or less continuous, and silica and other fluid constituents are leached as the host lithology is heated to above temperatures at which quartz will dissolve. These fluids migrate upward and emanate from seeps or faults. Seepage explains the paucity of cross-cutting root veins, minimal alteration, and district-wide carbonation. Ponding of dense saline silica-rich fluids (Chapter 9 and Turner and Gustavson, 1978) was ubiquitous. The rate of fluid flow and silica (gold, base metal) concentration in the brine and physical sedimentation determine the nature of the vein. Probably rapid outflow of supersaturated solution results in thick, unstructured mats of silica gel, in sea-floor shallow depression: the gel dehydrates to quartz. Slower precipitation from more dilute solutions in a distal environment results in laminated veins (QU). Bottom current strength and their nature play a significant role in determining vein types. If either fluid flow or bottom currents are significantly turbulent, silica will be dispersed. The nature of episodic influx of detrital material from turbidity currents and, in particular, the deposition environment further determines vein textures and significantly, the potential mining economics. It has been assumed in most studies that the veins are mineralized throughout their extent. Chataway (personal communication, 2015) investigated the Goldenville deposit in 1988–1989. He obtained old records that showed the location of day-to-day face sampling in the mine, which demonstated that grade control was done by observation of gold grains in the exposed face. After a few new faces failed to show gold, the miners moved to a new location. This work was confirmed by Hedley (1941). The importance of this historical information is that it shows the extent of the economic part of the vein and not the limit of the vein. A very important observation that demonstrates that looking at the veins alone does not show the total gold distribution, only what was economic at the time. The originals of this database are available in the Nova Scotia Department of Mines files in Stellarton, Nova Scotia.

Some cross-cutting veins are feeders of fluids to overlying strata. The complexity in vein textures increases with the intensity and duration of hydrothermal fluid flow. Some quartz textures reflect synsedimentary compaction and deformation. The degree to which metamorphism and deformation have affected vein textures is not entirely clear. Neither plays an essential role in the suggested sea-floor origin. Textures are primarily a reflection of complex system of primary sea-floor processes and are minimally affected by structure and deformation. Some veins are clearly later (Haynes, 1984;

Farqhuar and Haynes, 1986), and the scale of observation is important. Bob Chataway told the authors that all mining took place in bedding-parallel veins; cross-cutting veins were present, but did not contain sufficient gold to be exploited.

Folding of the Meguma is the result of regional shortening. Other areas of folded turbidite sequences do not host gold districts; therefore, deformation of turbidites is not causative of gold vein formation in itself. McBride visited the Coxheath Mine in 1989 and was able to photograph the gold-bearing vein where it crossed an anticline. The vein shows parasitic folds across the anticline; the axial plane slatey cleavage is visible in the slatey host, is axial planar to the minor folds, and can be seen in the folds except for the quartz; quartz is a poor recorder of deformation but may show undulose extinction, fragmentation, preferred lattice, and grain orientation (Spry, 1969, Plate XXVI). These basic structural geological relationship shows that the entire sequence was folded together by the single deforming event. The quartz vein clearly predates the deformation. The deformation does not do anything to the vein except fold it. There is not any evidence of remobilization (Figure 4.9).

Vein ages may reflect accumulation time of the Meguma during prolonged subsidence, increased lithospheric loading, and its formation in a rifted basin (Waldron et al., 2009). K/Ar and Re/Os ages are reset by granite intrusion and regional metamorphism. Temperatures derived from the arsenopyrite geothermometer are unrealistically high, judging by their textures. Some arsenopyrite is diagenetic or synsedimentary, and almost all crystals are zoned from S-rich centers to As-rich rims. Zoning reflects the availability of S and As, and growth kinetics. Isotopic disequilibrium between vein and wall rock reflects generation of fluids at depth, and discharge into a sea-floor with a different isotopic signature.

4.2 GOLD IN GREENSTONE BELTS

Conformable gold deposits are known in many greenstone belts in Canada and around the world. The authors have studied various gold deposits in Ontario, including the Swayze belt situated approximately 80 km north of Sudbury, Ontario. Chester Township and surrounding area has been the target of gold and some copper exploration for over 80 years. The results of these exploration projects form part of the following discussions. This greenstone belt differs from the average by having considerably more felsic material, especially in its eastern part, centered in Chester Township.

4.3 THE GEOLOGY OF CHESTER TOWNSHIP

4.3.1 Introduction

The geological relations among gold-bearing lode veins, felsic flows, and lamprophyres in Chester Township (Twp) are integral to understanding the synge- netic aspect of lode gold veins and represent key stepping stones for the main thesis of this book. Evidence for the following will be presented: (1) the Chester "granodiorite" or "trondhjemite laccolith" is a volcano- sedimentary complex; (2) graded bedding and facing direction indicators in tuffs, felsic epiclastics, and Gold Cycles within felsic "intrusions" permit unraveling of the regional stratigraphy and structure; and (3) of the two types of lamprophyre dikes described, the more common one represents Gold Cycle "E" (GCE) lithology; the other is a late-stage, cross-cutting biotite minette, which was subjected to deformation and greenschist facies metamorphism during regional deformation and isoclinal folding; (4) a very silica-rich rock described as "alaskite" may represent an early pulse of silica-rich hydrothermal fluid that presages and accompanies gold mineralization. Gold, minor base metals, and quartz are deposited from hydrothermal fluids that either accompany exsolution from felsic flows or emanate from contemporaneous sea-floor hydrothermal vents and seeps.

To place the discussion of lamprophyres and felsic volcanism into context, a brief overview of historical views of the "intrusion" and the general geology, structure, and gold mineralization in Chester Twp is warranted. Specific aspects of felsic volcanism as revealed by geological mapping, and whole rock and trace element analyses, are discussed later in Chapter 6. Chapter 5 deals with the significance of lamprophyres.

This exploration camp contains numerous prospects and several small mines with historical production including the Jack Rabbit, Chesbar, Gomac, and Young Shannon. The exploration was centered on small, high-grade veins. Recently, exploration was expanded to seek large open pit potential. Iamgold Inc. (www.iamgold.com) is currently actively investigating the area and Côté Lake. A multimillion-ounce gold deposit (minimum 131 million tonnes, grading 1 g Au/T) is being assessed for production.

4.3.2 Chester Complex, Chester Township, Ontario

The "Two Lamps" outcrop (Chapter 5) that shows the relationship of cross-cutting lamprophyres, lode gold vein,s and their host felsic flows and the Côté Gold deposit are all located in Chester Twp, Ontario, within a geological unit that the authors refer to as the Chester Complex (CC). Previous workers refer to it as Chester Intrusive Complex (Kontak et al., 2013), or Chester Granitoid Complex (Siragusa, 1993). The geology of Chester Twp, the location of the "Two Lamps" outcrop, and the other areas discussed are shown in the compilation (Figure 4.9). A detailed summary of previous work is warranted, since Kretchmar's mapping has shown that this unit consists of interbedded felsic and basalt–andesite flows and pyroclastics. It is not an intrusion (Kretschmar, 1999, 2011), a finding substantially at variance with published accounts. This claim is based on graded bedding in felsic pyroclastics, the fact that one can correlate lithological units and mineralization over hundreds of meters, and the presence of gold cycle and Gold Cycle "E" units (GCE), which permit unraveling of structure (McBride, 2002, 2010).

The Chester Complex is located in the Archean Swayze Greenstone Belt, parts of which were originally mapped by Laird (1932) and Furse (1934). More recently, the entire volcano-sedimentary rocks became known as the Swayze Belt (Heather and Shore, 1999). These rocks extend in an E–W direction for more than 60 km from Brunswick through Yeo, Chester, Osway, and Benton Twps and are part of the Abitibi Greenstone Belt. The Belt varies from about 6 to 1.5 km in width. It is bounded on the north by apparently younger volcanics of the Marion Group and on the south by granite-diorite complex. Northwest-trending Proterozoic faults and diabase dikes dissect it. Metamorphic grade is greenschist facies. The majority of numerous gold showings, several small gold mines, and the large low-grade Côté Lake deposit occur in Chester Twp between Clam and Cote Lakes.

In Osway Township, approximately 20 km to the west was the largest producer, the Jerome Mine in the Swayze Belt. This block is tilted, units dip 50–60°N, and, as will be seen from detailed core logs presented below, strata are overturned with facing directions to the SW.

On the original map of Laird (1932), the rocks are mapped as granite and diorite, granodiorite, alaskite facies, lamprophyre dikes, basic inclusions, and gold-bearing quartz veins. Laird noted variation in textures from granular to porphyritic that resemble meta-arkose or quartz porphyry and inclusions of older rocks. Siragusa (1981) called the same felsic rocks: batholithic intrusive trondhjemite and granite with some phases containing hornblende pods. He describes lithologies as medium- to coarse-grained potassic and sodic granite and fine-grained to aphanitic trondhjemite sheets. Lamprophyre dikes are mapped separately. Heather et al. (1996) describe the CC as a crudely stratified trondhjemite–diorite laccolith containing numerous screens and inclusions of mafic volcanic rocks. They regard it as a synvolcanic intrusion, emplaced along the Ridout Syncline, disrupting strata of the Chester Group. Subsequently, Heather and Shore (1999) described the CC as the Chester Granitoid Complex ranging in composition from biotite–trondhjemite to granodiorite. Heather (2001) describes lithologies as tonalite and noted that the volcanics to the north of the CC (Marion Group) have similar REE profiles. On OGS compilation Map 3511 of Ayer and Trowell (2002), the CC comprises tonalite, granodiorite, trondhjemite with intrusive diorite, gabbro, and melagabbro. U/Pb dating of zircon by van Breemen et al. (2006) from a leucocratic, quartz-rich biotite trondhjemite returned 2740 ± Ma. Berger (2011) mapped the eastern extension in Benneweis Twp and notes that the CC consists of biotite-bearing tonalite (<5%), diorite, and gabbro. He states that gabbro and diorite form elongate intrusions in the tonalite, showing complex textural relationships within the tonalite unit. Field relationships suggest commingling of units and coexistence of mafic and felsic magmas. Berger also notes the absence of a chilled margin to the CC.

Kontak et al. (2013), in a study of the Côté Gold deposit, describe tonalite as medium-grained, hypidiomorphic, and equigranular, with 40–50% quartz and 50–60% plagioclase. Fine-grained to slightly porphyritic textures with lesser quartz (<15%–20%) are rare. McBride worked extensively throughout the eastern part of the CC on the Jack Rabbit and Chesbar properties while

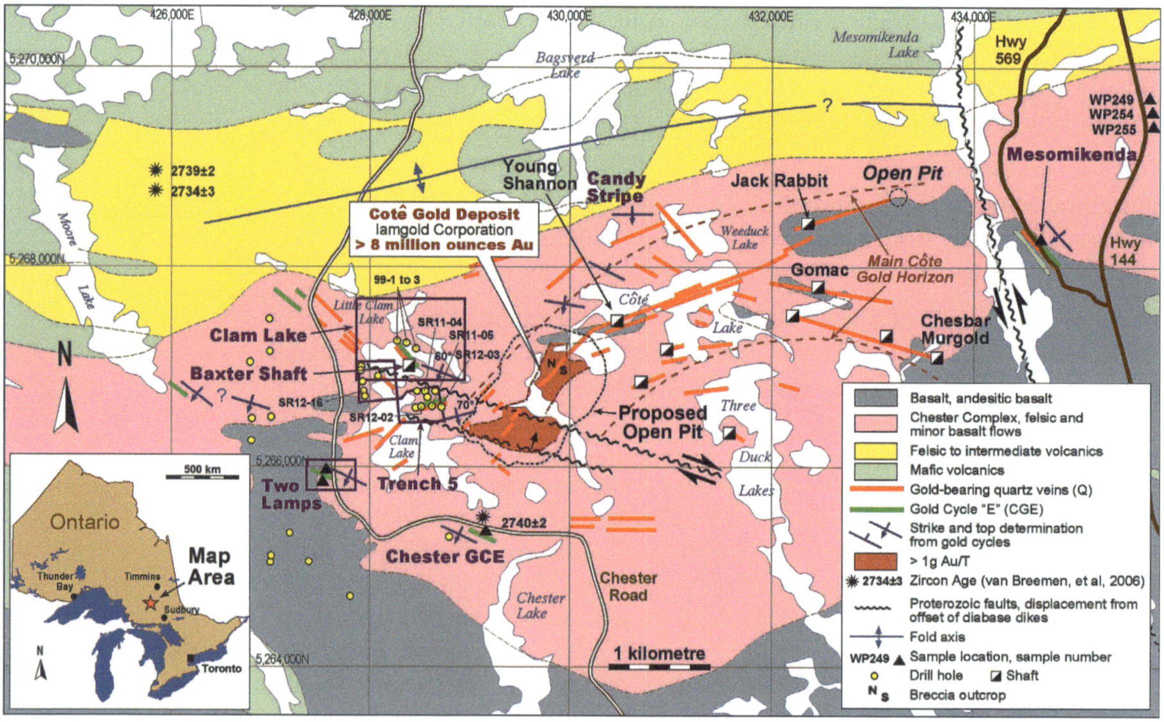

FIGURE 4.10

Compilation map of Chester Complex, Chester Twp, modified from OGS Map 3511 (*Ayer and Trowell, 2002*), data from Iamgold Corp. and the authors' mapping. Shown are locations referred to in the text for two lamprophyres, Chester, Mesomikenda GCE, Clam Lake, and Baxter Shaft drilling and Waypoint (WP 249, 254, 255) sampling on Hwy 144.

consulting with James Wade Engineering Ltd in the late 1980s and other clients in the early 1990s. He reviewed the exploration data for Northville Gold Corporation and Trelawney Mining and Exploration Inc McBride (2002, 2010).

4.3.3 Clam Lake Geology

Kretchmar's work involved geological mapping the central Chester Twp section of the CC, drilling, detailed core logging of 1100 m at the Baxter shaft at the north end of Clam Lake (Kretschmar, 1999 and Figure 4.8) and more recently, core from the east shore of Clam Lake (Figure 4.11), and core from various other locations in the CC. Some details of the 2012 work are summarized in Sanatana Resources Inc. news releases, accessible on the Sedar Website

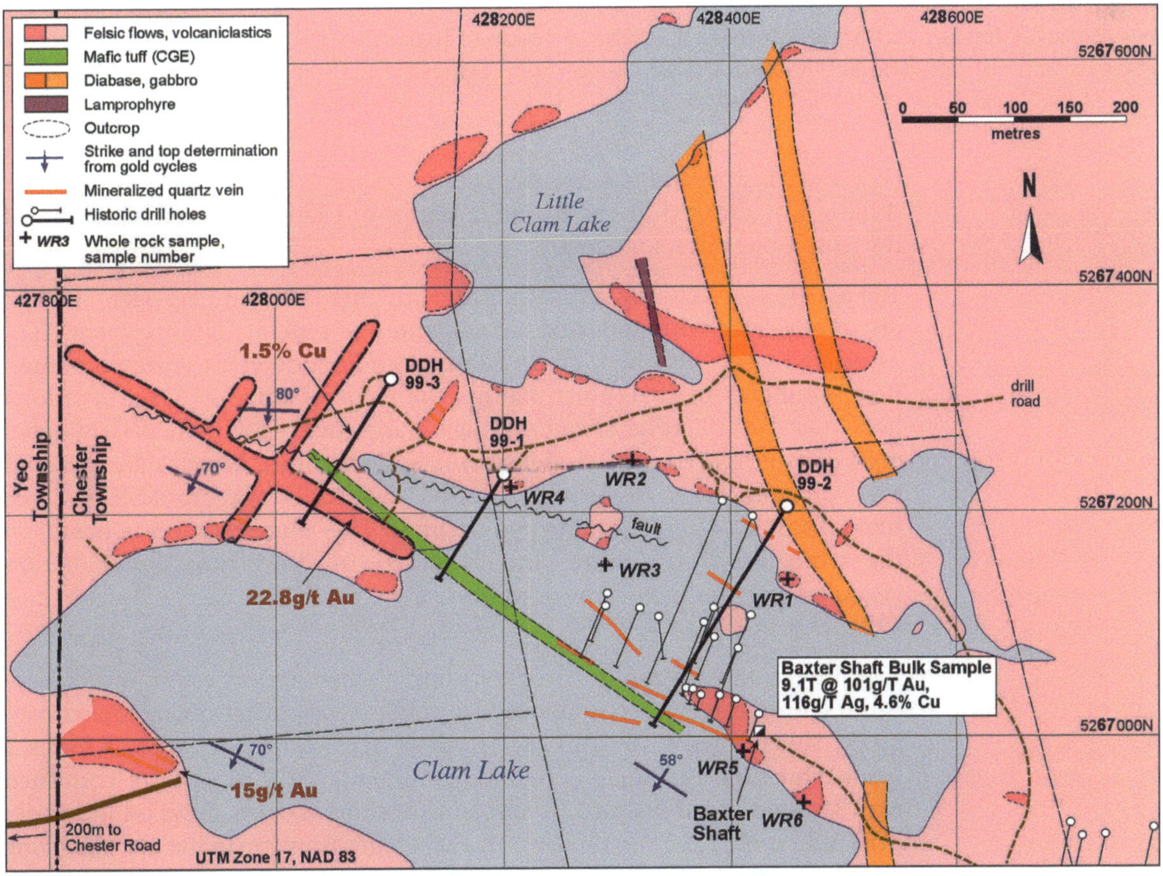

FIGURE 4.11

The mafic tuff (or chloritic GCE) and low-grade gold mineralization was traced for 450 m. Top determinations are from graded bedding in felsic epiclastics. Tops are to south and dips to north at 65°. The sequence is overturned. *Modified from Kretschmar, 1999.*

FIGURE 4.12

Baxter Peninsula, Clam Lake. Outcrop of felsic flows and GCE, quartz, and sulphides. Graded bedding in felsic units in drill core shows that sequence is overturned. High-grade gold from which a historic bulk sample (9.1 tons grading 101 g Au/T, 116 g Ag/T, and 4.6% Cu) was taken from quartz–sulfide vein in mafic interflow sediment (GCE).

(www.sedar.com) and www.sanatanaresources.com. The Sanatana 2011–2012 drilling was directed at the on-strike northward fault-displaced extension of the Côté Gold Mine stratigraphy.

The area is underlain by massive dacite–rhyolite flows and epiclastics. These are interbedded with andesitic basalts, and one type of "lamprophyre dike" associated with thin, high-grade quartz–sulfide lode veins represents interflow sediments or GCE units that are the focus of Chapter 5. Furthermore, host rocks to all mineralization (high-grade lode veins and disseminated gold and sulfides) are bi-modal basalts and rhyolites. Breccias are rare and have an autoclastic or peperite origin. The CC is not intrusive, but the remnant of a volcano–sedimentary edifice. This interpretation is based on the work described below and is based on Kretschmar's exploration (Figure 4.13). Felsic volcanism and the Côté Gold Deposit are described in Chapter 6.

The current analysis builds on 1999 drilling when andesitic tuff and graded-bedding epiclastics were encountered at the north end of Clam Lake (Kretschmar, 1999) and Figures 4.11–4.13. A 2-m-thick mafic quartz crystal tuff forms a marker horizon. Stratigraphy and discontinuous GCE units were correlated over 375 m. Graded bedding in tuffaceous sediments, felsic flows on the Baxter peninsula strike 120° and dip 65° NE. Facing directions are to the SW, signifying that the Baxter mineralization is on the south limb of an overturned anticline with its axis trending about 100°. Recent drilling confirms that the mineralization is stratabound and that units are overturned. Major lithologies are gabbroic

FIGURE 4.13
Clam Lake drill core from DH99-3-64 m. Typical felsic epiclastics and flows with rounded to subangular mafic clasts. Scale in centimeters. *Photo courtesy of Ken Reading.*

basalts, massive andesite, and intermediate tuff, felsic flows, and minor pink-to-red felsic feldspar porphyritic flows. Units are from 25 cm to 20 m thick and are separated by chlorite–carbonate GCE, thin intermediate and felsic tuffs, and "quartz units" or bedding-parallel veins, which vary from millimeters to 2 m thick. The GCE units are thin, discontinuous and difficult to correlate even over short distances between drill holes. However, each mineralized horizon has a specific precious metal, base metal, and magnetic signature, and shows clear evidence of bedding and systematic variation in gold content and sulfide mineralogy. A 400-m stratigraphic section was reconstructed.

The database for Hole SR12-03 comprises 429 multi-element chemical analyses and assays; scandium serves as an excellent inverse proxy for SiO_2 in instances when it is difficult to give fine-grained dark-green rocks a name during logging of the core. Scandium differentiates fine-grained dark green rocks; concentrations are 5–10 ppm for felsic flows, 20–30 ppm for intermediate flows and tuff, 30–45 ppm for massive medium to coarse grained basalt and gabbroic flows.

4.3.4 Jack Rabbit, Chesbar, Young Shannon, and Canorth Prospects

These prospects are situated in the CC to the east of Clam Lake and together outline a "V" pattern opening to the east with its apex near Clam Lake on the west (Figure 4.10). Exploration on the Jack Rabbit prospects in the 1980s followed up on previous pitting in the 1930s. Considerable stripping and drilling outlined three lenses within a zone, but none of this information is in the public realm. It is also unfortunate that the property was never geologically mapped. A summary is presented in McBride (2002, 2010) from numerous visits to the property. Mineralized zones are intermittently found along banded layering in of the host rocks. Individual lenses strike east–west, but their overall trend is closer to 60° and their dip approximately 60° to the south as a rusty band that step north in an en echelon pattern in granodioitic rocks. The banded strata extend beyond the property limits to host the 20 Zone of Chesbar and the C Zone of the Young–Shannon Mines. Zone 1's extension is the No. 20 Zone of Chesbar, which is well exposed and consists of bands of disseminated sulphides in a fine-grained felsic rock with angular blocks of a dark-gray quartz eye diorite. These zones consist of gold, pyrite, pyrrhotite, chalcopyrite, and quartz-carbonate in a light- to medium-gray granodiorite with blue quartz eyes.

The No. 3 Zone is considerably different; it occurs as a distinctive, highly schistose dark gray–to-black biotitic rock with carbonate and cherty quartz laminae carrying pyrite, pyrrhotite, chalcopyrite, and gold. This host rock is either a fine-grained mafic tuff or sediment; the strong schistosity indicates a high clay mineral content in the precursor rock.

4.3.5 Chesbar Property

The Chesbar Property is situated south of the Jack Rabbit in the southern arm of the "V" structure. No detailed geological map of the property exists. Most of the information comes from a 1990 study for Murgold Resources Inc. Two zones occur on this part of the property, the No. 1 and No. 3 zones. Gomac Mines exploited this vein in the 1930s with a small mill. It has since been probed by drilling, which outlined a steeply easterly raking 60-m lens along a structure that was traced for 800 m. South and parallel is the No. 3 vein, which lies in a band that has been traced by underground workings, drilling, and stripping for 1700 m. At the Bates Shaft at the west end of the Chesbar-Murgold system (Figure 4.10), this mineralization is parallel to compositional layering in the host rocks. This layering was observed to be infolded and was interpreted as tuffaceous. At the nearby decline, it is a schistose fine-grained pale gray rock with 5% clear-to-blue quartz eyes up to 3 mm in the band that is approximately 1 m wide. Pyrite and chalcopyrite are disseminated throughout the schistose host rock. It varies along strike down to 5 cm where the schistose host rock vanishes. At the Strathmore Shaft (most eastern), the mineralization is described as a schistose siliceous band with disseminated pyrite. Individual gold-bearing shoots plunge moderately to steeply west. The Strathmore mineralization seems to plunge moderately east.

4.3.6 Young–Shannon Property

Information on the Young–Shannon zones is taken from reports summarized in McBride, 2002 and a short visit to the property. Gold mineralization seems to follow both the Jack Rabbit and Chesbar mineralization directions and is suspected of being near the apex of the "V" structure. Most of the resources are under Three Duck Lake in a parallel series of south dipping veins, of which the C veins are to be the extension of the Jack Rabbit No. 2 and No. 3 Zones and the B veins are extensions of the Chesbar No. 1 and No. 3 veins. Cargill et al. (2009) state that the veins are composed mostly of the common sulphide minerals with gold, the limits of which grade into the wall rocks and are diffuse. Assaying had to be used to define the economic limits. Many parallel veins are found in the mine; they occur in both granodiorite and the underlying diorite, and are parallel to the contact between the two (Cargill and Gow, 2009, Figures 9.3 and 9.4). Quartz veining is usually absent.

4.3.7 Canorth Prospect

The Canorth prospect straddles Three Duck Lakes south of the Young–Shannon Mines and east of the Cote project structure and consists of gold-bearing sulphide bands in banded felsic and mafic rocks. It was well exposed by stripping a few years previously to McBride's evaluation in 1991. Gold mineralization is found east and west of the lake; it occurs as thin sulphide bands that can be followed intermittently across the stripped areas (Figure 4.14). These veins are parallel to layers rich in quartz eyes (Figure 4.15) or fragmental mafic volcanic rocks (Figure 4.16). Many parallel mafic and felsic bands are present in this area. The flattening of the fragments in the mafic bands suggests that this exposure is in the hinge of a fold. West of the lake, a large outcrop was investigated (Figure 4.17); it is a fine-grained, massive,

FIGURE 4.14

Thin bands of sulphide rich material defining the Canorth mineralization. The composition of the rock on the right of the photo is slightly different from that hosting the mineralization and on the left.

FIGURE 4.15
A quartz eye rock with many narrow bands containing connetrations of qauartz which are parallel to the sulphide bands, Canorth prospect.

featureless white felsic rock with 1-mm clear-quartz eyes. A combination of these elements indicates that this rock is a fine-grained featureless felsic tuff. Together these outcrops show the volcanic character of the Chester Complex and the disseminated nature of the sulphide bands making up the mineralization.

FIGURE 4.16
Contact between felsic fine-grained tuff and mafic volcanic agglomerate at gold showing on west side of Three Duck Lake.

FIGURE 4.17

Large outcrop west of Figure 4.16 which is composed of a very fine felsic matrix and 1 to 2 mm clear quartz eyes. This rock is very uniform and lacks any textures or banding.

4.3.8 Jerome Mine

The Jerome Mine lies approximately 20 km west northwest of the Chester Prospects in the Swayze Belt in a continuation of the same geology. It was the major producer in the area. Brown (1948) describes the vein as following the contact of footwall sedimentary rocks and the hanging-wall porphyry. It forms steeply west raking shoots and is limited by the depth of the porphyry that plunges approximately 30° to the east. Brown describes the mineralization as being composed of a bluish, cherty hanging wall and a white quartz-carbonate footwall. In 1989, McBride carried out a study of the underground workings; by this time, two mineralized zones had been defined: a south dipping north zone, and a north dipping south zone. Both had identical vein mineralogy, sedimentary footwalls, and porphyry hanging walls. The porphyry hanging wall is a felsic volcanic agglomerate sequence (Chapter 9, Figure 9.11). The importance of this agglomerate is obvious: rocks like this one were called porphyries in most of the older literature, when in fact they were volcanic tuffs and crystal tuffs. The mine geology is described in detail in Chapter 9.

4.3.9 Conclusions from Chester Area Field Studies

In conclusion, the Chester Area is part of the Swayze greenstone belt and seems to be composed of complex and discontinuous gold mineralizing systems. All the discussed prospects suffer from a lack of detailed geological mapping; thus, much field information must be gleaned from exploration records summarized in previous discussions (McBride, 2002, 2010). It is apparent from many of the observations that compositional banding has been mapped at all prospects. The rocks have too many volcanic characteristics to be considered multiple intrusions. This type of volcanic system contains many narrow horizons that are discontinuous but that are either too narrow or too low grade to have been economic by underground mining methods. They are not very common, but do form large low-grade gold mineralization systems in the Archean and Proterozoic of Ontario. McBride spent considerable time on a similar system called the Tower Mountain project west of Thunder Bay, Ontario and is familiar with the Dingman prospect in the Grenville of Southern Ontario. These prospects will not be discussed in this treatise, but much information is available in government files and on www.sedar.com for those who may be interested in more information.

4.4 GOLD VEINS IN THE MASKWA COMPLEX, MANITOBA

4.4.1 Setting and Location

The Little Bear Lake project is representative of early-stage exploration in the Canadian Archean Shield and lode veins in the Maskwa batholith, Manitoba. The work has revealed further insights into syngenetic aspects and leads to clear exploration guidelines for felsic "intrusion-hosted" deposits. Particularly significant is that the common mafic dikes, inclusions, and "lamprophyres" in the Maskwa batholith are GCE units (see Chapter 5). GC-facing direction determinations help in unraveling a major isoclinal fold structure, which in turn leads to questioning the nature of the host Maskwa "intrusion", Archean tonalitic (TTG) lithologies, and Archean crustal tectonics.

Kretschmar managed the exploration of a series of high-grade veins located in the Little Bear Lake area in the center of the Maskwa batholith in the southeast corner of Manitoba, near the center of an apparent TTG diapiric pluton, the Maskwa quartz diorite or Maskwa Lake Batholith (MLB), which separates the Cat Lake and Bird River greenstone belts (Figure 4.18) and underlies the Bird River volcanic terrain. It is located approximately 170 km northeast of Winnipeg, Manitoba. The geological context is provided by the work of Duguet et al. (2007, 2009), Gilbert (2013), and Yang and Gilbert (2014).

4.4.2 Geology of the Maskwa Lake Batholith and Its Gold Prospects

Its dimensions are a minimum of 60×30 km, with its western extension covered by Palaeozoic strata. The Maskwa Lake Batholith (MLB) is mapped as

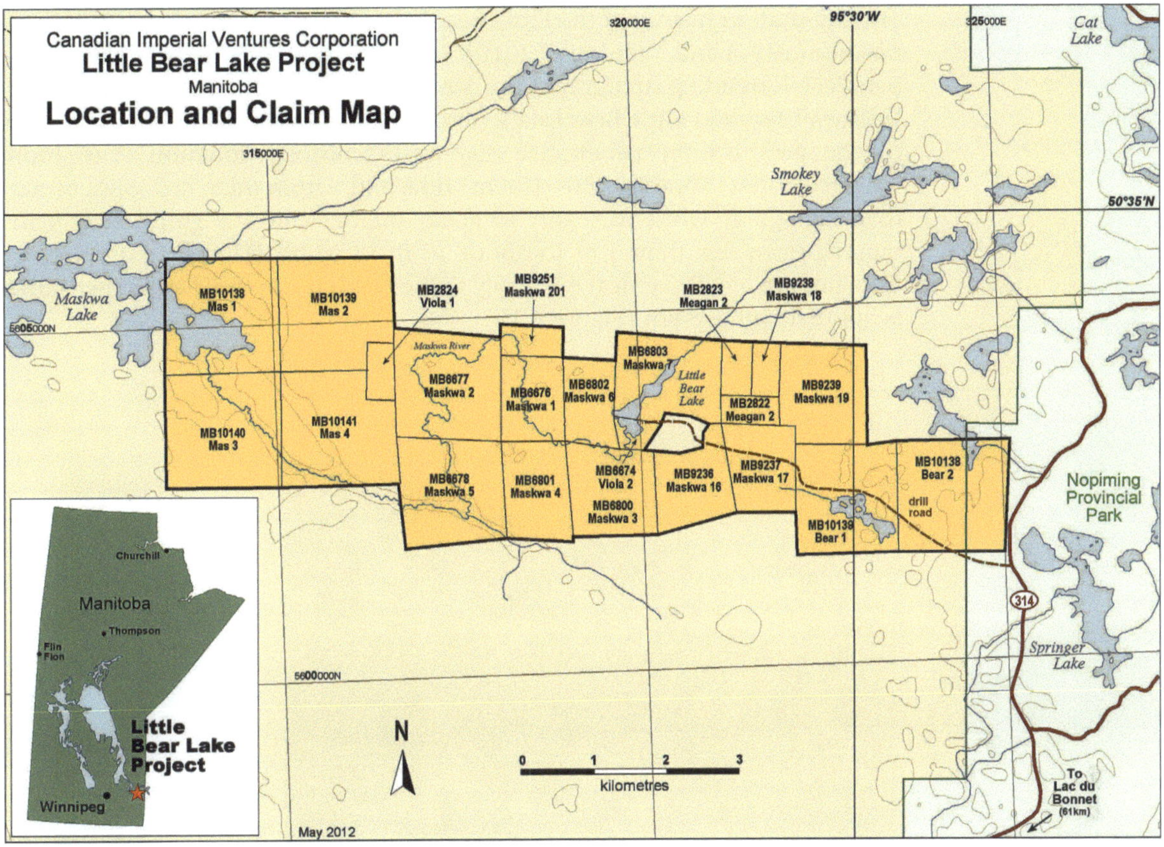

FIGURE 4.18

Location and claim map of Little Bear Lake Project.

white to pink, medium-grained microcline granite or tonalite with 1%–5% hornblende. Fine-grained mafic dikes (GCE of this study) are common. According to Duguet et al. (2007), new geochronological data on the MLB give a U-Pb zircon crystallization age of 2830 Ma, which brings into perspective its role during the collision between the North Caribou Terrain and the Winnipeg River Subprovince. This U-Pb zircon crystallization age represents the oldest lithology in the area. They suggest that the MLB acted as a small rigid microcontinent during this collision event and therefore it must be considered a distinct subprovince between the English River Subprovince and the Bird River Greenstone Belt. Preliminary map PMAP2013-7 (Yang, 2013) covers the Cat Lake–Euclid Lake area at the north margin of the MLB and mentions previous age dates of 2832 ± 0.9 Ma, 2844 ± 12 Ma, 2852 ± 1 Ma and 2737 ± 1 Ma for the MLB. There are no detailed maps to cover the entire MLB.

The exploration history of the Little Bear Lake area is well documented by Bernatchez (1997) and Therriault (2010). In the vicinity of Little Bear Lake, three parallel NE-trending stratigraphic horizons have been described (Stringer Lake, Maskwa Lake, and Little Bear Lake). According to Cerny et al. (1981), the rocks of the area exhibit penetrative east-west schistosity-foliation. The more "sheared zones" zones are 50–100 m wide, and within them are zones of narrow, high-grade white to sugary to glassy quartz–chlorite–gold–silver veins ranging from less than 1 m to about 18 m in width (Figure 4.20). Dips of the veins are mostly vertical or steeply to the north or south. Their known strike is approximately 8 km (Figure 4.19).

FIGURE 4.19

View (to east) of 1964 trench on Black Beaver Vein (person included for scale). Vein dips vertically; stratigraphic tops are to the left. *Photo courtesy of Ken Reading.*

Gold grades directly correlate with abundance of pyrite, galena, sphalerite, or chalcopyrite and pyrrhotite in the host quartz. Magnetite, molybdenite, and rare jamesonite (a sulfide of lead and antimony) have also been reported. Altered vein selvedges are commonly mineralized.

A detailed map of a small open pit at the west end of the Black Beaver Vein is shown as Figure 4.20.

This map shows the complex interplay of deposition of different kinds of quartz, pelagic-pelitic, and hydrothermal chlorite and sulfides. The units dip steeply, stratigraphic tops are to the north and a pink granitic unit (A3) at

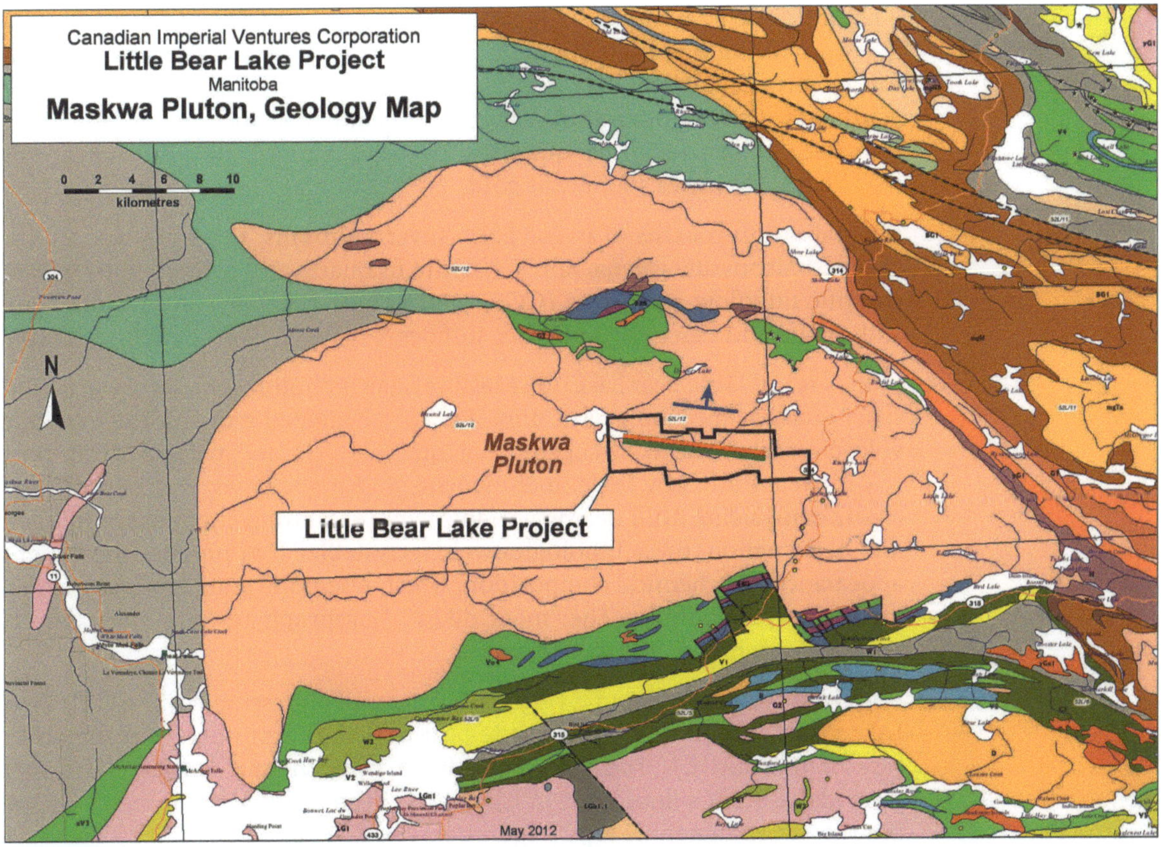

FIGURE 4.20

Little Bear Project. Maskwa Lake Batholith, Manitoba. Detailed geology in a trench at the east end of the Black Beaver (*Reid et al., 1986*). The vein is asymmetric. Quartz and sulfides (QU) occur preferentially on the north side of the vein and can be used for determination of facing directions and structural analysis. The mafic "E" division of the gold cycle (*Kretschmar, 2011; Chapter 3 and Figure 3.1*) is overlain by the next cycle of felsic crystal tuff or volcanic flow.

the bottom is a possible stratigraphic marker. The patterns demonstrate clearly that overall geological relations are not altered due to folding and metamorphism.

4.4.3 Discussion

Work in the Maskwa Lake Batholith reveals significant details about lode vein genesis and exploration potential for felsic "intrusion" hosted gold.

Numerous high-grade quartz–gold veins within narrow chlorite zones occur in more than 100 trenches and shallow exploration shafts. The veins, discovered in 1924, received sporadic exploration consisting of grab sampling, stripping, ground and airborne geophysics and airborne mag, EM, and limited drilling. The veins confirm existence of a high-grade vein corridor ("shear zone?") over a strike length of 8 km and a width of approximately 500 m. Drilling has been sporadic and shallow, and the line spacing used for geophysical surveys has not been suited to narrow, slightly conductive veins in a mafic host, within a highly resistant host rock of low magnetic response. In general, only quartz veins were assayed, despite indications that host rock carries significant gold.

Mineralization consists of native gold, silver, sulfosalts, tellurides, and minor pyrite, chalcopyrite, galena, and sphalerite in quartz and chlorite units within chloritic interflow sediments (GCE). Gold grade generally can be correlated with pyrite and percentage of other sulfides.

In the area of Little Bear Lake, the Maskwa Lake Batholith consists of thick feldspar–quartz–amphibole felsic units and coarse-grained probable intermediate volcanic rocks. The data suggest that quartz (QU) and pelagic-pelitic host (GCE) were buried by felsic, intermediate, and mafic flows, and minor pyroclastic equivalents from nearby volcanic vent (s). During later metamorphism and deformation, these lithologies were recrystallized, to produce the current appearance and chemical composition, which resembles granite, trondhjemite, tonalite, or granodiorite. The "mafic dikes", "lamprophyres", or "shear zones" represent original pelagic-pelitic units.

Both gold cycle types B and C (Figure 3.1) are well represented by the Little Bear Lake veins. A series of mainly pyrite-bearing quartz veins occur on the north side of the unit E1. These are considered hydrothermal quartz crusts that formed on the sea-floor (QU in Chapter 2). The resulting asymmetry can be used to tell stratigraphic tops (Figures 4.21 and 4.22).

The grade, thickness, and therefore economic potential of this type of mineralization is related to the interplay of at least three variables: the original gold concentration in the hydrothermal solution, the time interval available for its accumulation as quartz veins (QU), and the time between episodic deposition

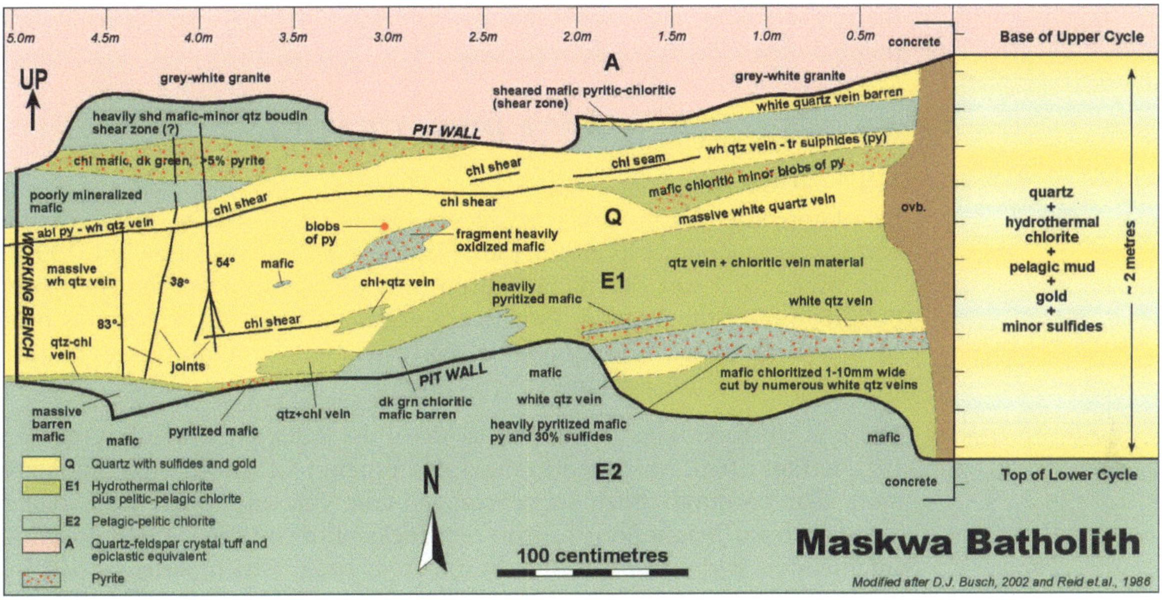

FIGURE 4.21

Drill Hole SF13-01 at 41 m. Upper rows: Unit 1. Bottom row: massive recrystallized sugary quartz to left of 41-m marker with pyrite overlies carbonate-chlorite GCE, with disseminated pyrite. Facing direction uphole from GCE asymmetry.

of clastic or epiclastic volcanic detritus (quartz–feldspar–hornblende crystal tuff in a turbidite-like unit or felsic flows by analogy with the Cote Gold deposit (Chapter 6).

A major implication of a syngenetic model of vein formation is that regional exploration can concentrate on favorable stratigraphic horizons and, within

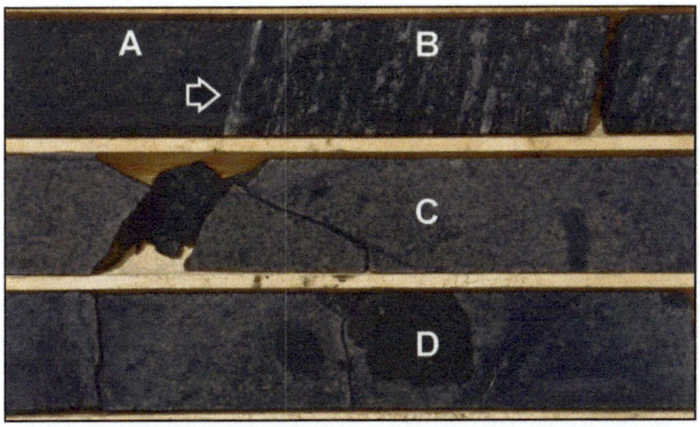

FIGURE 4.22

Drill Hole SF13-01 at 41 m. Upper rows. Unit 1. Bottom row, Massive recrystallized sugary quartz to left of 41 m marker with pyrite overlies carbonate-chlorite GCE, with disseminated pyrite. Facing direction is up hole from GCE asymmetry.

these, find thicker mafic strata and quartz veins (QU) that may reflect hydrothermal vent areas at sea floor fractures.

4.5 BEARDMORE AREA INVESTIGATIONS

4.5.1 Geology

Beardmore—Geraldton Ontario Greenstone Belt (GBGB) is typical of the Archean greenstone belts of the Canadian Shield (Card and Poulsen, 1998). The mineralized area belt occurs within the eastern portion of the Wabigoon Subprovince of the Superior Province. Since the mid-1930s, within this belt, the Beardmore-Geraldton Camps have produced 4.12 million ounces of gold from 19 past-producing mines.

These greenstone belts developed as the older sialic crust of the Onaman Lake terrain rifted, which led to the development of a volcanic—sedimentary cycle that produced thick sequences of mafic volcanic rocks. This system matured from magnesium to iron-rich rocks with the final tuffs becoming interbedded with iron carbonate exhalative beds, which were capped by clastic sedimentary rocks; locally felsic volcanic rocks were developed near subvolcanic intrusive centers. The major gold deposits show a special relationship to these centers and the final phases of mafic volcanicity and sedimentation (McBride, 1987). This discussion will include the work on the Hercules—Golden prospect, in Elmhirst and Rickaby Twps. They were mapped by MacKasey and Wallace (1978) and the Northern Empire mine area in McComber Twp. (McBride, 1987). Langford (1928) mapped the southern volcanic—sedimentary belts and observed that the sedimentary rocks decreased in grain size from the main volcanic center to the north. He also said that the sedimentary rocks seemed to be conformable with the underlying volcanic rocks, an observation shared by other government mappers in the area. Lafrance et al. (2004) present a different view of the regional geology. Barret and Fralick (1989) discuss sedimentation and environment of turbidite and iron formation. In the area, host volcanic rocks form two parallel units that strike east-west, one through Beardmore ad Geraldton and the second from north of Armstrong through Tashota. They are connected by a smaller belt trending NE from Beardmore. Significant gold mineralization occurs near Beardmore in the Northern Empire Mine, Leitch, Sand River, Sturgeon River, and Brookbank deposits. All lie within a volcanic—sedimentary sequence as conformable, folded, and boudinaged, and zoned mineralized bands that show a relationship to variations in host rock lithologies (McBride, 1987). The Hercules-Golden Mile prospects lie north of Beardmore and are associated with the Elmhirst "intrusion" and associated felsic volcanic rocks.

4.5.1.1 Mineralization of the Golden Mile—Hercules Gold Lode Veins in a Volcanic Environment

Extensive exploration focused on the high-grade Golden Mile vein in the Hercules Vein System hosted within the Elmhirst Intrusion (age 2734 Ma; Card and Poulsen, 1998; Kociumbas and Power-Fardy, 2010). Figure 4.23, compiled from their base map, Kretschmar's work, and that of Sage Gold Inc geologists (www.sagegoldinc.com). Veins in the Elmhirst intrusion are greenstone-hosted quartz-carbonate veins or mesothermal, orogenic (mesozonal and hypozonal), lode gold, shear-zone—related quartz-carbonate or "gold-only"

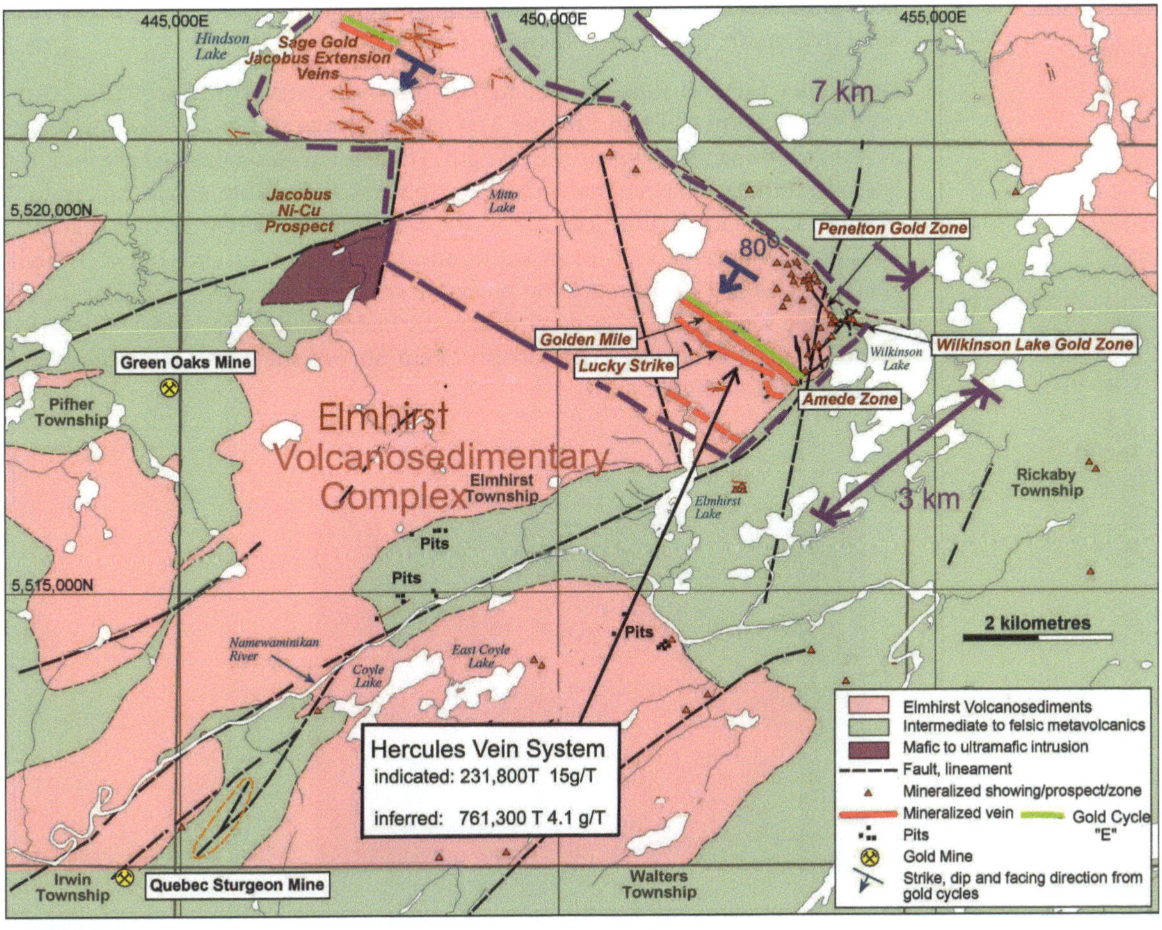

FIGURE 4.23

Vein arrays within the Elmhirst "intrusion", Elmhirst Twp, Beardmoree—Geraldton Greenstone Belt. Compilation from Kociumbas and Power-Fardy (2010), Sage Gold Inc., and Prodigy Gold (formerly Kodiak Exploration). Parallel veins strike NW—SE, dip steeply to the SW, and lie within a 7 × 3-km stratigraphic horizon.

deposits (Dubé and Gosselin, 2007), or "structurally controlled, complex epigenetic deposits that are hosted in deformed and metamorphosed terrains. They occur along major compressional to transtensional crustal-scale fault zones in deformed greenstone terrains of all ages." We shall see in Chapter 8 that gold is not found in the continental shear zones, but is much older and confined to the evolution of ocean floor rift systems.

4.5.1.2 Syngenetic Aspects

Kretschmar carried out considerable work on the property for Sage Gold. Based on lithic fragments, blue quartz eyes, incipient layering, and gold cycles, his observation is that the immediate host to the Golden Mile vein and Jacobus Extension veins as a feldspar–quartz–hornblende mafic and felsic flows with 5%–10% mafic pyroclastic fragments.

The Golden Mile vein was exposed in trenches and occurs at the top of a 2- to 3-m-thick GC "E" unit (refer to Chapter 3). Quartz veining shows striking differences in textures along its 1-km exposed length. There is a massive high-grade pod and lateral laminated extensions with open space and possible autoclastic breccia textures. GCE-facing directions are to the SW for all the veins Figures 4.24 and 4.25. Shoots average 1.75 m in width and the average distance between high-grade centres is 350 m, reflecting probable sea-floor fracture spacing or topography (see Kodiak Exp.company filings on Sedar www.sedar.com). Chapter 9 discusses vent geometry. The Jacobus Vein shows the same laminated fine-grained texture (Figure 4.26).

FIGURE 4.24

The "Golden Mile Vein" and GCE in the Elmhirst "intrusion". Complex laminated textures. View to SE. Q is 1.5 m thick, and E is 3 m thick. The central part of the vein is exposed for >1 km in a stripped trench. Facing direction to SW, dips steeply to SW.

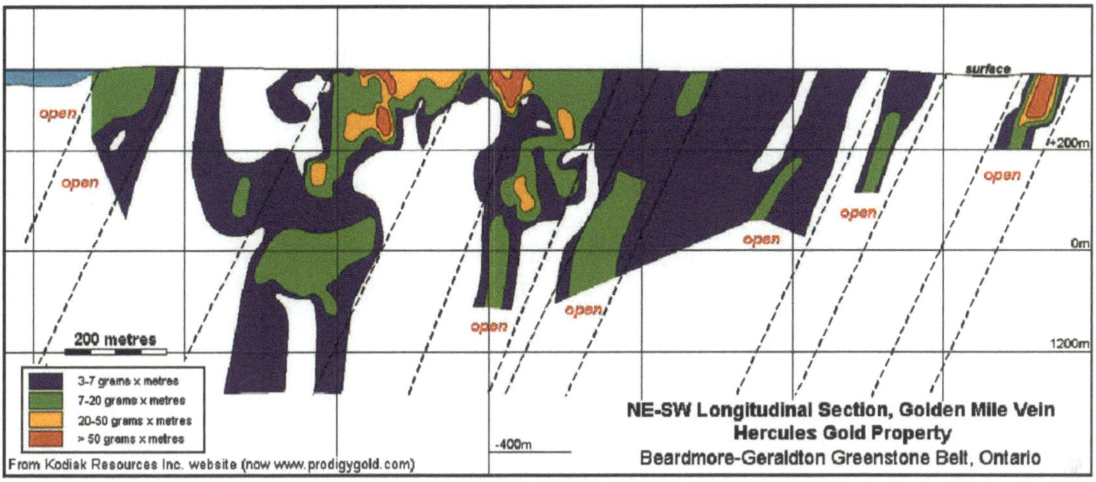

FIGURE 4.25

Longitudinal section of Golden Mile zone showing grade-thickness contours and a periodicity to ore shoots (see discussion of vent geometry).

FIGURE 4.26

Sage Gold, Jacobus Extension area. "Vein 03" and GCE in the Elmhirst "intrusion." Complex laminated textures. View to SE. Q is 1.0 m thick and E is 2 m thick. The central part of the vein is exposed for 200 m in a stripped trench. Facing direction to SW, dips steeply to SW. Veins are along strike from the Golden Mile vein.

4.5.2 Northen Empire Mine

McBride was responsible for a major evaluation of the Northern Empire mine and adjacent property in the fall and winter of 1981–1982. The first phase was the detailed mapping of the host volcanic–sedimentary belt from the mine for 10 km east to Ralph Lake (McBride, 1981); it was complemented by geophysical and humus geochemical and vein sampling programs across the entire property. The second program investigated the first two levels of the old mine and included geological mapping plus systematic chip sampling. None of the reports are in public files; however the results form a major part of McBride (1987).

Surface mapping added more detail; numerous parallel bands of volcanic flows tuffs and intervolcanic sediments were identified. Many of these included gold or sulphide mineralization, all of which was conformable and had been explored by extensive trenching in the 1930s. Sampling of these old exploration pits showed that the best gold mineralization was at the old mine. The underground program was even more revealing; it showed that the vein was bedding parallel and the host rock was coarser-grained in the economic sections of the vein relative to areas off the end of and between the ore shoots. These shoots defined steeply plunging bands that seem to apex below the deepest mine workings. Subsequent deformation rotated the rocks into a steep position, boudinaged the veins parallel to the fold limbs, and produced minor isoclinal folds in the cross-cutting ones (see Chapter 9). Gold in the vein was disseminated 2–4 mm arsenopyrite crystals and some fine-grained arsenopyrite bands. Of the gold, 60% could be recovered in a simple jig concentrate. McBride interpreted the veins as syngenetic chemical precipitates in the waning period of mafic iron-rich volcanism that were later folded into vertical ore shoots. Any faulting in the mine was post ore.

4.5.3 Discussion

The four diverse areas with significant gold mineralization, the Meguma, Swayzy, Maskwa, and Beardmore have been described. They show remarkably similar vein textures and basic geological relations. The difficulty and complexities associated with correlating lithologies where the limited drilling and lack of detailed geological information becomes clear. Government mapping is old and too widely spaced to really provide significant insights into the geological setting of the mineralization. More detailed mapping around the mineralized areas will greatly assist in understanding these gold deposit centers.

CHAPTER 5

Why Lamprophyres Have No Role in Lode Vein Genesis

CONTENTS

5.1 LAMPROPHYRES IN GOLD CAMPS

For a definitive treatise on lamprophyres, we are indebted to Nicholas Rock (Rock, 1991), who introduces them by noting that for many geologists, "lamprophyre" conjures up a negative image of rotten dikes forming minuscule, impersistent outcrops that are horribly altered and impossible to classify. He points out that this has delayed recognition that lamprophyres have a work confirms and reinforces Rock's prophetic statement—which is even truer today—by showing that in gold camps, frequently lamprophyres are GCE interflow sediments.

In the lode gold vein literature, there are many references to lamprophyres in gold camps (Figure 3.2 McLennan, 1915), describing a general model for their occurrence in the California Motherlode is one of the earliest. Moore (1940), in discussing the magmatic origin of gold deposits, stated that lamprophyres are simply indicators of the progressive steps in the process of differentiation toward a more alkalic and siliceous magma with simultaneous concentration of the metals. Dikes of these rocks impound the ore-bearing solutions and increase concentration of gold adjacent to them. More recent causal links with gold mineralization are included in work by McNeill and Kerrich (1986), Rock and Groves (1988), Wyman and Kerrich (1989), Rock et al. (1988, 1989), Kerrich and Wyman (1994), Wyman et al. (1995), Duggan and Jaques (1996), and Wyman et al. (1999).

Genetic links are attributed to favorable tectono-structural environments, e.g., as discussed by Wyman and Kerrich (1989), and Bierlein et al. (2001), who suggest that lamprophyre swarms represent the injection of volatile-rich magmas from deep sources along major structures associated with gold mineralization. They state that dikes have low gold contents except where they are directly overprinted by gold mineralization. On the other hand, a direct genetic connection due to high gold content in lamprophyre magmas has been suggested by Rock and Groves (1988), and a combination of these factors is suggested by Rock et al. (1989). Addressing the primitive nature of lamprophyre magma, Kerrich

The Metallogeny of Lode Gold Deposits. http://dx.doi.org/10.1016/B978-0-12-803222-0.00005-8

and Wyman (1994) discuss its significance for accretionary geodynamic settings, supercontinent cycles, and metallogenic processes, concepts that are then further developed by Wyman et al. (1999). Lamprophyres in the Yilgarn craton, studied by Taylor et al. (1994), and in the Laowangzhai gold deposit, studied by Huang et al. (2002), provide a wealth of geochemical data and a perspective on the issues surrounding the origin of lamprophyres in gold camps if their origin is magmatic, as is generally thought to be the case. A different perspective can be obtained from geological relations between cross-cutting mafic units in Chester Twp, Ontario. How this different perspective is arrived at is the focus of Chapter 5.

Relevant also is the debate on origin of alkaline rocks, in particular lamprophyres, addressed in reviews by Foley et al. (1987), Bergman (1987), Rock and Groves (1988) and others. Did they form by mantle or crustal processes? Classification is not straightforward and lamprophyres have only recently been included in the IUGS classification (Tappe et al., 2005). Nomenclature and genetic concepts are also addressed by Mitchell (1994), Wooley et al. (1996).

Important comparative studies of lamprophyre petrogenesis, from areas with no known gold, are by McCall et al. (1990) and Adams et al. (2005). The question becomes "what are lamprophyres and what relationship do they have to gold mineralization?" Kretschmar addresses this question with examples from his experience in the Chester Complex. McBride's experience with these rocks was as a student at the Haileybury School of Mines. The Professor, D.W. Atchison, pointed one out on a field trip because it had a contact metamorphic aureole with the adjacent Keewatin rocks. Subsequent exploration showed it to be a lamprolite and part of the local post Silurian intrusive system which was explored for diamonds.

5.2 THE "TWO LAMPS" OUTCROP, LAMPROPHYRES AND GCE IN CHESTER TWP

In outcrop, "lamprophyres" in the Chester Complex are dark to olive green, fine-grained, weather low, are from 50 cm to 3 m thick, up to 20+ m long, thin and thicken along their length (Figure 4.10). They are also described as inclusions, dikes, or mafic flows. Figure 5.1 shows photos of lamprophyres from two surface exposures, which are in fact GCE, as will be demonstrated. At the Chester showing, stratigraphy trends 120° dips steeply and facing direction is to SW. The mineralogy of the Chester GCE unit is predominantly penninite-magnetite. On the other hand, the Mesomikenda GCE from a blasted trench (Figure 5.1) consists mainly of crenulated, foliated, interbedded chlorite–carbonate and quartz. Photos of GCE and L from drill core are shown below, after description of the "Two Lamps" outcrop.

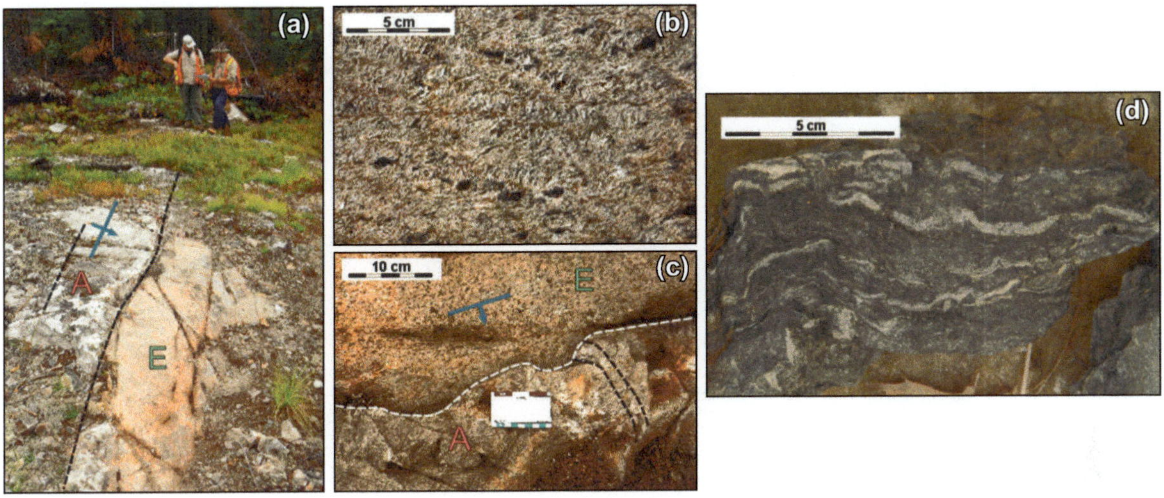

FIGURE 5.1

Photos of typical GCE in outcrop and hand specimens from Chester Twp. Frequently described in the literature as "lamprophyre." (a) Chester showing: regular basal contact of GCE (E) overlying felsic volcanic flow (A). Unit trends 120° dips steeply to north. Top determination is equivocal, but from regional considerations, stratigraphic tops are to SW. Location of grab sample E4. (b) Closeup of weathered GCE surface showing felted mass of penninite and magnetite. Field of view 15 cm. (c) Upper contact of Chester GCE (E) and irregular contact with felsic flow A, facing direction is to SE. Black dashed line outlines thin quartz vein or fracture filling in base of felsic flow. (d) Bedded chlorite-carbonate interflow sediment (GCE) from the Mesomikenda road trench.

5.2.1 The "Two Lamps" Outcrop

The Côté Lake prospect is an ideal place to study the lamprophyre problem, as the rocks of interest stand out because of color and grain size from the surrounding rocks. In the mafic volcanic setting like the Northern Empire Mine, the subtle differences are difficult to study. At the "Two Lamps" stripped area, providing a key outcrop, beside the Chester Road (Figure 5.2), differences between chlorite–carbonate interflow sediments (GCE) and lamprophyres (L) are very clear (Figure 5.3). Exposed in a 35 × 20 m area is an NW–SE trending volcanic sequence consisting of an intermediate feldspar–porphyritic flow, in contact with felsic flows separated by a thin (<1-m thick) GCE. The units are cut by an NE-trending lamprophyre (L). L is clearly later and contains felsic host rock xenoliths. It is dark green, dense, much harder, and has a uniform texture. In hand specimens, rock types TL2 and TL6 are virtually indistinguishable (Figure 5.3).

A mineralogical comparison of samples TL2 and TL6 from the Two Lamps outcrop, the Chester (E4) Showing (Figure 5.4(b) and (c)) and GCE from Mesomikenda (Figure 5.4(d)) are given in Table 5.1. Both lamprophyre and GCE from Two Lamps outcrop are biotite (or phlogopite)-bearing, but

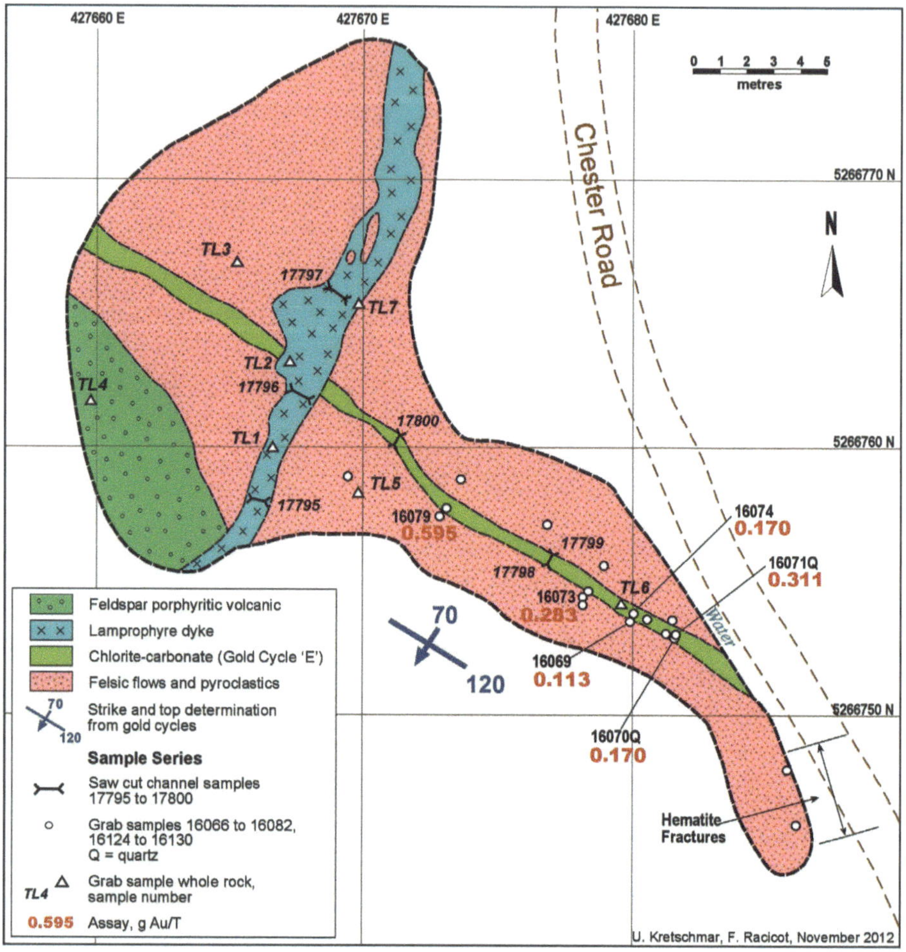

FIGURE 5.2

Geological and sample location map of Two Lamps outcrop, Chester Road, Km 12. Chester Twp. Gold assays >0.1 g/T from grab samples shown in red. TL series are grab samples.

Mesomikenda GCE has no biotite. The presence of plagioclase in GCE sample TL6 is consistent with an epiclastic or volcano-sedimentary origin (GCE) and lamprophyres (L) are very clear (Figure 5.3(b)).

Exposed in a 35 × 20 m area, is an NW–SE trending volcanic sequence consisting of an intermediate feldspar-porphyritic flow, in contact with felsic flows separated by a thin (<1-m thick) GCE. The units are cut by an NE-trending lamprophyre. It is clearly later, contains felsic host rock xenoliths, is dark green, dense, and much harder, and has a uniform texture. In hand specimens (Figure 5.3(b)), rock types TL2 and TL6 are virtually indistinguishable.

FIGURE 5.3

The "Two Lamps" outcrop. (a) Stripped outcrop exposing GCE (green dashed lines) and cross-cutting lamprophyre (L, blue dashed lines, and Figure 5). (b) Hand specimens of L (TL2) and GCE (TL6) are very similar in appearance, except that TL2 has biotite phenocrysts.

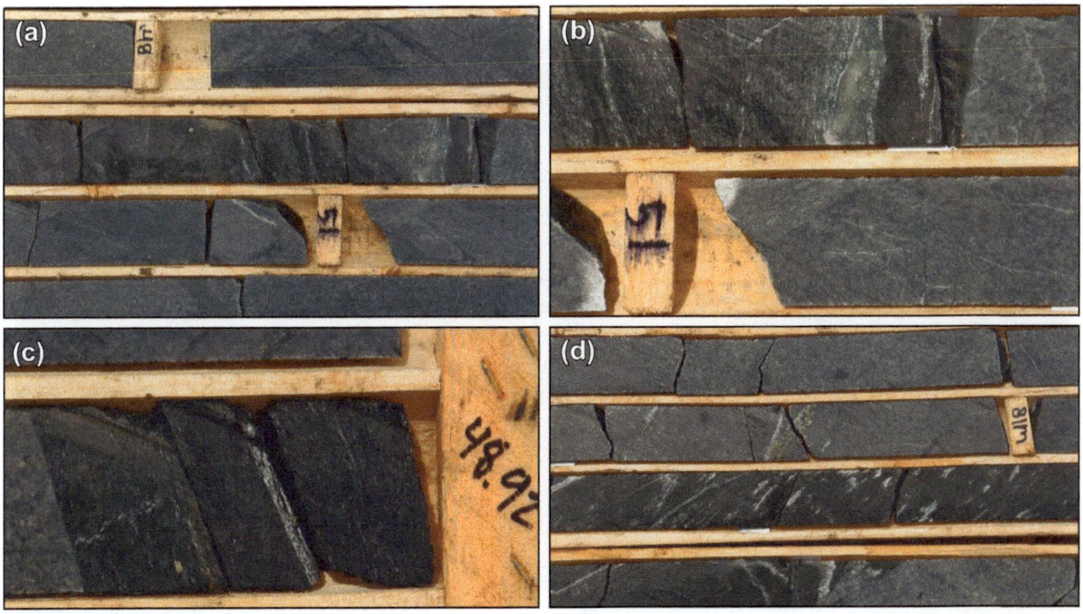

FIGURE 5.4

Typical GCE, common in every drill hole in thousands of meters of drilling within the CC; in the photos, the GCE are the fine-grained darker band, commonly with white carbonate.

Table 5.1 Mineralogical Comparison of Mesomikenda GCE, Two Lamps GCE, Lamprophyre, and Chester GCE. E4, Chester Twp. Point Counts Total 100%

Mesomikenda GCE	Mineral Count 105	TL6 Two Lamps GCE Analyses TL6, 17798, 17799, 17800 (Table 5.2)	Mineral Count 100	TL2 Lamprophyre Assays L750900 17795, 17796, 17797 (Tables 5.2 and 5.3)	Mineral Count 116	E4 Chester GCE Assay Tables	Mineral Count 100
Descriptive name: carbonate	Point count	None given	Point count %	Descriptive name: chlorite	Point count %	Descriptive name: penninite	Point count %
Chlorite green, feather-like	58	Plagioclase Some twins, altered, 1 mm or less	45	Chlorite Green	33	Groundmass <0.1 mm Penninite clusters	68
Carbonate up to 0.6 mm across, local layers 1.5 mm wide	14	Biotite Dark greenish brown, locally replaced by chlorite, 0.2 mm long	34	Calcite Crystalline, –0.5 mm across, twins	26	Carbonate 0.3 mm across	5
Opaque to 0.2 mm across, black in reflected light	10.5	Chlorite Green	13	Biotite Brown, 0.5 mm long	22	Quartz Sericite	22
Muscovite associated with carbonate	10.5	Calcite High birefringence, grains –0.5 mm across	4	Quartz 0.1 mm across, undulatory extinction	7	Feldspar	1
Quartz Sharp extinction (i.e., unstrained), 0.1 mm across	7	Epidote tiny grains, high relief, end order birefringence	4	Opaque	1	Opaque pale silvery yellow pyrite (?)	<1
		Opaque	Trace	Accessory mineral 0.02 mm, high relief, zircon, allanite	1		
The opaque is possibly magnetite, hematite, or chromite. The carbonate hosts tiny opaque black specks. The chlorite displays isoclinal to subisoclinal fads with some shearing along the axial surface. The quartz is recrystallized. Metamorphism to the greenschist facies was followed by microfolding. Prior to these events, the rock.		The foliation is defined by biotite and chlorite. The plagioclase appears brecciated. Metamorphosed to the greenschist facies, biotite zone, with some regression to the chlorite zone.		Calcite with extreme birefringence, which might be mistaken for a zeolite. Chlorite replaces biotite in part. The biotite and chlorite are aligned and the calcite is disposed in layers.		Most of the rock consists of a fine-grained, foliated mass of sericite, quartz, and feldspar, most of which is less than 0.05 mm. Penninite (chlorite group) has an anomalous blue interference color and occurs as elongate clusters and lenses.	

Description by D. Roussel, Laurentian University.

Within the Chester Complex, in outcrop, GCE generally weathers low, is foliated (commonly described as "shear" zone) and has a slightly lighter green hue. Lamprophyres weather higher, are rare and have only been identified in one other outcrop on a southeast bay of Little Clam Lake.

Within the Chester Complex, in outcrop generally GCE weather low, are foliated (commonly described as "shear" zone) and have a slightly lighter green hue. Lamprophyres weather higher, are rare, and have been identified only in one other outcrop on a southeast bay of Little Clam Lake. Photos of typical GCE from Clam Lake drill core and a single lamprophyre identified in core are shown in Figures 5.4–5.6.

Figure 5.4(a), (b), and (d) illustrate typical GCE, common in every drill hole in thousands of meters of drilling within the CC. Lamprophyres on the other hand (Figures 5.6(c) and 5.7(b)) are more difficult to identify visually and require additional chemical criteria, as discussed in Section 5.4.2.2. In Figure 5.4(a),

FIGURE 5.5

Photographs of GCE in drill core. (a) Drill hole SR11-01—501.5 m. Top of 1.5-m-thick GCE showing calcite and chlorite beds, hydrothermal chlorite, and quartz, stratigraphic tops (to right). (b) Drill hole SR12-01—345 m. Upper row shows folding or slumping in chlorite-carbonate GCE. This is overlain (to right) by gray felsic flow. Second and third rows shows pale pink, weakly schistose felsic rock, probably a tuff, with quartz veins or fracture filling and hairline hematite-filled fractures. Bottom row on far left, shows top of GCE with faint quartz eyes, in contact with overlying gray felsic rock.

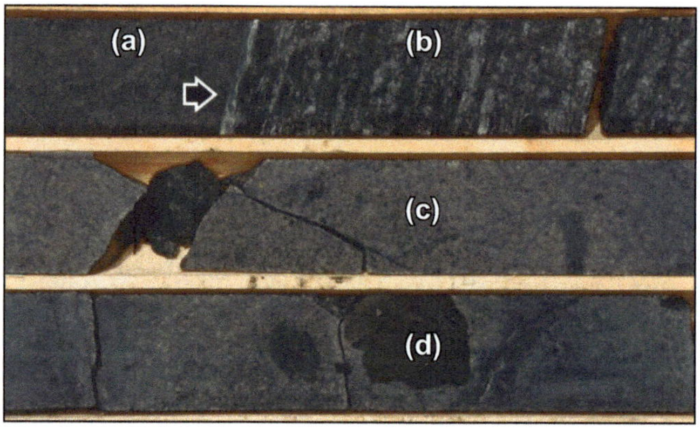

FIGURE 5.6
Photographs of sawed surface of core. Drill hole SR12-16, illustrating variable nature of GCE. (a) Top row: rock is provisionally classified as lamprophyre. Facing direction is to right, i.e., downhole; therefore (a) underlies (b). Arrow points to contact between samples. (c and d) Typical gray, medium-grained felsic flow, with mafic subangular breccia-sized volcaniclastic fragments in lower two rows.

the center of the image shows a 60-cm-thick GCE grading from incipient breccia texture, fining upward (to right) into bedded chlorite–carbonate and minor quartz. Figure 5.4(b) shows detail of (a), including upper contact of GCE with felsic flow. Chlorite–carbonate unit is bedded and has thin quartz–chlorite bed overlain by upward fining darker pelagic-pelitic material. Note uniform texture and absence of basal chilled margin in overlying felsic flow. (c) Black, medium-grained, foliated biotite (carbonate) lamprophyre with minor pyrite, knife-sharp, contact with felsic flow; foliated with carbonate band. (d) Foliated (bedded) GCE chlorite–carbonate–quartz interflow sediment overlain a very weakly schistose gray, pyritic felsic flow with angular mafic autobreccia or lapilli fragments. Irregular contact (center of bottom row) may be a small fold. Width of NQ core is 4.5 cm in diameter.

5.2.2 Chemical Differences: Lamprophyre or GCE?

Tables 5.1–5.3 present mineralogy and major and trace element analyses of lamprophyres and GCE. The samples are either surface grab or saw-cut channel samples from the Two Lamps outcrop (Figure 5.4), the Chester Showing and drill core from Clam Lake drilling (Figure 5.6). Table 5.2, presents all analyses from the Two Lamps outcrop. Shown are individual analyses of channel-cut, unweathered samples as well as grab samples. Averages and 1 standard deviation L_{avg} and GCE_{avg} combine grab and channel sample analyses for each rock type.

Petrogenetic ratios for the analyses in Table 5.2 are shown in Table 5.5. In an initial attempt to characterize GCE, it is compared to GLOSS, the average Global Subducting Sediment analysis of Plank and Langmuir (1989).

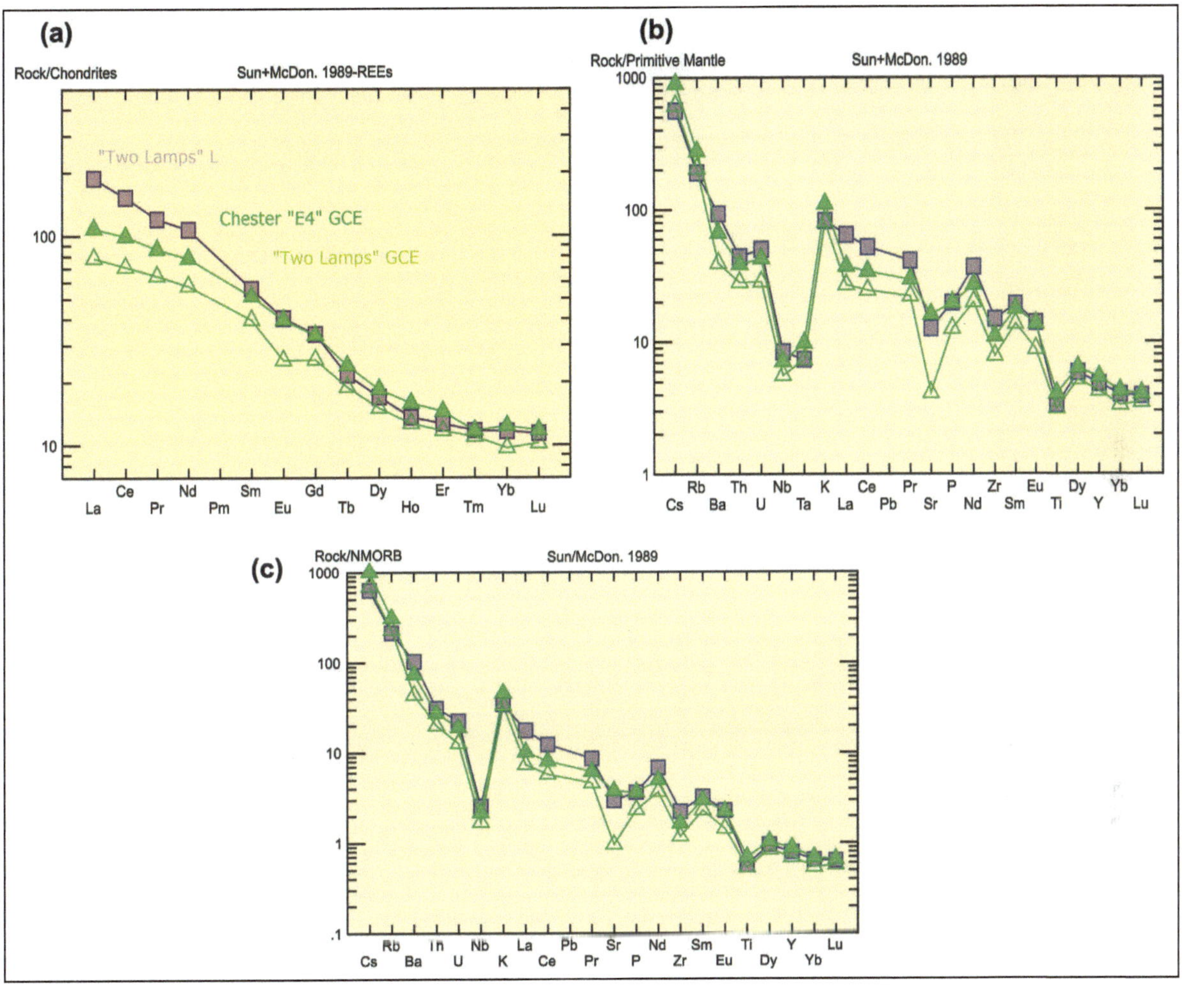

FIGURE 5.7

Petrographic plots from Two Lamps analyses. (a) Rock/chondrite. (b) Primitive mantle. (c) NMORB normalized trace element abundance diagrams (spidergrams) for Two Lamps lamprophyre (L) (blue square) and Gold Cycle interflow sediment (GCE) from analyses in Table 5.2. Blue square represents average of analyses L1, L2, and L3. Solid green triangle is average of analyses E1, E2, and E3. Open green triangle represents analysis Chester GCE sample E1, after Sun and McDonough (1989).

From Table 5.2, L_{avg} and GCE_{avg} whole rock analyses show significant differences. A simple comparison of whole-rock chemistry shows that SiO_2, Al_2O_3, Na_2O for GCE_{avg} is substantially lower than for L_{avg} and MgO is higher. Concentrations of the other major oxides are similar. Comparing GCE_{avg} and GLOSS of Plank and Langmuir (1998) the following are higher: SiO_2, P_2O_5, FeO, and K_2O. LOI is approximately the same, CaO and P_2O_5 are lower, and MgO is much lower. Twenty petrogenetic element ratios (Table 5.5) permit more detailed comparisons.

Table 5.2 Whole Rock and Trace Element Analyses

	L1b	L2b	L3b	TL1a	TL2a	TL7a			E1b	E2b	E3b	TL6a	E97	E98	E4				
Lithology	L	L	L	L	L	L	L_{avg}	1σ	GCE	GCE	GCE	GCE	GCE	GCE	GCE_{avg}	1σ	GCE	GLOSS	1σ
Analysis Element	17795	17796	17797	TL1	TL2	TL7		17798	17799	17800		TL6	750897a	750898a			50896	a	
Au[5] (ppb)%	<0.001	2.11	1.05	0.002	0.001	0.001	0.527	0.80	3.425	6.606	12.99	0.002			7.675				
SiO_2	52.6	52.75	52.9	55.15	53.45	54.17	53.50	0.90	45.14	46.42	48.29	44.4	50.4	54.1	48.13	3.33	57.4	58.57	2.49
Al_2O_3	15.89	15.7	15.92	17.44	16.46	16.42	16.31	0.58	9.34	11.46	12.74	8.37	13.75	16.1	11.96	2.61	15.3	11.91	0.94
Fe_2O_3	8.65	8.36	8.13	8.78	8.58	8.47	8.50	0.21	8.56	8.78	8.97	8.36	9.29	8.35	8.72	0.34	5.54	$FeO = 5.2$	0.42
CaO	5.83	5.33	5.56	2.64	4.79	3.92	4.68	1.10	9.6	9.28	8.25	10.3	7.36	5.31	8.35	1.66	5.31	5.95	1.75
MgO	4.22	4.32	4.04	4.16	4.22	4.17	4.19	0.08	11.23	8.44	6.74	10.93	6.32	4.29	7.99	2.50	2.49	2.48	0.16
Na_2O	4.14	4.81	4.35	5.19	4.56	4.36	4.57	0.35	0.05	1.47	2.51	0.05	3.03	4.99	2.02	1.74	4.09	2.43	0.20
K_2O	2.56	2.02	2.89	2.67	2.9	2.8	2.64	0.30	2.43	3.48	3.19	1.91	2.75	1.83	2.60	0.61	3.49	2.04	0.16
Cr_2O_3	<0.01	<0.01	<0.01	<0.01	0.01	0.01	0.00	0.00	0.11	0.06	0.04	0.12	0.03	0.01	0.06	0.04	0.01	–	
TiO_2	0.73	0.72	0.7	0.77	0.77	0.77	0.74	0.03	0.68	0.83	0.92	0.66	1	0.75	0.81	0.12	0.56	0.62	0.04
MnO	0.11	0.11	0.11	0.09	0.12	0.11	0.11	0.01	0.15	0.14	0.14	0.14	0.14	0.12	0.14	0.01	0.13	0.32	0.13
P_2O_5	0.447	0.418	0.434	0.44	0.481	0.442	0.44	0.02	0.275	0.395	0.465	0.264	0.53	0.45	0.40	0.10	0.35	0.19	0.05
SrO	0.03	0.02	0.03	0.03	0.04	0.04	0.03	0.01	0.01	0.02	0.05	0.01	0.06	0.03	0.03	0.02	0.03	–	
BaO	0.06	0.04	0.07	0.06	0.06	0.06	0.06	0.01	0.02	0.04	0.04	0.01	0.05	0.05	0.04	0.01	0.12	–	
LOI[4]	4.09	4.39	4	2.39	3.55	3.09	3.59	0.68	11.55	8.61	6.5	12.7	5.79	4.76	8.32	2.95	5.32	$CO_2 = 7.29$	1.44
H_2O																		7.29	0.41
Sum[4]	99.36	98.99	99.14	99.81	99.99	98.83	99.35	0.42	99.14	99.43	98.84	98.23	100.5	101.14	99.55	0.99	100.14		
Ag[6]	<0.5	<0.5	<0.5	<0.01	0.01	<0.01	0.00	0.00		<0.5	<0.5			<1	0.00	0.00	1		
As[6]	<5	<5	<5	1.5	0.8	1.2	0.58	0.62		<5	<5	<0.2			0.00	0.00			
Ba	680	495	782	600	650	700	651	88.8	278	482	456	220	434	448	386	100	1045	776	137.1
Be				1.38	1.33	1.4	1.37	0.03				1.26			1.26	0			
Bi				0.07	0.06	0.07	0.07	0.00				0.01			0.01	0			
Cd[6]	<0.5	<0.5	<0.5					0.00		<0.5	<0.5	0.21			0.07	0.10			
Ce	93.3	91.1	93.5	78.2	73.4	77.3	84.47	8.33	43.6	56.1	65.6	39.5	68.3	95.7	61.5	18.6	92.1	57.3	10.3
Co	23.9	27.2	27.2	34	21.9	23.7	26.32	3.93	41.4	38.4	35.3	39.7	35.6	24.6	35.8	5.5	18	21.9	9.48
Cr	70	70	60	42	48	53	57.17	10.5	880	490	330	581	230	80	432	259	40	78.9	7.06
Cs	4.44	3.5	5.41	4.48	4.77	4.67	4.55	0.57	4.87	7.11	7.04	3.73	6.34	2.62	5.29	1.69	2.26	3.48	0.50
Cu	<5	14	13	9.9	1.7	3.9	7.08	5.48	19	62	93	8.7	101	10	49	38	<5	75	16.07
Dy	4.06	4.28	4.61				4.32	0.23	3.83	4.37	5.09		5.47	4.18	4.59	0.60	3.41	4.99	1.86
Er	1.96	2.09	2.24				2.10	0.11	1.94	2.23	2.54		2.46	2.01	2.24	0.24	1.7	2.92	1.06
Eu	2.26	2.3	2.47				2.34	0.09	1.48	2.16	2.49		2.57	2.29	2.20	0.39	1.71	1.31	0.44
Ga	20	19.8	20.6	19.35	19.9	20.9	20.09	0.52	13	16.2	18.1	0.11	17.6	20.4	14.2	6.7	18.6	–	
Ge				0.29	0.28	0.28	0.28	0.00				0.23			0.23	0			
Gd	6.66	6.9	7.32				6.96	0.27	5.28	6.32	7.42		7.7	6.49	6.64	0.86	4.86	5.26	2.04
Hf	4.2	4.2	4.4	3.6	3.5	3.6	3.92	0.36	2.4	3.1	3.6	1.8	3.7	4.2	3.1	0.8	4.6	4.06	0.30

Ho	0.71	0.79	0.82				0.77	0.05	0.72	0.84	0.95		0.99	0.78	0.86	0.10	0.64	–	
In				0.031	0.046	0.049	0.04	0.01				0.05			0.05	0			
La	44.4	43.8	45.3	34.1	29.3	31	37.98	6.68	18.6	23.3	28.1	16.4	29.1	43.4	26.48	8.85	45.3	28.8	6.8
Li				34.8	35.6	35.3	35.33	0.38				39.2			39.20	0			
Lu	0.28	0.29	0.29				0.29	0.00	0.26	0.3	0.33		0.35	0.31	0.31	0.03	0.28	0.413	0.133
Mn				612	851	829	764	108				1080			1080	0			
Mo	<2	<2	<2	0.1	0.08	0.17	0.06	0.06	<2	<2	<2	0.05	<2	<2	0.01	0.018	<2	–	
Nb	6.2	6.1	6.4	6.1	6.5	6.6	6.32	0.20	4	4.7	5.5	1.6	5.6	6.4	4.63	1.55	6.6	8.94	0.94
Nd	49.8	48.8	49.9				49.50	0.50	27	33.7	39.3		39.9	47	37.38	6.69	39	27	8.3
Ni	25	26	23	20.8	20.3	23.7	23.13	2.06	203	114	50	185	39	23	102	71	16	70.54	14.73
Pb	<5	<5	<5	2.7	2.7	2.7	1.35	1.35	<5	<5	<5	1.5	<5	<5	0.25	0.56	<5	19.9	5.4
Pr	11.4	11.25	11.5				11.38	0.10	6.13	7.62	8.86		9.22	11.65	8.70	1.83	10.45	–	
Re				<0.002	<0.00	<0.00	0.00	0.00				<0.00			0.00	0			
Rb	121	98.2	143	83.4	86.3	83	102	22	130	183	168.5	106.5	136	84.6	135.0	33.7	120.5	57.2	6.66
S				0.14	0.11	0.08	0.11	0.02				<0.01			0.00	0			
Sb				0.08	0.08	0.1	0.09	0.01				0.07			0.07	0			
Sc[6]				14	13.6	14.2	13.9	0.2				33.7			33.7	0		13.1	1.03
Se				2	2	2	2	0				1			1	0			
Sm	8.58	8.38	8.76				8.57	0.16	6.08	7.3	8.46		9.07	8.71	7.92	1.10	6.67	5.78	1.83
Sn	1	<1	1	0.9	1.1	1.2	0.87	0.40	1	1	1	0.8	1	1	0.97	0.07	1	–	
Sr	264	245	293	223	273	291	265	25	87.2	231	448	72.4	545	238	270	174	283	327	53.8
Ta	0.3	0.3	0.4	0.34	0.35	0.36	0.34	0.03	0.3	0.3	0.4	0.13	0.4	0.3	0.31	0.09	0.4	0.63	0.06
Te				<0.05	<0.05	<0.05		0.00				<0.05							
Tb	0.76	0.8	0.88				0.81	0.05	0.71	0.83	0.96		1.09	0.81	0.88	0.13	0.65	–	
Th	3.73	3.67	3.87	3.2	2.7	2.9	3.35	0.44	2.41	2.94	3.66	2.2	3.35	3.96	3.09	0.64	5.03	6.91	0.80
Tl	0.5	<0.5	0.6	0.57	0.64	0.58	0.48	0.22	0.5	0.7	0.7	0.42	0.5	<0.5	0.47	0.23	<0.5	–	
Tm	0.29	0.3	0.32				0.30	0.01	0.28	0.32	0.37		0.36	0.29	0.32	0.04	0.25	–	
U	1.08	1.03	1.07	0.9	0.7	0.8	0.93	0.14	0.6	0.75	0.95	0.5	1.01	0.98	0.80	0.20	1.26	1.68	0.18
V	199	201	204	126	129	137	166	36	169	206	248	153	235	154	194	38	93	110	10.7
W	<1	<1	<1	0.4	0.5	0.4	0.22	0.22	<1	<1	1	0.2	1	<1	0.37	0.45	1	–	
Y	19.9	22.5	23.7	20.5	17.3	20.9	20.80	2.02	19.4	22.8	26.4	15.2	29.2	22.8	22.6	4.5	20.8	29.8	9.92
Yb	1.91	1.97	2.06				1.98	0.06	1.65	1.92	2.2		2.27	1.95	2.00	0.22	1.69	2.76	0.88
Zn	110	114	111	112	113	115	113	2	134	123	113	145	111	119	124	12	75	86.4	8.88
Zr	164	164	171	137.5	126.5	134.5	150	17.2	88	111	134	65.5	140	170	118	35	200	130	8.5

Notes: Analyses by SGS, Sudbury, Ontario. Major elements, %, trace elements, ppm. Method codes: 1, ME-ICP06; 2, OA-GRA05; 3, ME-MS81; 4, ME-XRF06; 5, Au-ICP21; 6, ME-4ACD81. Sample types: a, surface grab; b, channel-cut; c, core. Locations (see Figure 5.2): L1, L2, L3, E5, E6 Two Lamps outcrop. GLOSS = Global Subducting Sediment of Plank and Langmuir (1998). Two Lamps GCE, Lamprophyre and Comparison to GLOSS.

Table 5.3 Trace Element Composition of GCE from Clam Lake Core

Analysis No.		E17501	E17512	E17513	E17511	E17509	E17510
		SR11-04	SR11-04	SR11-05	SR12-02	SR12-02	SR12-02
Hole and Depth		43.6–43.95	299.57–300.15	380.25–380.65	261.9–262.54	271.35–271.92	284.5–285.35
Rock Type		Daisy Dio	GCE	GCE	GCE	GCE	GCE
Ag	ppm	0.02	0.1	0.02	0.26	0.05	1.17
Al	%	7.05	7.93	7.08	7.61	6.92	7.93
As	ppm	1.7	42.5	1.4	4.5	2.5	4
Ba	ppm	180	290	130	30	170	490
Be	ppm	0.66	0.4	0.79	1.25	0.44	0.5
Bi	ppm	0.06	1.07	0.06	0.34	0.1	0.95
Ca	%	6.4	4.75	3.86	1.98	6.13	3.13
Cd	ppm	<0.02	0.56	0.08	<0.02	<0.02	0.04
Ce	ppm	8.52	20.3	58.1	20.3	8.46	29.2
Co	ppm	23.3	35.5	20.1	38.8	41.8	25.2
Cr	ppm	176	128	133	75	163	221
Cs	ppm	0.42	1.09	0.4	0.27	0.2	1.72
Cu	ppm	46.8	19.9	24.8	168.5	124	1610
Fe	%	7.34	7.93	5.68	9.87	7.25	7.54
Ga	ppm	14.3	19.15	18.55	22.1	13.6	20
Ge	ppm	0.11	0.13	0.14	0.13	0.12	0.14
Hf	ppm	1.4	1.2	3.1	2.6	0.7	1.5
In	ppm	0.063	0.078	0.078	0.092	0.068	0.121
K	%	0.67	1.78	0.47	0.15	0.84	1.69
La	ppm	2.7	7.4	24.3	8.1	3.2	8.7
Li	ppm	50.2	54.2	28.8	38.5	39.7	89.2
Mg	%	4.18	4.56	2.73	4.17	4.05	4.3
Mn	ppm	890	1460	785	1120	1140	1160
Mo	ppm	0.34	1.92	1.02	0.61	0.18	0.2
Na	%	1.12	0.95	2.44	3.61	1.9	2.03
Nb	ppm	2.6	4.3	6.9	12.3	1.8	4.6
Ni	ppm	89.7	156.5	45.8	65.3	76.9	120.5
P	ppm	570	530	950	750	280	460
Pb	ppm	0.9	5.2	1.6	2.7	1.3	19.3
Rb	ppm	22.9	69.3	15.6	1.8	35.8	43.5
Re	ppm	0.002	0.002	0.002	0.002	0.002	0.002
S	%	0.05	0.05	0.06	0.08	0.14	0.29
Sb	ppm	0.16	0.33	0.12	0.18	0.28	0.2
Sc	ppm	29.7	21.2	20.2	37.6	42.6	16.9
Se	ppm	<1	<1	1	1	1	2
Sn	ppm	1.2	5.2	3	2	1.9	4.9
Sr	ppm	82.2	245	162.5	35.3	122.5	194.5

Table 5.3 Trace Element Composition of GCE from Clam Lake Core continued

Analysis No.		E17501	E17512	E17513	E17511	E17509	E17510
		SR11-04	SR11-04	SR11-05	SR12-02	SR12-02	SR12-02
Hole and Depth		43.6–43.95	299.57–300.15	380.25–380.65	261.9–262.54	271.35–271.92	284.5–285.35
Rock Type		Daisy Dio	GCE	GCE	GCE	GCE	GCE
Ta	ppm	0.16	0.26	0.52	0.74	0.12	0.23
Te	ppm	<0.05	<0.05	<0.05	0.1	<0.05	0.44
Th	ppm	0.7	0.7	3.3	1.6	0.3	0.4
Ti	%	0.416	0.755	0.526	0.842	0.411	0.547
Tl	ppm	0.07	0.31	0.06	0.03	0.1	0.32
U	ppm	0.2	0.2	0.7	0.5	0.1	0.3
V	ppm	182	153	145	245	232	145
W	ppm	2.7	0.8	0.4	1.6	0.2	0.8
Y	ppm	21	20.1	23.4	30.2	16	13.7
Zn	ppm	75	134	86	174	114	132
Zr	ppm	48.8	40.4	110.5	92	24.1	50.1
Nb	ppm	2.6	4.3	6.9	12.3	1.8	4.6
Pb	ppm	0.9	5.2	1.6	2.7	1.3	19.3

Notes: Location of drill holes on Figure 5.1. Samples are GCE chlorite–silica–carbonate interflow sediments. E17511 has vague breccia texture. E17510 is folded and displays foliation parallel to core axis.

Trace element analyses of GCE from core (locations on Figure 4.10) are presented in Table 5.3.

Both primitive mantle and NMORB normalized plots (Figure 5.7) show significant depletion of Nb, Ta, and Ti, which is a tectonic signature indicating an island arc environment with magma derived from a metasomatized mantle wedge source (John Ayer, 2012, written communication). Figure 5.7 also shows that L samples have very similar overall REE concentrations, and although they are somewhat similar to the overall patterns of the GCE samples, they have higher concentrations of LREE. The GCE show progressively more depletion in LREE relative to each other for samples with more primitive compositions (higher Mg, Cr, and Ni). Sample (17798, E1) also shows a slight Eu depletion. This sample has the highest LOI (11.5%). An accepted interpretation (John Ayer, 2012, written communication) is that this represents alteration rather than original magmatic signature, since plagioclase fractionation typically evident in rocks with Eu depletion is unlikely to have occurred at mantle wedge source depths (>70 km). Kretschmar's interpretation is that since it is an interflow sediment, its composition reflects sedimentation conditions. E1 also stratigraphically overlies E2. The following figures show the Two Lamps analyses on various classification plots.

L and GCE have broadly similar chemical compositions and distinctly calc-alkaline affinity. There is a high degree of fractionation of the REE and trace element patterns with elevated LREE and incompatible trace elements.

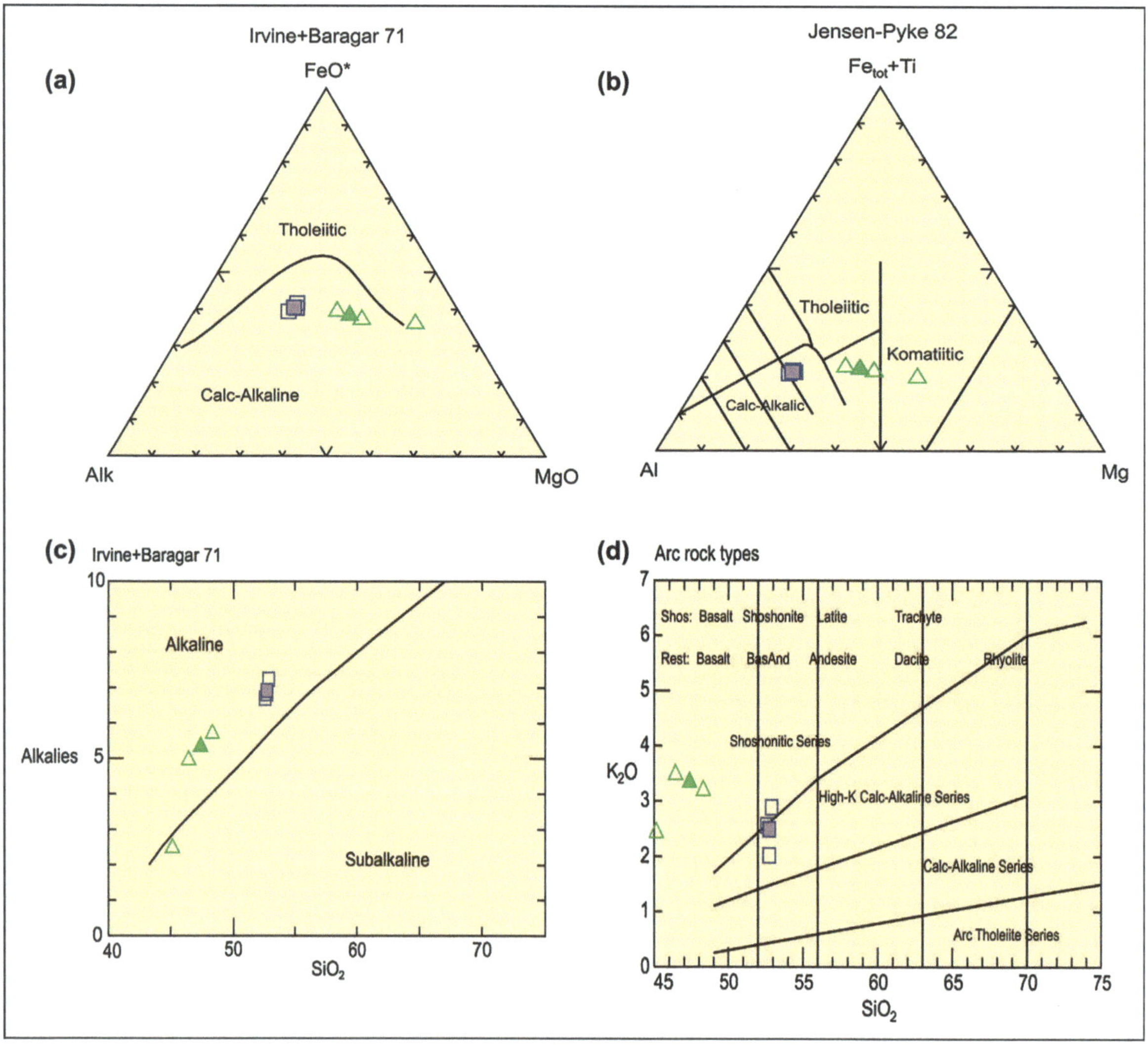

FIGURE 5.8

Two Lamps lamprophyre and GCE analyses (Table 5.3) plotted on petrographic classification diagrams. (a and c) Irvine and Baragar (1971). (b) AFM diagram of Jensen (1976). (d) Arc rock types of Peccerillo and Taylor (1976). Symbols as in Figure 5.7 with the addition of open blue squares (L1, L2 and L3). Solid blue square is average of L1, L2 and L3.

The Two Lamp GCE has higher concentrations of Mg, Cr, Ni, and Ca and lower concentrations of Si, Al, Na, and P series fields.

Classification plots also illustrate the dominantly calc-alkaline affinity of both samples. On the Jensen diagram (Jensen, 1976), GCE samples fall into the tholeiitic and komatiitic fields, probably because the plot does not include alkali elements. On the Irvine and Barrager plot of alkalis versus SiO_2 (Irvine and Barragar, 1971) shows that most of the samples plots in the alkaline field, but the more comprehensive subdivision of arc rocks by Peccerillo and Taylor (1976), illustrates that both L and GCE plot in the shonshonitic to high K calc-alkaline.

Petrographic ratios (Table 5.5) show a significant difference between L00 and E99: V/Cr (6.7 vs 2.7), Cr/Ni (0.27 vs 0.01), and Mg (40.2 vs 47.0). Pb concentrations are <5 ppm, and Nb/Pb* was calculated assuming 5 ppm Pb for purposes of plotting on Figure 5.9.

5.2.3 Whole-Rock and Trace Element Discrimination

Several trace element discrimination plots were developed for lamprophyres. Figure 5.9 shows analyses from Tables 5.4, 5.5, 5.8, and 5.9 plotted on the V/Cr versus Nb/Pb diagram originated by Rock (1991, Figure 5.7) and used by Ashley et al. (1994). Both lamprophyres and GCE units plot in the field of calc-alkaline shoshonitic lamprophyres. This ratio diagram does not readily separate lamprophyres from GCE—a possible reason why the distinctive nature of GCEs has not been fully appreciated. According to Rock (1991), no other petrographic ratio diagram has been found suitable. Analyses in Figure 5.9 are further discussed in Section 5.4.2.2.

5.2.4 Trace Element and Petrographic Classification

Analyses of superior, fresh, saw-cut lamprophyre (samples L1, L2, L3, analyses 17795, 17796, and 17797 respectively) and GCE (samples E1, E2, E3, analyses 17798, 17799, and 17800 respectively) from Table 5.2 are plotted in Figures 5.12–5.14. L1, L2, and L3 have closely similar composition. GCE sample E1 (analysis 17798) has a more "primitive" composition (higher MgO content) and is shown separately as open green triangle.

Lithological classification diagrams in Figure 5.10 place the GCE samples on the boundary between alkaline and subalkaline basaltic and the L samples as andesites. The Le Bas et al. (1986) diagram indicates the lamprophyres are trachy-andesites, whereas the GCE samples are trachy-basalt and sample E1 is picritic. The samples also plot within the calc-alkaline island arc basalt fields on tectonic discrimination diagrams. Further trace element plots to illustrate the difference

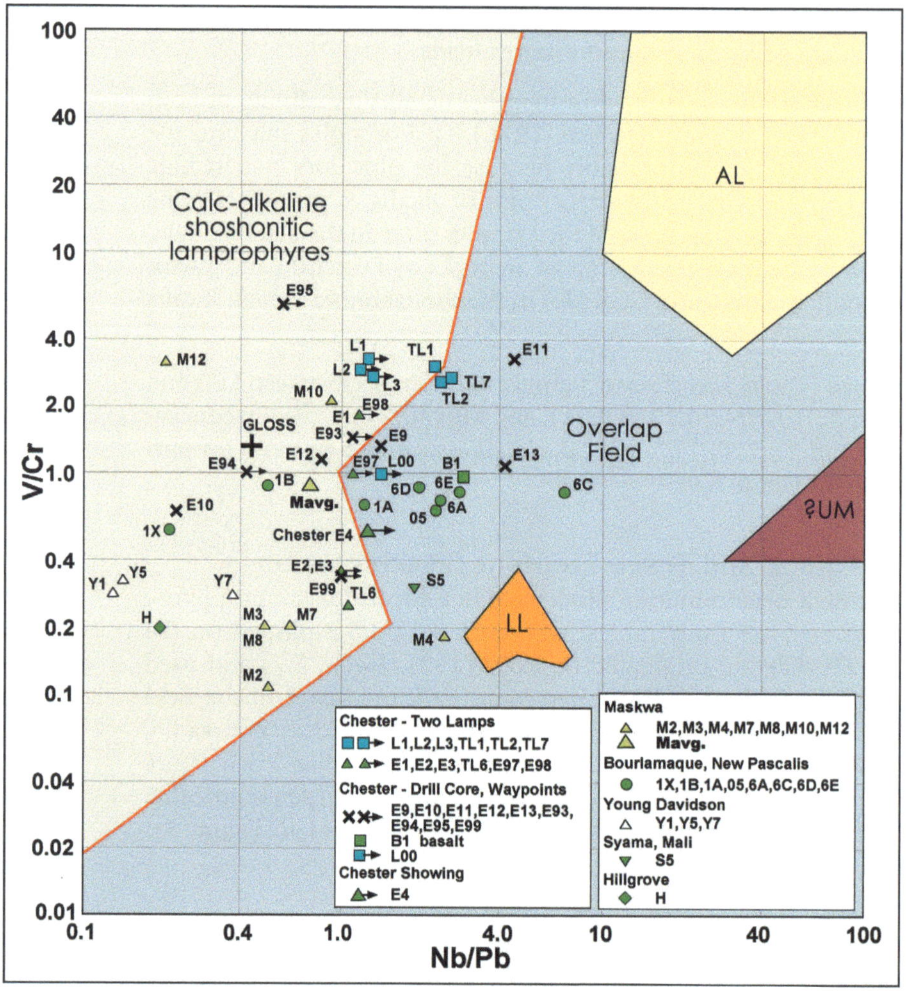

FIGURE 5.9

Whole-rock chemical analyses plotted on V/Cr vs Nb/Pb trace element ratio diagram of Ashley et al. (1994). Tables 5.5 and 5.6: L (L1, L2, L3, TL1,2,7 and L00); GCE (E1, E2, E3, TL6 E97, E98) grab sample Chester Twp, Two Lamps outcrop, GLOSS × Chester Twp, GCE from core and Waypoints (E9-E12, E93-95, E99); B1, "Daisy Diorite" basalt; Table 5.6: M02-M12, Maskwa pluton, Manitoba; Table 5.7: Y1, Y5, Y7, Young-Davidson, Matheson; 1X, 1B, 1A, 05, 6A, 6X, 6D, 6E New Mine, batholith S5, Syama gold deposit, Mali; Hillgrove. Symbols with arrows are upper limit values calculated for Pb detection limit of 5 ppb. If Pb > 5 ppb, points will lie to left.

and the extent of separation of Two Lamps GCE and L and Chester GCE (Tables 5.2 and 5.3) are shown in Figure 5.12.

The discriminant plots point out that although there is a clear difference between L and GCE, significant chemical variability for GCE is in keeping

Table 5.4 Chester Complex

	WP249	WP254	WP255	SR12-16: 41.2 m	SR12-16: 41.5 m
	L75089	L750894	L750895	L750900	L750899
Rock Type	GCE	GCE	GCE	L	GCE
SiO_2	53	45.3	47.9	47.7	47.7
Al_2O_3	14.45	14.05	12.15	12.5	9.34
Fe_2O_3	13.35	11.85	14.85	9.37	9.67
CaO	2.69	10.1	8.29	8.23	10.9
MgO	6.27	7.29	4.41	6.3	8.58
Na_2O	3.55	1.47	2.28	2.85	0.06
K_2O	1.2	0.55	0.59	4.31	3.31
Cr_2O_3	0.03	0.04	0.01	0.03	0.06
TiO_2	1.33	0.92	1.31	0.88	0.71
MnO	0.16	0.18	0.22	0.13	0.16
P_2O_5	0.5	0.09	0.11	0.79	0.48
SrO	0.01	0.01	0.01	0.02	0.01
BaO	0.02	0.02	0.01	0.11	0.06
LOI**	4.19	9.82	9.14	6.82	10.4
Total	100.75	101.69	101.28	100.04	101.44
Element					
Ba	183.5	132.5	76.8	914	510
Ce	67.2	7.6	10.7	124.5	81.5
Co	37.9	220	0.52	21	1.67
Cr	220	290	0.24	16.1	0.75
Cs	0.52	0.24	0.55	17.6	0.97
Cu	94	143	97	18.6	1.71
Dy	4.86	3.23	4.19	17.6	2.57
Er	2.49	1.97	2.62	20.4	2.29
Eu	1.67	0.75	0.97	13.3	1.89
Ga	21	16.1	17.6	17.9	13.3
Gd	5.71	2.56	3.54	9.11	6.45
Hf	3.9	1.5	2.3	5	3.5
Ho	0.96	0.69	0.91	0.91	0.76
La	26.3	2.8	4	56.5	37.1
Lu	0.37	0.32	0.43	0.29	0.23
Mo	<2	<2	<2	<2	<2
Nb	6	2.1	2.9	6.9	4.9
Nd	29.2	5.9	7.9	63.6	42.3
Ni	55	145	39	60	98
Pb	<5	<5	<5	<5	<5

Continued

Table 5.4 Chester Complex continued

	WP249	WP254	WP255	SR12-16: 41.2 m	SR12-16: 41.5 m
	L75089	L750894	L750895	L750900	L750899
Rock Type	GCE	GCE	GCE	L	GCE
Pr	7.09	1.15	1.59	15.45	10.2
Rb	29.9	20.6	25.8	202	163.5
Sm	6.31	1.92	2.55	12.35	8.6
Sn	2	1	<1	2	1
Sr	76.2	116.5	95	133.5	93.9
Ta	0.4	0.1	0.2	0.4	0.3
Tb	0.85	0.47	0.65	1.08	0.81
Th	5.06	0.21	0.39	6.93	4.45
Tl	<0.5	<0.5	<0.5	0.5	0.5
Tm	0.36	0.29	0.4	0.3	0.24
U	1.26	0.07	0.12	2.16	1.42
V	307	290	346	206	164
W	1	1	<1	<1	<1
Y	27	20.1	25.3	26.3	22.9
Yb	2.18	1.96	2.73	26.3	1.5
Zn	137	89	118	129	162
Zr	140	50	80	190	130

*Notes: Analyses by SGS Lab, Sudbury. Analytical Methods: * ME-ICP06, ** OA-GRA05, *** ME-MS81. Waypoint and drill hole locations in Figure 5.1. Whole-Rock Analyses. GCE and lamprophyre from outcrops and drill core. Sample locations in Figure 4.9. Figure 5.1 shows photos of L750900 (lamprophyre) and L750899 (GCE).*

with its origin as a sediment and in fact invalidates interpretations based on chemistry alone. At the Two Lamps outcrop E1 overlies E2, whereas E3 is a 1-m composite sample encompassing the entire width taken to the NW, along strike from E1 and E2 (Figure 5.9). On the other hand E4, a grab sample of GCE from the Chester Showing has a significantly different composition and plots in separate areas on discrimination plots in Figure 5.9 and Figure 5.11.

Petrographic ratios (Tables 5.5 and 5.6), the significance of which is discussed below, are calculated from the analyses in Tables 5.2–5.4. Those analyses for which V/Cr and Nb/Pb ratios can be calculated are plotted on Figure 5.9. These ratios further illustrate the significant differences between GCE and L at the Two Lamps outcrop.

The purported role of lamprophyres in the genesis of gold deposits are addressed in Section 5.4, but first it is worthwhile to briefly examine the extent to which lamprophyres in other gold camps may be GCE.

Table 5.5 Element Ratios for Two Lamps Lamprophyre, Two Lamps and Chester GCE Analyses, and Comparison to GLOSS

	L1	L2	L3	TL1	TL2	TL7			E1	E2	E3	TL6	E97	E98			E4	GLOSS Chester
Rock Type	**L**	**L**	**L**	**L**	**L**	**L**	**L_{avg}**	**1σ**	**GCE**	**GCE**	**GCE**	**GCE**	**GCE**	**GCE**	**GCE_{avg}**	**1σ**	**GCE**	**Sediment**
Sam	17795	17796	17797	TL1	TL2	TL7			17	17799	1	TL6	750897	750898			75	
Al_2O	21.77	21.81	22.74	22.60	21.40	21.30	21.94	0.55	13	13.81	1	12.70	13.80	21.50	14.90	2.98	27	17.1
Ba/	2.58	2.02	2.67	2.69	2.38	2.41	2.46	0.23	3.	2.09	1	3.03	0.80	1.88	2.00	0.90	1.	2.54
Cr/N	2.80	2.69	2.61	2.00	2.40	2.20	2.45	0.28	4.	4.30	6	3.10	0.10	0.10	3.09	2.35	8.	1.12
10	17.58	17.18	16.79	26.60	27.90	27.80	22.31	5.15	15	15.78	1	14.90	16.80	18.00	16.11	1.01	24	29.6
K/B	3.13	3.41	3.07	3.69	3.70	3.32	3.39	0.25	7.	5.99	5	7.21	5.26	3.39	5.82	1.31	2.	2.18
La/	7.16	7.18	7.08	5.59	4.51	4.70	6.04	1.15	4.	4.96	5	10.25	5.20	6.78	6.16	1.95	6.	3.22
Nb/	Pb < 5			Pb < 2			0.00	0.00	Pb									0.45
Rb/	0.46	0.40	0.49	0.37	0.32	0.29	0.39	0.07	1.	0.79	0	1.47	0.25	0.36	0.79	0.52	0.	0.17
Sm/	0.17	0.17	0.17						0.	0.22	0							0.21
Sr/N		5.30	5.02	5.87	No Nd	No Nd	2.70	2.66	3.	6.85	1	No Nd	13.70	5.10	6.71	4.67	7.	12.1
Th/L	0.08	0.08	0.09	0.09	0.09	0.09	0.09	0.00	0.	0.13	0	0.13	0.12	0.09	0.12	0.01	0.	0.24
Th/	3.45	3.56	3.62	4.00	4.00	4.00	3.77	0.23	4.	3.92	3	4.00	3.30	4.00	3.85	0.25	4.	4.11
10	22.74	22.38	22.63	23.27	21.34	21.56	22.32	0.67	27	26.49	2	33.59	23.93	23.29	27.00	3.34	25	53.2
Ti/V	3.61	3.52	3.29	3.61	3.52	3.29	3.47	0.13	2.	2.41	2	2.55	2.50	4.00	2.68	0.60	5.	3.37
Ti/Z	2.67	2.63	2.45	3.00	3.40	3.20	2.89	0.34	4.	4.48	4	5.30	4.30	3.60	4.40	0.52	3.	2.86
V/Cr	2.84	2.87	3.40	3.00	2.69	2.58	2.90	0.26	0.	0.42	0	0.26	72.76	78.17	25.43	35.42	0.	1.39
Zr/N	26.50	26.90	27.60	22.50	19.50	20.40	23.90	3.24	22	23.60	2	40.90	25.00	26.60	27.08	6.33	30	14.5
Zr/S	13.24	12.38	13.83	9.80	9.30	9.30	11.34	1.86	6.	6.93	7	1.90	No Sc	No Sc	3.73	3.17	No	No
Zr/Y	8.24	7.29	7.22	6.70	7.30	6.40	7.19	0.58	4.	4.87	5	4.30	4.80	7.50	5.18	1.07	9.	4.36
Mg#	52.10	53.50	52.50	51.29	52.22	52.25	52.31	0.65	74	68.10	6	74.39	60.19	53.31	65.52	7.66	50	32.2

Notes: Throughout, mg# = wt% (100(MgO/MgO + FeO + 0.889 Fe_2O_3)).*

Table 5.6 Element Ratios From Core and Waypoint Samples

	E17501	E17512	E17513	E17511	E17509	E17510	WP249	WP254	WP255		
	SR11-04	SR11-04	SR11-05	SR12-	SR12-02	SR12-02	L750893	L750894	L750895	L750900	L750899
	43.6-	299.57 -	380.2502-	261.9 -	271.5-	285.35				SR12-	SR12-16
							GCE	GCE	GCE	L	GCE
Al_2O_3/TiO_2	No WR						10.86	15.27	9.27	14.20	13.15
Ba/Sr	2.19	1.18	0.80	0.85	1.39	2.52	2.41	1.14	0.81	6.85	5.43
Cr/Ni	1.96	0.82	2.90	1.15	2.12	1.83	4.00	2.00	0.01	0.27	0.01
K/Rb	29.26	25.69	30.13	83.33	23.46	38.85	33.31	22.16	18.98	17.71	16.80
K/Ba	3.72	6.14	3.62	5.00	4.94	3.45	5.43	3.45	6.38	3.91	5.39
La/Nb	1.04	1.72	3.52	0.66	1.78	1.89	4.38	1.33	1.38	8.19	7.57
Nb/Pb	2.89	0.83	4.31	4.56	1.38	0.24	Pb < 5				
Rb/Sr	0.28	0.28	0.10	0.05	0.29	0.22	0.39	0.18	0.27	1.51	1.74
Sm/Nd	No Nd						0.22	0.33	0.32	0.19	0.20
Sr/Nd	No Nd						2.61	19.75	12.03	2.10	2.22
Th/La	0.26	0.09	0.14	0.20	0.09	0.05	0.192	0.075	0.098	0.123	0.120
Th/U	3.50	3.50	4.71	3.20	3.00	1.33	4.02	3.00	3.25	3.21	3.13
1000Th/Zr	14.34	17.33	29.86	17.39	12.45	7.98	36.14	4.20	4.88	36.47	34.23
Ti/V	2.29	4.93	3.63	3.44	1.77	3.77	2.60	1.90	2.27	2.56	2.59
Ti/Zr	8.52	18.69	4.76	9.15	17.05	10.92	5.69	11.02	9.81	2.77	3.27
V/Cr	1.03	1.20	1.09	3.27	1.42	0.66	10.01	7.13	32.40	6.71	2.70
Zr/Nb	18.77	9.40	16.01	7.48	13.39	10.89	23.33	23.81	27.59	27.54	26.53
Zr/Sc	1.64	1.91	5.47	2.45	0.57	2.96	No Sc				
Zr/Y	2.32	2.01	4.72	3.05	1.51	3.66	5.19	2.49	3.16	7.22	5.68
Mg#	No WR						51.1	57.8	39.8	59.9	66.4

Table 5.7 Maskwa Pluton, Manitoba. Analyses of GCE from Treasure and Silver Fox veins

Analysis		**E8**	**E9**	**E10**	**E11**	**E12**	**E13**	**E14**
Sample		**306802**	**306803**	**306804**	**306807**	**306808**	**306810**	**306812**
Waypoint		**WP 037**	**WP 037**	**WP 021**	**WP 018**	**WP 005**	**WP 041**	**WP 055**
Location		**TV**	**TV**	**SF**	**SF**	**SF**	**TV**	**TV**
Detection limit								
Analyte								
Au[3] ppb	5	<5	<5	444	196	1080	49	<5
SiO_2		53.6	53.4	45.5	42	55	55.7	50.8
Al_2O_3		13.1	13	4.89	4.66	12.1	16.5	16.1
Fe_2O_3		7.39	7.26	15.2	14.5	7.12	9.3	7.51
CaO		7.19	7.22	11.3	12.8	4.01	6.4	7.13
MgO		9.84	9.9	10.9	12.2	10.2	3.6	6.35
Na_2O		2.6	2.61	0.12	0.14	2.75	3.92	4.22
K_2O		3.26	3.28	2.52	2.93	3.48	1.94	3.06
Cr_2O_3		0.1	0.1	0.11	0.14	0.13	<0.01	<0.01
TiO_2		0.55	0.56	0.85	0.82	0.41	0.62	0.69
MnO		0.13	0.13	0.29	0.26	0.17	0.16	0.13
P_2O_5		0.27	0.29	0.05	0.04	0.13	0.16	0.34
LOI		1.91	1.9	6.44	4.87	3.45	1.5	2.99
Total		99.9	99.6	98.1	95.5	99	99.9	99.4
Element[1]								
Al%	0.01	6.71	6.67	2.55	2.43	6.18	8.46	8.12
Ba ppm	0.5	1500	1550	133	140	1040	212	1130
Be ppm	5	<5	<5	<5	<5	<5	<5	<5
Ca%	0.1	5.1	5.2	7.8	9	2.9	4.5	5
Cr ppm	10	700	690	830	1010	810	50	60
Cu ppm	5	17	13	434	305	72	29	16
Fe%	0.01	5.09	5.04	10.3	10.1	4.97	6.41	5.1
K%	0.1	2.9	2.9	2.2	2.7	3	1.7	2.7
Li ppm	10	80	80	140	150	120	80	120
Mg%	0.01	5.9	5.96	6.5	7.36	6.13	2.18	3.76
Mn ppm	10	1000	1000	2230	2040	1160	1270	910
Ni ppm	5	277	296	413	490	242	47	50
P%	0.01	0.13	0.14	0.03	0.03	0.07	0.08	0.16
Sc ppm	5	17	16	22	23	21	49	18
Sr ppm	0.1	577	586	339	376	447	265	1260
Ti%	0.01	0.32	0.33	0.5	0.48	0.23	0.37	0.4
V ppm	5	139	136	149	205	182	109	180
Zn ppm	5	86	82	216	194	998	128	82

Continued

Table 5.7 Maskwa Pluton, Manitoba. Analyses of GCE from Treasure and Silver Fox veins continued

Analysis		E8	E9	E10	E11	E12	E13	E14
Sample		306802	306803	306804	306807	306808	306810	306812
Waypoint		WP 037	WP 037	WP 021	WP 018	WP 005	WP 041	WP 055
Location		TV	TV	SF	SF	SF	TV	TV
All ppm								
Ag	1	<1	<1	<1	<1	6	<1	<1
As	5	<5	<5	<5	<5	<5	<5	<5
Bi	0.1	0.1	<0.1	0.3	1.6	1.8	0.2	<0.1
Cd	0.2	<0.2	<0.2	0.3	0.3	6.4	<0.2	<0.2
Ce	0.1	112	107	24.2	23.4	118	86.4	162
Co	0.5	33.9	31.3	66	73	39.2	22.6	28.9
Cs	0.1	3.3	3.1	7.3	7.3	6.6	2.2	4.1
Dy	0.05	3.12	2.91	3.06	2.51	2.98	13.4	3.86
Er	0.05	1.49	1.62	1.57	1.28	1.72	7.61	1.9
Eu	0.05	1.74	1.81	1.46	1.07	1.69	1.68	2.88
Ga	1	15	15	16	9	19	21	19
Gd	0.05	4.65	4.46	3.4	2.98	4.8	13.7	6.89
Ge	1	1	1	2	2	2	2	1
Hf	1	3	3	4	1	2	8	4
Ho	0.05	0.6	0.57	0.55	0.44	0.58	2.65	0.67
In	0.2	<0.2	<0.2	<0.2	<0.2	<0.2	<0.2	<0.2
La	0.1	62	57.3	10.4	10.3	63.3	30.7	90
Lu	0.05	0.25	0.22	0.22	0.17	0.27	0.82	0.31
Mo	2	2	<2	<2	<2	<2	<2	<2
Nb	1	4	4	15	5	4	11	4
Nd	0.1	50.4	46.9	14.9	14.7	48.4	57.7	71.8
Pb	5	8	8	6	8	258	12	20
Pr	0.05	12.8	12.1	3.39	3.37	13.2	12.8	19.1
Rb	0.2	76.1	73.2	109	127	95.4	79.4	107
Sb	0.1	0.2	0.1	0.2	0.2	0.4	0.1	0.5
Sm	0.1	7.2	6.8	3.5	3.3	7.5	13.1	10.2
Sn	1	2	1	2	2	2	2	2
Ta	0.5	<0.5	<0.5	<0.5	<0.5	<0.5	<0.5	<0.5
Tb	0.05	0.59	0.56	0.53	0.44	0.59	2.3	0.82
Th	0.1	10.7	9.5	1	0.7	20	2.8	17.3
Tl	0.5	<0.5	<0.5	0.7	0.7	0.7	<0.5	0.7
Tm	0.05	0.21	0.18	0.19	0.14	0.22	0.89	0.2
U	0.05	2.17	2	1.54	0.76	3.65	1	3.07
W	1	<1	<1	13	4	3	1	<1

Table 5.7 Maskwa Pluton, Manitoba. Analyses of GCE from Treasure and Silver Fox veins continued

Analysis		E8	E9	E10	E11	E12	E13	E14
Sample		306802	306803	306804	306807	306808	306810	306812
Waypoint		WP 037	WP 037	WP 021	WP 018	WP 005	WP 041	WP 055
Location		TV	TV	SF	SF	SF	TV	TV
Y	0.5	14.6	14.8	14.2	11.6	16	64.1	17.6
Yb	0.1	1.3	1.3	1.3	1	1.4	6.1	1.5
Zr	0.5	121	116	135	49.6	77.6	274	160

Notes: Analytical Method: 1, ICM90A; 2, XRF76C detection limit, 0.01%; 3, Au (FAI323), detection limit, 5 ppb.
Abbreviations: TV, Treasure Vein; SF, Silver Fox Vein; UTM locations of Waypoints: NAD 83 Zone 17. WP 05, 0318564/5603686; WP 18, 0318770/5603643; WP 021, 0318816/5603638; WP 37, 003210153/5603409; WP 41, 0320771/5603284; WP 55, 0321193/5603128.

5.3 LAMPROPHYRES FROM OTHER GOLD CAMPS

5.3.1 "Lamprophyres" in The Maskwa Lake Batholith (MLB)

Chemical analyses of GCE from MLB are presented in Table 5.7; data are from Kretschmar's geological mapping and sampling. Background information on geology and mineralization is provided by Kretschmar (2011).

Within the Little Bear Lake zone, field mapping and logging of 2500 m of core lode shows that vein quartz occurs consistently on the north side of "chlorite dikes" (GCE units). These rocks are fine-grained, dark green, and commonly foliated (schistose). They are Maskwa GCE interflow sediments (Kretschmar, 2011). The seven whole-rock analyses in Table 5.7 are representative of GCE from the Silver Fox (SF) and Treasure Vein (TV). A distinguishing feature of these analyses is their consistently lower LOI relative to the Two Lamps GCE, which may be attributed to metamorphic devolatilization. Element ratios for these analyses and average (M_{avg}) are shown in Table 5.8. Plotted on V/Cr versus Nb/Pb diagram Figure 5.9, they show a wide scatter of values. The analyses fall into two groups, one of which includes Chester core TL6 and Chester sample E4 (Table 5.2). The other includes Young-Davidson, Syama, and Hillgrove, as discussed below.

The Maskwa GCE also plot in the field of mafic enclaves, xenoliths, mafic dikes, and inclusions on petrographic plots of TTG rocks, e.g., Whalen et al. (2002). This has important implications for the origin of the TTG suite and early Archean crustal processes, which are briefly addressed in Chapter 6.

Comparing Maskwa average (M_{avg}) to GCE_{avg}, L_{avg} and GLOSS, the difficulty of using average whole rock analyses becomes apparent because of the variable

Table 5.8 Element Ratios for Chemical Analyses of GCE Units from the Maskwa Pluton

	Analysis No.	30680	306803	306804	30680	306808	306810	30681		
	Waypoint	**WP 037**	**WP 037**	**WP 021**	**WP 018**	**WP 005**	**WP 041**	**WP 055**	**Avg**	**1σ**
1	Al_2O_3/TiO_2	23.82	23.21	5.75	5.68	29.51	26.61	23.33	19.70	9.08
2	Ba/Sr	2.60	2.65	0.39	0.37	2.33	0.80	0.90	1.43	0.97
3	Cr/Ni	2.53	2.33	2.01	2.06	3.35	1.06	1.20	2.08	0.73
4	1000 K/Rb	35.6	37.2	19.2	19.1	30.3	20.3	23.7	26.5	7.2
5	K/Ba	1.93	1.87	16.54	19.29	2.88	8.02	2.39	7.56	6.88
6	La/Nb	1.23	1.22	0.70	0.70	1.31	0.53	1.25	0.99	0.31
7	Rb/Sr	0.132	0.125	0.321	0.338	0.216	0.300	0.0849	0.217	0.09
8	Nb/Pb	0.500	0.500	2.500	0.625	0.016	0.917	0.200	0.751	0.76
9	Sm/Nd	0.143	0.145	0.235	0.224	0.155	0.227	0.142	0.182	0.04
10	Sr/Nd	11.45	12.49	22.75	25.58	9.24	4.59	17.55	14.81	6.96
11	Th/La	0.173	0.166	0.096	0.068	0.316	0.091	0.192	0.157	0.07
12	Th/U	4.93	4.75	0.65	0.92	5.48	2.80	5.64	3.60	1.97
13	1000Th/Zr	88.43	81.90	7.41	14.11	257.73	10.22	108.13	81.13	81.8
14	Ti/V	2.30	2.43	3.36	2.34	1.26	3.39	2.22	2.47	0.68
15	Ti/Zr	2.64	2.84	3.70	9.68	2.96	1.35	2.50	3.67	2.54
16	V/Cr	27.25	26.15	19.10	22.78	62.76	24.22	36.00	31.18	13.7
17	Zr/Sc	7.12	7.25	6.14	2.16	3.70	5.59	8.89	5.83	2.11
18	Zr/Y	8.29	7.84	9.51	4.28	4.85	4.27	9.09	6.87	2.15
19	Zr/Nb	30.3	29.0	9.0	9.9	19.4	24.9	40.0	23.2	10.4
20	mg#	74.8	75.8	61.5	65.2	76.1	46.3	65.3	66.4	9.9

Table 5.9 Select Whole Rock and Trace Element GCE Analyses from Gold Deposit Literature Described as Lamprophyre

Sam	D01	D52	D20	D60	S5	S1	Y1	Y7	Y5	H	1Xb	1B	>1A	>05	>6A	6C	6D	>E6	N1	N9	N13	N21	N23
Oxide																							
SiO_2	51.0	45.81	45.06	44.8	40.34	44.54	43.39	45.55	43.38	57.99	39.6	43.4	44.3	51.9	47.6	53.8	50.3	63.3	39.3	40.2	30.2	55.3	44.8
TiO_2	0.58	0.56	0.69	0.54	0.63	1	1.09		1.066	0.46	0.83	0.71	0.83	0.88	0.9	0.99	0.76	0.22	0.74	0.87	0.43	0.51	0.51
$A1_2O_3$	13.5	12.61	13.07	12.3	9.16	11.84	12.46	10.02	13.16	12.89	8.88	9.85	9.18	9.18	11.5	10.5	11.5	8.98	8.9	11.3	8	15	15.8
Fe_2O_3	1.53	2.35	1.96	2.72	10.35	9	11.94	6.17	10.28	0.31	7.17	5.65	6.41	12.6	14.5	12.4	13.2	9.82	9.14	9.22	9.12	4.94	7.5
FeO	3.76	3.47	3.91	3.40	–	–	6.63	4.32	–	3.8													
MnO	0.11	0.17	0.10	0.15	0.33	0.29	–	0.11	0.162	0.08	0.12	0.15	0.13	0.12	0.12	0.11	0.12	0.08	0.17	0.18	0.2	0.08	0.16
MgO	5.50	7.50	10.90	9.50	8.59	5.32	7.79	7.87	6.71	3.64	8.37	4.74	5.45	10.4	12.9	11.8	12.3	8.51	11.4	7.67	14.7	4.02	6.55
CaO	5.30	7.20	9.20	7.70	9.49	8.6	7.56	8.66	6.64	3.87	11.5	16.7	14.8	4.77	2.72	2.11	2.36	1.88	11.4	10.6	15.3	5.86	6.76
Na_2O	1.46	1.28	1.60	1.45	0.28	1.98	3.81	0.44	4.29	2.2	3.11	3.12	3.7	<0.0	0.01	<0.0	0.35	<0.0	0.65	1.1	<0.0	4.5	2.48
K_2O	5.30	5.57	3.69	4.50	1.55	1.22	0.23	6.27	2.13	4.05	2.09	0.42	0.21	0.11	0.06	0.2	0.1	0.45	0.6	1.67	0.07	1.18	2.4
P_2O_5	0.44	0.56	0.78	0.52	0.37	0.24	0.45	–	0.36	0.1	1.13	1.16	1.28	1.19	1.29	1.35	1.3	1.06	0.63	0.44	0.12	0.2	0.07
H_2O^+	1.72	3.30	2.67	0.97	S = 0.1	0.06	<0.01	0.06	2.5	0.22									3.1	2.7	3.9	1.7	2.5
H_2O^-	0.01	0.11	0.59	0.01	$C_{organic}$	0.11	n.d.	n.d.	n.d.	n.d.									0.1	0.1	0.1	0.1	0.1
CO_2	8.83	8.70	5.10	10.4	C_{total}	3.93	n.d.	n.d.	n.d.	$H_2O = 1$									15.9	14.6	20.2	8.28	10.4
				LOI	13.94	13.0	7.92	12.3	10.48	$CO_2 = 8$	16	13.5	7.23	4.7	6.7	5.7	6.08	4.39					
Total	99.1	99.1	99.3	99.0	99.6	101.	95.35	89.4	90.67	99.73	98.8	99.1	95.4	95.9	98.4	99.1	98.7	99.3	102.	100.	102.	101.	100
Au(p)					0.01	0.02	0.975			0.268	0.62	0.21	0.014	2.7	0.96	0.06	0.38	0.51	<1	2 ppb	<1pp	<1	<1
Ag	n.d.	n.d.	n.d.	n.d.	0.5	0.3	0.9	n.d.	n.d.	0.7									0.15	0.17	0.25	0.19	0.14
As	n.d.	n.d.	n.d.	n.d.	<10	<10	5.9	3.1	15.7	819	<1		<1	2	<1	<1	1	<1	<0.5	<0.5	<0.5	<0.5	<0.5
Ba	1594	2009	2602	1459	430	120	947.3	>174	527	601	2162	102	252	95	27	75	187	63	3.3	8.5	<0.5	12	44
Be							2.32	7.69	2														
Bi							<0.15	0.37	1.5										0.1	0.1	<0.1	<0.1	<0.1
Cd							0.15	0.15															
Ce	76.6	52.38	98.52	44.4	49	9	83.52	294.6	81.3	49									116.6	56.41	87.6	56.0	10.4
Co					45	40	43.76	33.3	60	12	35	34	42	68	99	96	86	78	43.7	31.9	45.8	17	30.8
Cr	349	292	521	565	505	215	870	429	604	325	242	218	254	294	390	354	333	313	890	381	1073	290	239
							42.37	19.92			1.2		0.6	0.5	0.6	0.6	0.5	0.5	n.d.	nd	nd	nd	nd
Cu	12.8	34.3	37.3	35.1	93	81	11.5	18.9		22	52	36	54	52	48	60	48	47	<1	29	<1	3	<1
Dy	3.63	4.13	4.93	3.55			4.61	9.13	4.47														
Er	1.73	1.94	1.96	1.87			2.37	2.78	2.32														
Eu	1.71	1.51	2.00	1.33			2	7.01	2.11											1.66	1.56	1.01	0.66
Ga					25	23	18.21	15.54	18	16									2.88				
Gd	5.35	5.37	5.92	4.59			5.95	19.29	5.89														
Hf	4.92	4.33	4.82	3.69			3.94	8.42	3.3	n.d	4	4	5	6	5	6	8	4	4.1	3.2	1.3	2.8	1.2
Ho	0.67	0.75	0.74	0.70			0.85	1.27	0.86														
In							0.08	0.06															
La	35.8	25.95	50.07	21			48.17	139.1		29									50.7	23.82	37.68	26.6	3.96

Continued

Table 5.9 Select Whole Rock and Trace Element GCE Analyses from Gold Deposit Literature Described as Lamprophyre continued

Sam	D01	D52	D20	D60	S5	S1	Y1	Y7	Y5	H	1Xb	1B	>1A	>05	>6A	6C	6D	>E6	N1	N9	N13	N21	N23
Li							78.8	47.3											19	15	18	12	16
Lu	0.22	0.25	0.22	0.25			0.29	0.23											0.28	0.35	0.13	0.13	0.23
Mo							2.57	0.22			2		2	12	<26		2	9	0.9	1.6	3	3.2	1.3
Nb	7.64	7.81	10.3	6.59	17	15	5.37	10.4	5.3	8	12	13	10	37	25	28	24	36					
Nd	34.2	28.33	48.75	23.8			38.01	153.9	37.7										57.69	31.8	44.07	24.2	5.67
Ni	78.5	74.8	365	263	345	56	294.2	265.2		32	182	171	206	167	215	194	179	154	252	77	338	91	117
P					807	524																	
Pb					9	<2	42.3	28.2	37	37	54	25	8	16	10	4	12	13	2.5	1.6	1.4	2.4	2.2
Pr	8.93	7.20	12.65	5.75			9.58	37.55	9.37														
Rb	214	259	140	185	39	17	197.0	206.2	81	14.5	50	7		22	10	12	9	26	12	36	<10	27	53
Sb					20	<5	1.59	3.13	2.6	123	0.3		0.3	0.5	0.6	0.6	0.4	0.5	0.1	0.1	<0.1	0.1	<0.1
Sc	22.5	20.2	27.3	26			35.6	16.7	33.6	15	13	12	12	13	14	12	15	10	25.89	31.48	25.26	12.1	38.4
Se							<0.4	<0.4											<0.1	<0.1	<0.1	<0.1	<0.1
Sm	6.45	6.20	8.44	5.43			7.3	28.88	7.65										11.98	6.78	6.88	4.14	1.77
Sn							1.43	2.27	2														
Sr	609	935	1049	783	225	114	444.5	>156	392	287	1794	456	308	182	150	200	206	200	345	255	762	304	204
Ta	0.55	0.58	0.73	0.49			0.32	0.45	0.32		0.7		1.3	1.2	0.8	0.9	0.9	1	0.31	0.28	0.3	0.27	0.13
Tb	0.67	0.67	0.81	0.67			0.81	2.05	0.85										1.19	0.68	0.47	0.33	0.33
Te							0.03	<0.0															
Th	11.3	16.7	11.1	7.26			5.29	15.76	3.93	11	24	28	29	32	28	35	41	28	8.2	3	3.5	4.1	0.6
Ti							6591	4053	1066														
Tl							2.77	1.42	1.85														
Tm	0.26	0.23	0.27	0.26			0.32	0.31	0.332														
U	3.6	6.43	3.65	3.14			2.18	2.21	1.32	3	5		7	9	7	6	13	7	1.29	0.74	0.32	0.92	nd
V	123	139	113	112	154	158	247.2	120.2	204	77	130	194	182	201	292	288	298	243	168	210	133	96	186
W							11.37	3.43	36.2										4.8	2.9	3.7	<0.5	<0.5
Y	17.8	20.00	19.13	20.9	10.3	3.1	23.71	36.22	22.5		53	53	11	24	17	19	15	20	28.9	27	14	11	13
Yb	1.50	1.67	1.56	1.70			1.97	1.7	2.06										1.5	2.25	0.73	0.67	1.44
Zn	36.2	65.4	74.2	51.1	37	22	323	102	110		109	68	84	191	211	207	200	143	118	64	88	56	58
Zr	121	174	109	80.4			150	368	138		267	285	146	199	166	209	226	182	175	133	77	128	46

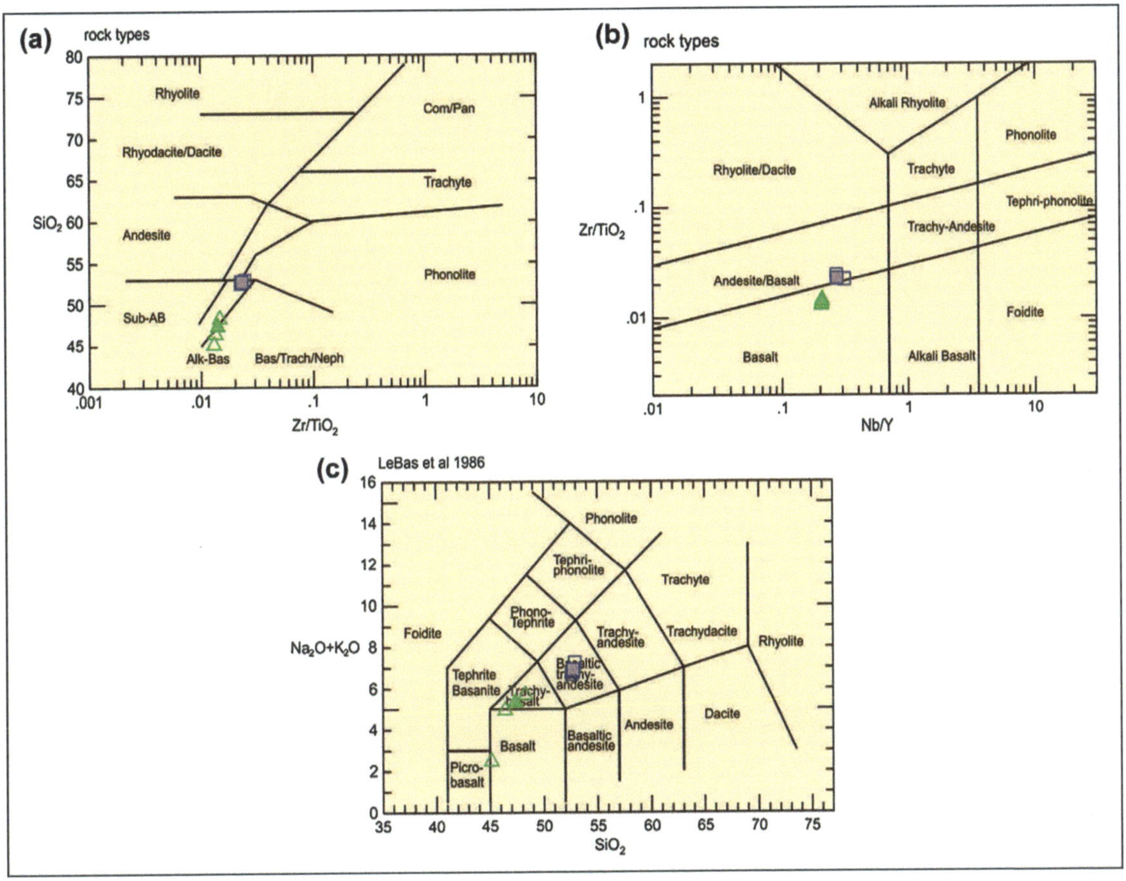

FIGURE 5.10

Two Lamps analyses (Table 5.2) plotted on lithological classification diagrams of Le Bas et al. (1986). (a) SiO_2 versus Zr/TiO_2. (b) Zr/TiO_2 versus Nb/Y. (c) $Na_2O + K_2O$ versus SiO_2.

composition of GCE. Metamorphism may further complicate the chemistry. Eleven of 18 ratios or 61% (no Nb/Pb and Sm/Nd) for M_{avg} falls within 1σ of GCE_{avg}. Comparing M_{avg} to L_{avg} and GLOSS (Global Ocean Subducting Sediment of Plank and Langmuir) (1989), discussed in Section 5.4.2), the correspondence is 44.4% and 36.8% respectively (Table 5.8).

5.3.2 The Bourlamaque Batholith

The use of GCE top determinations to unravel the structure of the Bourlamaque gold camp was briefly addressed by Kretschmar (2011). Of direct interest in this section on lamprophyres is to compare chemical analyses of "lamprophyres" and "mafic dikes" in the Bourlamaque to data from the "Two Lamps" outcrop.

The 10 × 25 km, 2711 Ma (Wong et al., 1991) Bourlamaque batholith in the Val d'Or camp hosts eight mines. According to Campiglio and Darling

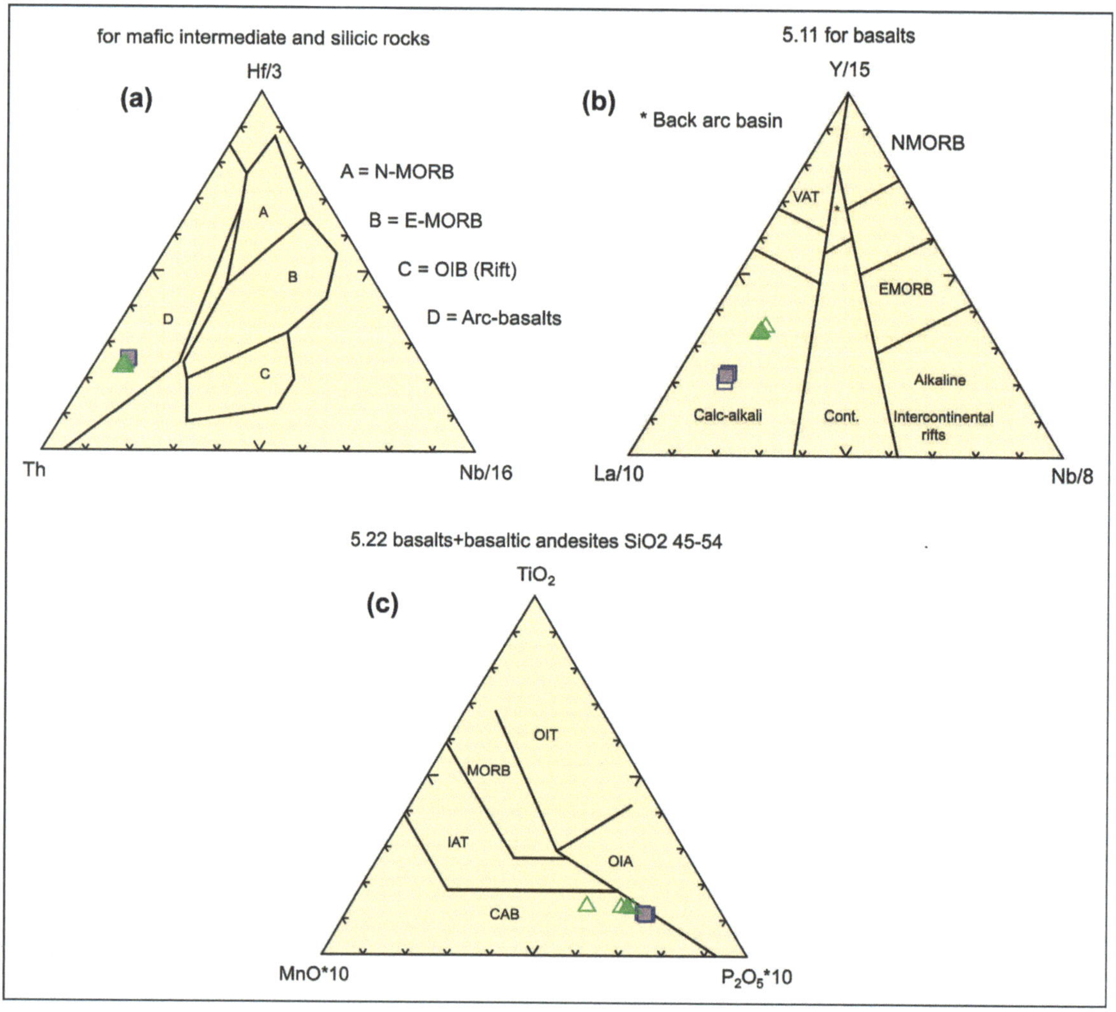

FIGURE 5.11

Two Lamps data plotted on various Tectonic Discrimination diagrams; symbols as in Figure 5.8. (a) Ta-Th-Hf plot after Wood (1980). (b) Nb-La-Y discriminant diagram for arc, MORB/WP, and intermediate contaminated rocks, after Cabanis and Lecolle (1989). (c) TiO_2-MnO*10-P_2O_5*10, dividers after Mullen (1983). Fields labeled as follows: CAB, calc-alkaline basalt; IAT, island arc tholeiites; MORB, mid-ocean ridge basalt; OIT, ocean island tholeiites; OIA, ocean island alkaline rocks.

(1976), it is a synvolcanic quartz tonalite. In a gravity study, Jebrak et al. (1991) point out that it resembles a dike rather than a sill, and confirm a distinct density difference between the northern and southern parts. Claoue-Long et al. (1990) found multiple periods of zircon growth in veins hosted by the Bourlamaque batholith. Belkabir et al. (1993) discuss the relation between dikes and auriferous shear zones, and indicate that complex shear zone and vein patterns within the pluton reflect the influence of diorite dikes. They consider that these

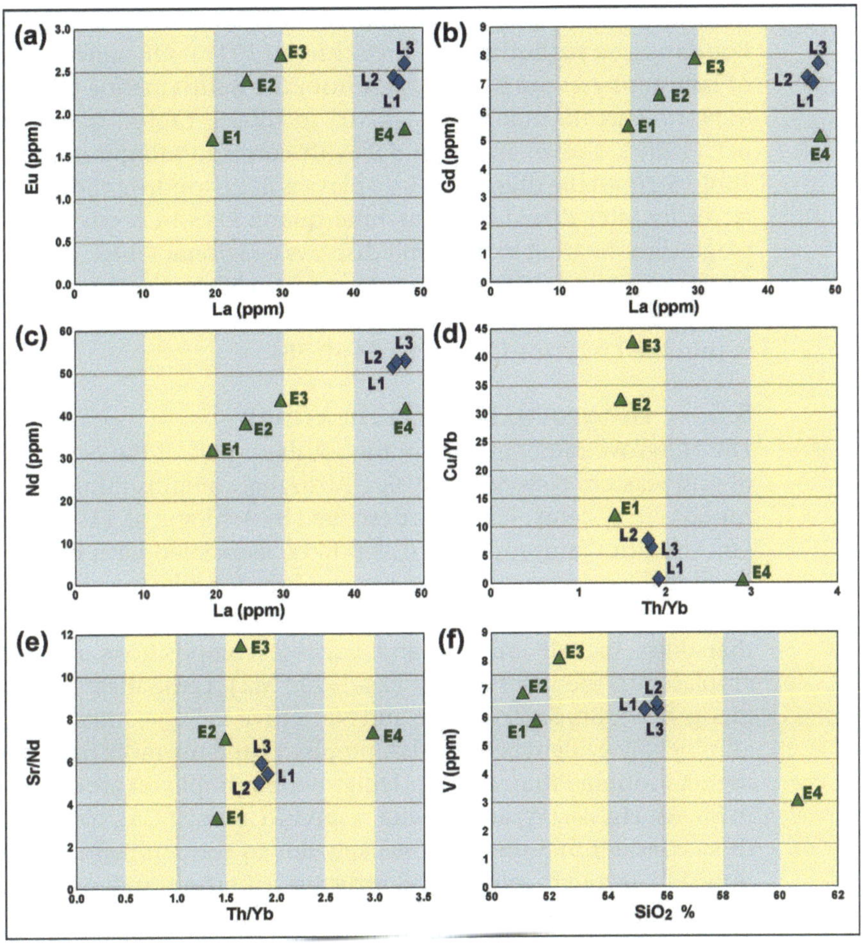

FIGURE 5.12

Chester Twp. Trace element plots to illustrate chemical differences between Two Lamps GCE and L and Chester GCE. GCE(E1[17798], E2[17799], E3[17800]), L (L1[17795], L2[17796], L3[17797]), and Chester GCE(E4[750896]). Original data in Tables 5.3 and 5.5. Recalculated to 100% volatile-free basis. (a) Eu versus La. (b) Gd versus La. (c) Nd versus La. (d) Cu/Yb versus Th/Yb. (e) Sr/Nd versus Th/Yb. (f) V versus SiO_2.

acted as weak layers that were activated during subsequent deformation and thus illustrate the importance of layer anisotropy in auriferous shear zone development. The Beaufort mine was studied by Tremblay (2001), the Ferderber mine by Vu et al. (1987), and the gold potential of the Bourlamaque batholith by Taner and Trudel (1989). Recent dating and mapping of veins by Lemarchand et al. (2010) show the spatial relationship of gold-bearing quartz veins to dikes. Daigneault and Gaboury (2007) present a compilation of these studies. In particular, their mine cross-sections show a consistent spatial relationship between lode veins and mafic dikes, which the authors interpreted in terms of Gold Cycles, and then used GCE asymmetry to infer folding of the Bourlamaque batholith, e.g. Kretschmar (2011). Of significance to the study of lamprophyres are five analyses from the Bourlamaque batholith identified as North Pascalis mafic dikes, NP-01 to NP-23, Table 1 of Burrows and Spooner (1989), shown in Table 5.9 with corresponding petrographic ratios in Table 5.10. Mafic dike NP-23 displays a flat chondrite tholeiitic REE pattern, is cut by later dikes, contains blue quartz eyes in a coarse chlorite-altered, plagioclase-bearing matrix, and displays a regional fabric, a description which conforms well to GCE. Because there are no Nb analyses for NP samples, these cannot be plotted in Figure 5.9. V/Cr ratios (0.12–0.78, Table 5.10) are within the range for GCE_{avg} (0.73–0.58).

5.3.3 Hillgrove, New South Wales

The Hillgrove mineral field, in the southern part of the New England Orogen of northeastern New South Wales, Australia, contains numerous mesothermal Au-Sb vein systems. These are described by Ashley et al. (1994) as calc-alkaline (shoshonitic) lamprophyre (CAL) dikes, associated with mineralization with dilational lode structures acting as conduits for dike intrusion, which occurred before and after major quartz–stibnite veining. Their descriptions further stated that dikes include minette and vogesite compositions and that they were emplaced in the late Permian (247–255 Ma). Least-altered dikes are enriched in Mg, K, Ba, Rb, Sr, Zr, Th, Cr, and Ni relative to I-type intrusions, and chemical affinities are evident between lamprophyres and the more mafic members of the high-K Moonbi Plutonic Suite. Hillgrove lamprophyres are commonly enriched in Sb, As, Hg, Au, W, and Bi with respect to average CAL compositions. The authors consider that this is most likely due to contamination of magma during intrusion through mineralized structures. Carbon and oxygen isotopic data from carbonates in least-altered, post-lode lamprophyres are indistinguishable from carbonate in altered dikes and veins, and the authors suggest that hydrothermal interaction continued after dike intrusion. Dike emplacement and hydrothermal mineralization are contemporaneous; fresh and altered dikes have the same isotopic composition. The authors suggest the fluid source is either metamorphic or magmatic, and that carbon isotopes suggest an organic, sedimentary origin. In Figure 5.9 of the present study, Hillgrove lamprophyre analysis R64069 (H) plots with Young-Davidson, Syama, and some Maskwa GCE and a similar volcanic–sedimentary origin may be inferred. The stated organic, sedimentary source for the vein fluids also is permissible evidence that the lamprophyres are GCE interflow sediments.

5.3.4 Laowangzhai Lamprophyres

A thorough study of lamprophyres was carried out for the Laowangzhai area by Huang et al. (2002), from which work whole rock and trace analyses of four minettes are shown in Tables 5.9 and 5.10. V/Cr ratios in the range of

Table 5.10 Element Ratios for Select GCE ("Lamprophyre") Whole Rock and Trace Element Analyses from Gold Deposit Literature

	Ratio	YD-B-1	YD-52	YD-20	YD-60	SYD55	SYD51	YD06-10-	YD07–33A	YD07-52	HillgrovE	1X	1B	1A	05	6A	6C	6D
1	Al_2O_3/TiO_2	23.40	22.52	18.94	22.83	14.54	11.84	11.43	No	12.35	28.0	11	14	13	10	13	11	12
2	Ba/Sr	2.62	2.15	2.48	1.86	1.91	1.05	2.13	1.12	1.34	2.1	1.2	0.2	0.8	0.5	0.2	0.4	0.9
3	Cr/Ni	4.45	3.90	1.43	2.15	1.46	3.84	2.96	1.62	No Ni	10.2	1.3	1.3	1.2	1.8	1.8	1.8	1.9
4	K/Rb	20.56	17.85	21.88	20.19	32.99	59.56	0.97	25.23	21.83	231.8	347	498	–	42	50	138	92
5	K/Ba	2.76	2.30	1.18	2.56	2.99	8.44	0.20	2.99	3.35	5.6	8	33	7	10	18	22	4
6	La/Nb	4.69	3.32	4.86	3.19	No La	–	8.97	13.37	No La	3.6	No La						
7	Nb/Pb	No Pb	–	–	–	1.89	Pb < 2	0.13	0.37	0.14	0.2	0.22	0.52	1.25	2.31	2.50	7.00	2.00
8	Sm/Nd	0.28	0.22	0.17	0.23			0.19	0.19	0.20	No Sm	No Sm						
9	Sr/Nd	17.77	33.00	21.52	32.78	No Nd	–	11.69	no Sr	10.40	No Nd	No Nd						
10	Th/La	0.32	0.64	0.22	0.35	No Th	–	0.11	0.11	no La	0.4							
11	Th/U	3.14	2.60	3.04	2.31	No Th	–	2.43	7.13	2.98	3.7	5		4	4	4	6	3
12	1000Th/Z	93.39	95.98	101.83	90.30	No th	–	35.27	42.83	28.48	No Th	89.9	98.2	198.6	160.8	168.7	167.5	181.4
13	Ti/V	no Ti	–	–	–	No Ti	–	26.66	33.72	52.25	No Ti	3.82	2.19	2.73	2.62	1.85	2.06	1.53
14	Ti/Zr	no Ti	–	–	–	No Ti	–	43.94	11.01	77.25	No Ti, Zr	19	15	34	27	33	28	26
15	V/Cr	0.35	0.48	0.22	0.20	0.30	0.73	0.28	0.28	0.34	0.2	0.54	0.89	0.72	0.68	0.75	0.81	0.89
16	Zr/Nb	15.8	22.3	10.6	12.3	No Zr	No Zr	27.9	35.4	31.7	No Zr	22.3	21.9	14.6	5.4	6.6	7.5	9.4
17	Zr/Sc	5.38	8.61	3.99	3.09	No Zr	–	4.21	22.04	4.11	No Zr	21	24	12	15	12	17	15
18	Zr/Y	6.79	8.70	5.70	3.84	No Zr	–	6.33	10.16	6.13	No Zr	21	24	12	8	10	11	15

Notes for Tables 5.9 and 5.10. D. Lamprophyres from Laowangzhai gold, Yunnan, Dongualin Province, China. Huang et al. (2002). Rock-type minette. D01: YD-B-1; D52: YD-52; D20: YD-20; D60:YD-60; S. Syama Gold deposit, Mali. Analyses from Olson et al. (1992). Table 5.1, p. 327. Altered andesitic-lamprophyric intrusions. Analyses carried out by BHP Minerals International Sunnyvale Minerals laboratory. Major elements by wet chemistry and trace elements by ICP; S5: SYD55; S1: SYD51; Y. Young-Davidson Gold Deposit. Sample description by Martin (2012): Mine Cross Section 23,240 m E. Y1:YD06-10-558.2, analysis 802811, sample of biotite-rich lower lamprophyre, sheared. Y7:YD07-33A-1463.5, analysis 802847, geochemistry of the different syenite; lamprophyre dike. Y5:YD07-52-629.1, analysis 803325. Described as sediment and also lamprophyre. Samples were analyzed at Geoscience Laboratories (GeoLabs) in Sudbury, Ontario; H. Hillgrove Au-Sb Veins. Analysis from Ashley et al. (1994). Additional elements: F = 0.15%; less O = S, 0.11%; less O = F2, 0.06%. H, analysis R64069; Canadian Arrow, Matheson, analyses from McNeill and Kerrich (1986). Symbols and analyses: Xb, grab sample, location not noted, 1B, 1A, 05, 6A, 6C, 6D, 6E; N. New Pascalis Mine in the Bourlamaque Batholith from Burrows and Spooner (1989), Table 5.1. S < 20 ppm, B = 21 ppm. N1: NP-01; N9: NP-09; N13: NP-13; N21: NP-21; N23: NP-23. Laowangzhai minettes, Syama, Mali; Young-Davidson, Matheson; Hillgrove NSW, Australia; Canadian Arrow, Matheson; New Pascalis Mine, Bourlamaque Batholith, Quebec.

0.2–0.5 are similar to those in Young-Davidson, Hillsdale, some Maskwa, and TL6. Cr/Ni and Sm/Nd correspond exactly to GCE_{avg}. Ba/Sr, La/Nb, Zr/Y are within 1σ of GCE_{avg}. However, K/Rb (20.2) and 1000 Th/Zr (95) are significantly higher than either GCE_{avg} (16.1 ± 1.0 and 27.0 ± 3.0, respectively) or L_{avg}. It is apparent that chemical data without geological control cannot be used to identify GCE. This observation is not surprising, since there are so many variable in the deposition of intervolcanic sedimentary rocks.

5.3.5 Abitibi Lamprophyres

The geologic setting, widespread distribution, petrography, geochemistry, and origin of lamprophyres in the Superior Province were studied by Wyman and Kerrich (1989). All lamprophyres examined by them occur as narrow dikes, generally 0.5 m to 1.0 wide and are rarely continuous for more than 20 m along strike. They are steeply dipping and commonly possess sheared margins composed of 1–2 cm foliated chloritized material. Chill features in nonfoliated margins are extremely rare, and no contact effects were detected in host rocks. Fresh dikes have SiO_2 (47–51%), TiO_2 (0.55–0.60%), P_2O_5 (0.36–1.58%), K_2O/Na_2O 1.5–2.3, and Mg# (78–80). Unaltered dikes are characterized by extreme enrichments of the lithophile elements K, Rb, Ba, Cs, Th, U, and light rare earth elements (LREE), as well as enhanced abundances of Cr (650–920 ppm), Co (37–42), Ni (264–355), and Sc (22–28). MORB normalized plots display the following: consistent enrichments of K, Rb, and Ba relative to Sr and LREE; variable enrichments of Rh and Ba relative to K; and troughs at Ta-Nb and Ti. Compared with fresh rocks, altered samples have systematically lower K, Rh, Ba, and Sr contents, but similar REE distributions and profiles from Th-Yb and Sc-Ni on normalized diagrams. Suites of post-Archean ultramafic lamprophyres in the Superior Province are stated to be enriched in large ion lithophile elements (LILE), LREE, and compatible elements, relative to their Archean counterparts and lack troughs at Ta-Nb and Ti. Age relations are based on the following: cross-cutting by dated dikes (e.g., Matachewan diabase 2643 Ma); the authors' interpretation that lamprophyres postdate the main ductile deformation; and the temporal and spatial association with gold mineralization that systematically developed late in the tectonic-magmatic evolution of each subprovince: the gold mineralization event has either been directly dated in several areas, or alternatively is closely bracketed by independent high-precision U-Pb ages on intrusive rocks. The authors conclude that alkaline magmatism is constrained to the terminal stages of alkaline tectonic-magmatic evolution of each subprovince during diachronous transpressive events. Comparing GCE_{avg} to their analyses, SiO_2, Cr, Co, and Ni correspond closely. Ti is slightly higher; P_2O_5 is much lower; and K_2O/Na_2O of GCE_{avg} falls below the Abitibi range. On an MgO vs Ni plot (Wyman and Kerrich, 1989, Figure 12), L_{avg}, GCE_{avg} and M_{avg} correspond with their L analyses from Red Lake, Shebandowan, Schreiber-Marathon, but Ni and MgO values for Wawa and Pickle Lake are higher than all three averages. The chemistry and mode of occurrence strongly suggests the described dikes are GCE of the current study.

5.3.6 Canadian Arrow, Matheson, Ontario

In an otherwise excellent geochemical study of two lamprophyres in the Canadian Arrow mine that are exposed at the edge of an open pit, the geological map (McNeill and Kerrich, 1986, Figure 1(b)) does not reveal whether they are parallel to bedding or cross-cutting, so it is not obvious from their mapping whether they are GCEs or lamprophyres. The host rock is described as trondhjemitic intrusive. Analyses from the open pit are presented in Table 5.9 and petrographic ratios in Table 5.10. The freshest, most undeformed dike Sample 1X is considered the parent by the authors and the other analyses are compared to it. Sample 1A is the least altered and the remainder (05, 6A, 6C, 6D, 6E) taken across strike, are stated to represent progressive alteration.

In the V/Cr vs Nb/Pb diagram, Figure 5.12, the analyses span a wide range of Nb/Pb and a more restricted range of V/Cr. None of the analyses plot near Two Lamps L (L1, L2, L3, Table 5.5) but the Canadian Arrow dikes span a compositional range similar to Maskwa GCE. However, Zr/Sc (15″ is similar to L (11.3″ La 2.5″) average 1.9) for GCE_{avg} is 6.2 and 2.0 for L_{avg} 6.1 ± 1.2″, values that are similar to the lowest Canadian.

Arrow dike (samples not identified in Figure 8, p. 336 of McNeill and Kerrich (1986)).

Again, the chemical data are permissive but do not constitute proof that the Abitibi lamprophyres are GCE of this study.

5.3.7 Young-Davidson Syenite-Hosted Gold Deposit, Matachewan, Ontario

Three chemical analyses of lamprophyres (Tables 5.9 and 5.10) and the following summary are taken from a study by Martin (2012). The Young-Davidson (YD) gold deposit is located southwest of Kirkland Lake, in the southern Abitibi greenstone belt, Canada. It is currently the subject of study by Linnen et al. (2012). Historical gold production was mainly from quartz veins and disseminated pyrite hosted in altered syenite. Underground reserves and resources stated by AuRico Gold Corporation (formerly Northgate Minerals Co.) are 14.6 million tons (measured and indicated) at an average grade of 2.03 g Au/T and an underground reserve of 39.1 million tons (proven and probable) at an average grade of 2.79 g Au/T (Edmunds, 2009). Three generations of veins are identified by (Martin, 2012) from cross-cutting relationships at an underground ore crosscut: V1 boudinaged iron–carbonate veins, V2 folded quartz–pyrite veinlets, and V3 planar quartz-carbonate veins. Structural characterization indicates that V2 and V3 extensional vein arrays developed during the D2 deformation, which generated steeply dipping, east–west striking penetrative S2 foliation. Pyrite mapping combined with alteration assemblages, mineralization characteristics, and mineral chemistry suggests that the syenite-hosted mineralization formed from interaction between multiple fluids with distinct physicochemical conditions. The formation model proposed for the Young-Davidson deposit by Martin (2012) is of mixing between magmatic and metamorphic fluids during deformation that generated the extensive veining systems at Young-Davidson.

Analyses Y1, Y5, and Y7 (Tables 5.9 and 5.10) are from drill core from the Lower Boundary Zone, with locations shown in mine cross section 23,240 mE (Martin, 2012, analyses 802811:Y1, 803325:Y5, and 802847:Y7), sections that are also shown in Linnen et al. (2012, p. 28). Samples were identified as biotite-rich lamprophyre or sediment. Y1 is hematite–potassic altered, Y5 is described as sediment, and Y7 described as fresh lamprophyre. Y1 and Y5 show distinctively low Nb/Pb (Figure 5.9), but all three have V/Cr similar to Two Lamps E2 and E3 and on this basis and their clearly bedding-parallel occurrence (mine cross-section 23,240 m E) may be considered GCE interflow sediments.

5.3.8 Malartic, Quebec

"Chloritic seams" in the Malartic mine, described by Derry (1939), have mineralogy of hornblende, biotite, chlorite and feldspar in variable proportion. They are 1–5 ft wide, are dark green, in places schistose, with 40% plagioclase and orthoclase feldspar, hornblende, biotite, and tourmaline. Derry states that they are not true dikes and do not have chilled borders, even when they are not "unsheared". They show considerable variation of texture and composition, with some biotite as the main mineral, but chlorite and hornblende are more common. They are cut by later pegmatites. No chemical analyses are presented. The mine cross-sections and descriptions permit a tentative identification of the chloritic seams as GCE. The Malartic Mine as well as the mafic rocks are currently the subject of extensive study (Perrouty et al., 2014).

5.3.9 Tennant Creek Inlier, NT, Australia

The lamprophyres described by Duggan and Jacques (1996) and references therein, from the Tennant Creek Inlier, which hosts major gold deposits, vary from volumetrically minor, strongly mafic, to mica-rich rocks (minettes), to leucocratic types dominated by quartz and orthoclase. Ubiquitous alteration ranges from hematite staining of orthoclase and minor development of chlorite, through to extensive alteration of the primary igneous assemblage with the development of abundant chlorite, hematite, quartz, carbonate, and clays. The dikes were dated by phlogophite K-Ar, zircon U-Pb methods at

1700–1660 Ma, which is younger than host mineralization 1830–1825 Ma. Although stated to be indistinguishable in petrography and major element abundances, the lamprophyres are divisible into two distinct groups on the basis of trace element geochemistry. Group 1 (Zr/Nb > 20) is more variable, with generally higher Sr, Ba, and LREE as well as lower U, Th, and HREE as compared with Group 2 (Zr/Nb < 12). The two groups are substantially spatially distinct, with the Group 2 rocks restricted mostly to a small part of the southeastern outcrop area of the Warramunga Group.

For the Tennant Ck data, averages of V/Cr for the high and low Zr/Nb groups from Duggan and Jaques (1996, Table 3) are 0.34 and 1.59, respectively, whereas Nb/Pb are 0.79 and 1.23. Comparing the data with Two Lamps analyses (Table 5.2), all analyses except 71060183 may be GCE. For this latter analysis, V/Cr is 6.03 due to low (32-ppm) Cr, whereas Nb/Pb is 0.2, which plots near E95 in Figure 5.7 and suggests that it may be a true lamprophyre. In summary:

1. The presence of quartz and feldspar in leucocratic dikes is compatible with a GCE origin,
2. Zr/Nb of GCE_{avg} and L_{avg} overlap, and this ratio does not discriminate between them,
3. Zr and Nb concentrations of both Two Lamps L and GCE are lower than the Tennant Creek samples and do not clearly fall into either the high or low Zr/Nb groups of Duggan and Jacques (1996, Figure 3).
4. Based on published analyses alone, the majority of Tennant Creek Inlier lamprophyres may be classed as GCE units. Confirmation would require field mapping and specific dating of dikes and host rock. The conclusions of Duggan and Jacques are that lamprophyres are localized along the same structural channels that control flow of the mineralized fluid, or association of lamprophyres and mineralization is the artifact location of drill sites closer to mineralization. Both of these conclusions are compatible with a proposed syndepositional origin.

5.4 DISCUSSION OF LAMPROPHYRES AND GCE

5.4.1 Dating and the Relative Age of Lamprophyres

According to Rock (1991, p.153 and Table 9.2), wherever detailed field and/or geochronological evidence is available, lamprophyres overlap gold mineralization in both time and space, and are often the only rock to have been emplaced precisely at the same time as the gold. He also points out that the space–time connection between lamprophyres and gold deposits is close to a scientific "fact," with both quantitative and statistical data as support. He cites Hodgson and Troop (1988) who, in turn, state that both true lamprophyres and potassically altered mafic dikes of all types show a statistically significant correlation with economic status in the deposits of the Abitibi Greenstone Belt.

An example in which the gold deposits, lamprophyres, and host granites are from different locations and therefore a stated >100 My age difference does not appear to relate to gold genesis is provided by lamprophyres described by Duggan and Jacques (1996). The dikes are considered postmineralization because muscovite K-Ar ages in the gold deposits east of Tennant Creek, where dikes also occur, gave older ages of 1830–1825 Ma (Compston and McDougall, 1994). However, lamprophyres in the Warramunga group and the Warrego granite itself give a different age (1663–1700 Ma). The relationship of the lamprophyre dikes to the mineralization is not clear from published accounts.

5.4.2 Distinguishing Lamprophyres (L) from Gold Cycle E (GCE) Interflow Sediments

5.4.2.1 Field Criteria

GCE units are bedding parallel, are dark green to black in color, weather low, are generally laminated or bedded, and are almost always foliated (so-called "shear zones"). Biotite–phlogopite is a common primary mineral in lamprophyres (minette, vogesite, kersantite?) but does not uniquely define a lamprophyre. Biotite is not a primary mineral in GCE, but it may form from chlorite at higher metamorphic grades or during contact metamorphism, adding another level of difficulty to distinguishing between L and GCE in higher-grade metamorphic areas. GCE generally are less magnetic than L (but Mesomikenda GCE has modal magnetite). Major differences are summarized in Table 5.11.

GCE units are generally discontinuous and weather recessively, and the timing relationship is difficult to determine. Thermal and thermo-mechanical erosion and felsic magma contamination, analogous to such processes described for interflow sediments and Kambalda komatiites (Williams et al., 1998), may contribute to the discontinuous nature of GCE. In higher-grade metamorphic terrains (e.g., Maskwa batholith, Chapter 4) GCE are amphibolites and, if disrupted or faulted, may be identified as xenoliths.

5.4.2.2 Chemical Distinction

The close similarity in field occurrence and chemistry of the two rock types is one reason why the distinction between them has not been fully appreciated in either the lode gold or lamprophyre literature. For lamprophyres, using element ratios for identification is complicated by possible assimilation of host rock and sampling bias. A wide range in major element composition may be due to differentiation, high volatile content, modal xenoliths, macrocryst, and primary mineral variation. For GCE, variable bulk chemistry results

Table 5.11 Field Criteria for Distinguishing Between Interflow Sediment (GCE) and Intrusive Lamprophyre (L)

Gold Cycle "E" (GCE)	Lamprophyre[a]
Olive green to dark green, black if carbonaceous or high-grade metamorphic	Dark green to black
Soft, weathers low	Dense and hard, weathers high
Biotite in higher metamorphic grade	Porphyritic phlogopite (biotite), pyroxene, amphibole, olivine, groundmass chlorite, carbonate
LOI higher than L, up to 11% or 12%, but lower (4–5%) in higher-grade metamorphic terrain	LOI less than about 5%. Calcite present if intruded into CO_2-rich environment.
Bedded (foliated); may show upward fining, chemical gradients, asymmetry	Foliated if metamorphosed, flowage differentiation, chilled margins
Main minerals: chlorite–carbonate–quartz	Magnetite, pyrite accessories
Au variable and high if close to hydrothermal vent, base metals	Au low to zero, base metals low to absent, but elevated if gold-bearing
Significant and variable	Stratigraphy traversed
Possible epiclastic–detrital feldspar, pyroxene (amphibole) or quartz, possible lithic clasts	Xenoliths and inclusions of country rock
Nonmagnetic or low-magnetic susceptibility	Higher-magnetic susceptibility

[a]*Note: For further details on lamprophyre (L) characteristics, see Rock (1984, 1991) and references in the present study, e.g., McCall et al. (1990) and Adams et al. (2005).*

from interplay of terrigenous or volcaniclastic sedimentation and variable hydrothermal fluid composition. Metamorphism may result in further chemical and physical variation.

Chemical analyses of GCE and L from the Two Lamps outcrop show overlap in both V/Cr and Nb/Pb for grab samples (Figure 5.9). Similarly large variation is shown by the channel samples. In GCE, asymmetry may be displayed as differences in foliation (upward fining) or differences in carbonate, sulfide, or quartz content across strike. As stated, this reflects changes in contemporaneous volcanism, sedimentation, and/or accompanying hydrothermal activity that is reflected in the variable trace element chemistry. E1 and E2 (Table 5.2, Figures 5.9 and 5.11) are 30-cm-long, sawed channel samples taken from the center of Two Lamps GCE. From graded bedding in drill core and regional considerations, facing directions are to southwest, and E1 stratigraphically overlies E2. Figure 5.9 and other discriminating plots (Figures 5.8 to 5.12) show that there is a decrease in all elements and ratios between samples E1 and E2. The most extreme separation is shown on the LeBas alkali–silica diagram (Figure 5.8(c)). E1 plots on the boundary of basalt and picro-basalt while GCE_{avg} and E2 fall in the field of basaltic trachy-andesite.

In trying to further characterize GCE, for E1, E2, and E3, the Th/La average of 0.12 is within upper range of compositions for 2.75 Ga Archean continental crust estimate by Plank (2005, Figure 9) after Taylor and McLennan (1985), but the Th/La for Chester GCE (E4, 0.11) falls below the range. Apparently, average Th/La is very sensitive to the proportions of the different pelagic units. This line of investigation would appear to be worth pursuing, since Th/La is significantly different for L_{avg} (0.09) and GCE_{avg} (0.12). The Zr/Y (>7) for L places it in the calc-alkaline magma clan of Barrett and MacLean (1994), whereas most Two Lamps GCE (Zr/Y = 4.5−7.5) are placed in the transitional magma clan (Zr/Y = 4.5−7).

Wyman and Kerrich (1993), quoting Goldstein (1988), state that the decoupling of Rb-Sr and Sm-Nd systematics is common in clastic sediments. Sr is lost during weathering, and clastic sediments typically have Sr/Nd ~ 3, compared with Sr/Nd ~ 40 in volcanogenic detritus. Addition of juvenile volcanogenic material buffers the Sr isotopic composition of clastic sediments, but does not strongly affect the Nd isotopic composition. The Sr/Nd ratio varies between 3.2 (E1) and 11.4 (E3), (Table 5.5). For L_{avg} it is 0.186. For M_{avg} (Tables 5.7 and 5.9) the following ratios are within 1σ for GCE_{avg}: Th/U, Ti/V. Ti/Zr, Zr/Sc, Zr/Y.

In the lamprophyre literature, Rock (1984) has identified three lamprophyre petrogenetic associations: (1) with calc-alkaline granitoid plutons, (2) with shoshonitic volcanic and subvolcanic suites, and (3) with appinite-breccia pipe complexes. GCE of the current study are unlikely to belong to (3), but could belong to either or both of (1) and (2).

Taylor et al. (1994) found that variable mobile lithophile element (K, Rb, Ba, Sr) concentrations suggest that the F, S, and CO_2 contents of the Yilgarn lamprophyres can effectively discriminate mineralized lamprophyres from nonmineralized samples. S, Au, CO_2, and Au/Pd content of lamprophyres appears to be a function of proximity to gold deposits. They find that a direct genetic link between gold mineralization and shoshonitic lamprophyres is unlikely, and that correlations among Au abundance, geochemical alteration, carbonation, proximity to gold deposits, and abundance of chalcophile and Au-pathfinder elements suggest that relative Au enrichments and depletions are largely the result of external fluid infiltration following lamprophyre emplacement. They also propose (Taylor et al., 1994, Figures 10 and 11) three classes of lamprophyres on the basis of S and F/CO_2 contents and ratio plots. Based on findings of the current study, any one or all three could be GCE. More importantly, the present work suggests that elevated Au, base metal, trace element, and CO_2 and H_2O content need not necessarily result from alteration or contamination of magmas, as suggested by several authors, e.g., Wyman and Kerrich (1989) for Greenstone Belt metallogeny, Harris et al. (2000) for Swayze lithogeochemical data, Huang et al. (2002) for Laowangzhai lamprophyres, and Ashley et al. (1994) for the Hillgrove lamprophyres. If these lamprophyres are indeed GCE units, their variable composition reflects primary sedimentation conditions. It is, however, noteworthy that at the Campbell Mine, Red Lake, a true lamprophyre cross-cuts gold-bearing carbonate veins and bleached basalts in a surface outcrop, displays chilled margins, and has assimilated significant gold, e.g., Dubé et al. (2004).

Three other examples further demonstrate the difficulty of identifying GCE from gold deposit literature. An excellent geological cross-section (Olson et al., 1992, Figure 1) of the Syama gold deposit in Mali and good sample descriptions suggest that it is highly likely that analyses SYD55 and SYD51 (Tables 5.9 and 5.10 and S1 and S2 on Figure 5.11) represent GCE. Al_2O_3/TiO_2 (13.2 vs 13.8 for GCE_{avg}) and V/Cr (0.52 vs 0.45 for GCE_{avg}) correspond well, but K/Rb (15.02 vs 46.3 for GCE_{avg}) does not. No other ratios are available for comparison. For the Young-Davidson deposit (Tables 5.9 and 5.10 and Y1,Y5, Y7 on Figure 5.9), while good geological cross sections and brief descriptions, Martin (2012) suggests that the samples are GCE, only Sr/Nd, and V/Cr fall "1σ. In Figure 5.7, both Syama and Young-Davidson analyses plot within GCE_{avg} on the low V/Cr end of the diagram either within the field of Shoshonitic Calc-Alkaline lamprophyres or in the overlap field. For Olympic Dam, six mafic to ultramafic dikes analyzed by Johnson and McCulloch (1995, Table 1) have Sm/Nd ranging from 0.13 (MX4548) to (0.89 (RX4539), which encompasses both GCE_{avg} σ and L_{avg}" 1σ. No other ratios are available, and the published geological descriptions do not permit classification as L or GCE.

To conclude, the variability in chemical composition of GCE is a reflection of its origin and points to both the difficulty of using purely chemical criteria to differentiate between L and GCE and the need to develop more specific criteria. In keeping with the syngenetic theme of this book, the authors interpret variable volatile and trace element content in GCE as largely a primary feature, related to the interplay of hydrothermal and submarine volcanic-sedimentary processes unless proven otherwise. It is not necessarily due to alteration. Clearly: (1) field relations or detailed drill core logging must be used to discriminate between a sediment or a dike; (2) detailed chemical, mineralogical, and thin section study is required before specific causes can be attributed to the variation; and (3) high gold and base metal concentrations of true lamprophyre may indeed provide a possible vector to mineralization as envisaged by Taylor et al. (1994), not because the gold concentration of the lamprophyre magma is elevated but because the magma was contaminated by gold-bearing strata it traversed.

5.4.3 Mafic Enclaves and the Trondhjemite-Tonalite-Granodiorite (TTG) Suite

A link between felsic rocks, the TTG suite, and lode gold Au deposits has been drawn by Burrows et al. (1986), Jemelieta and Spooner (1989), and Spooner (1991), among others. When one follows this link and the debates surrounding the origin of TTG suite lithologies (e.g., Condie, 2014), the presence of mafic dikes, enclaves, screens, cognate mafic to ultramafic inclusions, or mafic intrusions/rafts are frequently noted (Whalen et al., 2002). In their study, mafic group B inclusions and, to a lesser extent, group A are stated to be compatible with mafic components of andesite–dacite–rhyolite (ADR) arc-type suites (Whalen et al., 2002, p. 1560) with the stated implications about their petrogenesis under conditions of low pressure, dry melting, and fluxing by hydrous fluids derived from subducted lithosphere.

In the present study, this line of investigation cannot be pursued in detail, except to note that in Harker and other variation diagrams (e.g., Figures 4–6 of Whalen et al., 2002), redrawn as Figure 5.11, some Maskwa samples (Tables 5.7 and 5.8) and GCE analyses from Two Lamps and Chester (Tables 5.2–5.4) fall in the fields A or B of mafic intrusions/rafts (Figure 5.11).

Maskwa GCE show a wide range of Al_2O_3, SiO_2, Mg#, and Rb/Sr but M_{avg} (11.9%, 53.2%, 47.8, and 0.22, respectively) falls within or close to group B. Two Lamps GCE (E1, E2, E3) analyses fall within mafic group B in Figure 5.11(a), within and at higher Mg# in Figure 5.11(b), and at significantly higher Rb/Sr ratios in Figure 5.11(c). Two Lamps L (L1, L2, L3) have higher SiO_2. Chester GCE (E4) has significantly higher SiO_2, due to an admixture of hydrothermal quartz, in keeping with its volcanic-sedimentary origin.

These results suggest that mafic enclaves and "cognate mafic to ultramafic inclusions" in some TTG lithologies are GCE units. When the felsic volcanic pile is dismembered during deformation and metamorphism, these GCE division mafic units become virtually unrecognizable. This has substantial implications for the genesis of TTG rocks and early crustal processes. Also corroborated are the conclusions of Burrows and Jemielita (1989), that lithophile-element systematics in ubiquitous Archean TTG suites do not rule out a magmatic origin for Archean Au fluids.

5.5 CONCLUSIONS

In the quest to understand lode gold veins, it has been necessary to stray into the distant territory of lamprophyres and TTG rocks, a terra incognita for the exploration geologist. The reward, however, has been substantial, because the authors hope it has been become evident that: (1) there are "lamprophyres" in gold camps that are GCE interflow sediments, and this means that they must have formed subaqueously; and (2) if they formed subaqueously, their host felsic rocks formed subaqueously and therefore cannot be felsic intrusions. A further question then arises as to the nature of this felsic volcanism and its relationship to gold. To understand this better, Kretschmar turns again, in Chapter 6, to the Chester Complex in Ontario, an excellent exposure of a GCE-bearing unconformity in Trench 5 and the information provided from drilling along stratigraphic strike extension of the Côté Gold deposit.

CHAPTER 6

Felsic Volcanism Associated with Mineralization in the Chester Complex—Type Gold Deposits

CONTENTS

6.1 GEOLOGY OF THE CHESTER COMPLEX

The Chester Complex (CC) hosts numerous prospects and several small mines with historical production (Chapter 4). The authors' work has covered all of the known deposits and much of the complex and associated rocks; there is a large database of historical and government assessment work, and descriptions have been published by Cook and Roscoe (2011), Kontak et al. (2013), and the IMG Website (www.iamgold.com). The current chapter builds on the published company data posted on their Website and is followed by our evidence from field work that the ore originates from combined sea-floor hydrothermal fluids and fluids exsolved from felsic and mafic volcanic flows.

Chester Township is located in the Swayze greenstone belt in the southwestern extension of the Abitibi greenstone belt of the Superior Province. The Swayze belt contains a diversity of extrusive and intrusive rock types ranging from ultramafic through felsic in composition, as well as both chemical and clastic sedimentary rocks. Igneous rocks predominate and include both volcanic and plutonic rocks.

Locally the area overlies a narrow greenstone belt assemblage that separates the Kenogamissi granitoid complex to the north from the Ramsey—Algoma granitoid complex to the south, a portion of the northern edge of which is called the Chester Granitoid Complex (CGC). Trelawney's Chester Project (now Iamgold's Côté Gold Project) is underlain by calc-alkalic pyroclastic metavolcanics of felsic to intermediate composition (Chester Group), felsic to intermediate granitoid intrusive rocks, namely tonalite, trondhjemite, granodiorite, and diorite of the CGC and related migmatites (?). The granitoid rocks are very heterogeneous, reflecting a number of primary igneous intrusive phases, migmatization, and assimilation of older country rocks and local rafts and screens of the intruded supracrustal lithologies.

Major lithologies in the Chester Complex are basalt-diorite (B), felsic flows mapped as "granodiorite, syenite, or trondhjemite" (F), a variety of pyroclastics and tuff, and variable Gold Cycle pelagic-pelitic Gold Cycle "E" interflow

The Metallogeny of Lode Gold Deposits. http://dx.doi.org/10.1016/B978-0-12-803222-0.00006-X

sediments (GCE). Typical lithologies from outcrop and core are illustrated in Figures 6.1–6.9.

Figure 6.7 is an interesting sample. It shows the character of the end of a flow breccia, and Figure 6.8 shows the same texture in drill core. An additional agglomeritic breccia is shown in Figure 6.9. It should be noted that the smallest fragments in these photographs are angular and not partially reabsorbed, which precludes an intrusive origin. According to Kontak (2013), these breccia textures are stated to be characteristic of the rocks in the vicinity of Côté Lake mineralization; they also occur in several core intervals both on the Clam Lake claim and elsewhere on the property.

FIGURE 6.1
Angular bomb-sized heterolithic volcanics in a fine-grained felsic matrix. Outcrop is close to the Iamgold-Sanatana eastern boundary of claim 3011820.

FIGURE 6.2
Typical massive felsic flow with rounded mafic tuff to bomb-sized volcaniclastic fragments. Clam Lake, North Shore.

FIGURE 6.3
Felsic flow with heterolithic fragments. Chester Road, Km11.

FIGURE 6.4
Typical basalt and diorite. (a) Coarse-grained diorite with occasional mafic xenoliths. Hole SR11-04—27.5 m. (b) Hematite altered "Daisy diorite." (c) Basalt with possible unmixing texture.

FIGURE 6.5
Crystal tuff showing a weakly schistose matrix. Chester Road.

FIGURE 6.6
Essentially monolithic felsic peperite agglomerate with rimmed fragments and a chloritic matrix.

FIGURE 6.7 AND 6.8
Flow top or autobreccia. Angular to subrounded pink felsic fragments in dark pelitic-pelagic chloritic matrix. Graded bedding confirms facing direction to south in overturned flow sequence that dips to north.

FIGURE 6.9

Côté Gold Peperite Breccias. (a) North breccia. (b) South breccia. Both breccias are essentially monolithic blocky angular unaltered fragments in chloritic matrix, but south breccia has a small proportion of fragments of heterolithic origin.

The key relationships in the gold cycle are demonstrated by trenching and include a felsic flow sequence (Figure 6.10). This photograph shows a series of parallel bands that make up the gold cycle. Figure 6.11 illustrates the detail of the typical section across the outcrop.

FIGURE 6.10

Photograph of Trench 5 with geological contacts. "Gold Cycle" or interflow sediment is about 10 m thick. View to north. Facing directions to south. *Photograph of U. Kretschmar. Detail and units are described on Figure 6.11 from trench 5, which exposes the Gold Cycle or interflow sediment at the contact between gabbroic flows and overlying felsic units.*

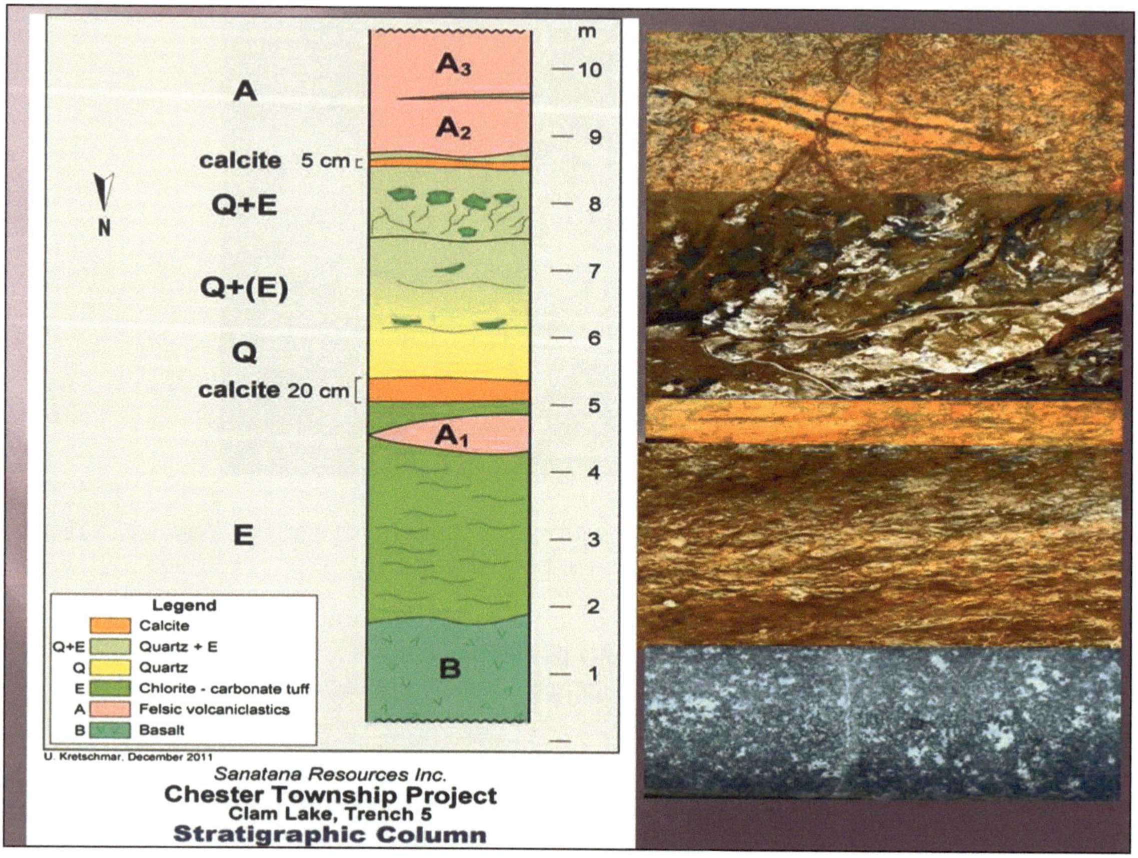

FIGURE 6.11

Stratigraphic column with "Gold Cycle" or interflow chlorite—quartz—carbonate unit exposed in trench shown in Figure 6.10. Right-hand column shows representative photographs of lithologies. Notable are two pure calcite beds, an upper one that is 5 cm thick within mixed Q + E lithology, and a lower 20-cm-thick bed that signals the onset of silica precipitation, from mainly quartz—carbonate—chlorite pelagic-pelitic sedimentation.

Drilling showed that the sequence is overturned. From field measurements, units dip to the north at about 65°. Assays and whole rock analyses are from channel-sawed chip samples averaging 30 cm in length and 3 cm in width. The highest-grade assays 3.0 g Au/T occur in chloritic calcareous tuff and quartz. The chemistry of the flows is discussed in Section 6.7. Additional characteristic of the felsic volcanism is seen in the "Candy Strip" outcrop (Figure 6.12 and 6.13).

Felsic massive and pyroclastic flows may be from <1 to 20 m thick. The thicker ones are uniform, massive, have the appearance of fine-grained granite, except for the presence of angular mafic pyroclastic fragments, which suggests that the

FIGURE 6.12

Chester Complex. Southeast striking, steeply NE dipping sequence of felsic flows from the "Candy Stripe" stripped outcrop. The area displayed in the photo-panel is an approx. 8 × 10-m outcrop. Red fiduciary marks are approximately 1 m apart. Darker areas are weathered surfaces. North is at bottom and stratigraphic tops at top of photo. Flows display incipient autoclastic fractures resulting from cooling during flowage, and appear to have been deposited into sulfide-silica—rich brine pool, which is reflected by very thin chlorite—pyrite alteration selvages and interstitial sulfide "veins." These sulphide veins show minimal penetration into fractures. Direction of flowage is interpreted as left to right.

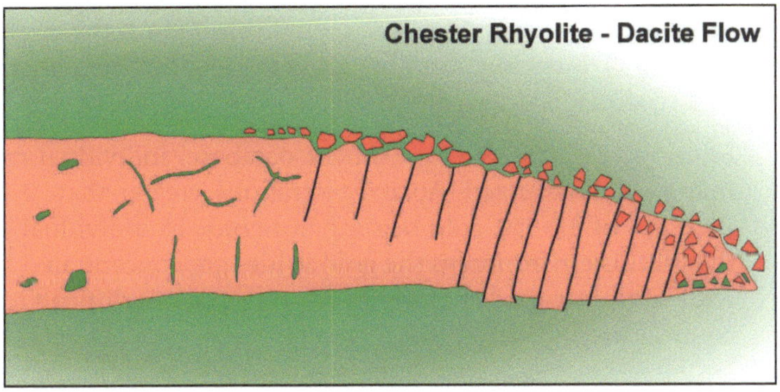

FIGURE 6.13

Schematic felsic flow geometry, typical of silica-rich flows in Chester Twp and Côté Gold—type felsic hosted gold deposits, based on "Candy Stripe" stripped area and drill core.

rocks are not intrusive but are volcanic crystal tuffs. They may show subtle indications of graded bedding. These silica-rich rocks break up and brecciate as they flow, often showing brecciated flow tops and fragments hosted by a pelagic mud matrix. Common are distinct angular fragments that appear to be stacked, evidence that the shape is primary. (i.e., not due to later folding and deformation). The breccia outcrops in Figures 6.7 and 6.8 are interpreted as peperite and also show this angular fragmentation.

FIGURE 6.14
Quartz band which shows massive white fine-grained quartz adjacent to dark grey material with angular matrix-supported quartz clasts.

Silica precipitation in interflow environment shows a brecciation mechanism in operation (Figure 6.14). This texture and asymmetry are uncommon, but may reflect bottom currents that were generated by the next felsic flow.

6.2 MINERALIZATION

The ultimate objective, one might say the only objective, is to understand the mineralizing process in this environment and to use this knowledge to find more ore. Mineralization is scattered throughout the rocks in very low values (<0.5 g/T), but some short higher-grade sections can enhance the average grade of a section.

From the assay and trace element database, individual mineralized horizons may be characterized. Au concentrations greater than 0.10 g Au/T show the presence of 10–50 gold-bearing horizons, in individual drill holes. It seems that in this environment the gold values are concentrated in high-grade shoots with large, low-grade haloes. The overall grade is difficult to predict because the controls are illusive.

6.3 TOWARD A MODEL FOR FELSIC-HOSTED GOLD DEPOSITS BASED ON CHESTER TWP GEOLOGY AND MINERALIZATION

6.3.1 Lithology

In the Chester Complex, major lithologies consist of a sequence of interbedded felsic and basaltic flows, trending roughly 100° and dipping north at 50–60°. The felsic flows are 1- to 20-m thick quartz-eye dacite and rhyolite. Several percent of subangular to rounded heterolithic mafic pyroclastic fragments are common. A fine-grained buff color and peppery appearance are common on weathered and glaciated outcrops. The felsic rocks form short, stubby flows that are laterally discontinuous and change thickness. The thick ones are massive, uniform, and resemble granite. Distinctive felsic units include thin flesh or red hematite–stained, porous textured monomineralic plagioclase–quartz feldspar flows. These have variable silica content that may range from syenite to rhyolite. A white plagioclase–quartz unit has 74.9% SiO_2.

Interbedded are gabbroic basalts and massive flows of approximately similar thickness. The basalts are dark green and may be coarse with a gabbroic texture, or medium to fine grained. A distinctive "daisy diorite" texture, possibly indicative of bi-modal magma unmixing, characterizes some coarse flows. Upward fining shows consistent stratigraphic tops to the south. QEA asymmetry is common and consists of a massive basal unit (felsic flow or A), which is overlain by a chlorite–carbonate unit (generally well-bedded GCE), which in turn may be overlain by a quartz–sulfide–gold unit (the "vein" or QU).

Flows may be interbedded with thin chlorite–carbonate interflow sediments or tuffs also described as "lamprophyres" or "Gold Cycles E" lithologies. These range from 1 cm to about 1 m in thickness. Their composition is mainly chlorite, probably derived from devitrified glass. Thinly interbedded hydrothermal carbonate, silica, and sulfide layers are common. These GCE units weather low and are generally discontinuous, dismembered, and difficult to map or characterize. Thicker units may encompass quartz units ("veins"), chlorite, ankerite, or calcite. Interbedded sulfides are mainly pyrite but may include sulfosalts and tellurides.

Minor lithologies consist of pyroclastics including lapilli to bomb-sized heterolithic clasts, tuffs, feldspar crystal tuffs, and felsic feldspar porphyritic flows.

There are more mafic flows higher in the stratigraphy as basaltic volcanism starts to predominate over felsic volcanism.

6.3.2 Textures

In most outcrops, felsic flows are massive and structureless. Less common are felsic pyroclastics, flows that are silica-rich, and incorporated mafic flow lithologies, the result of autobrecciation and fragmentation during flowage. Brecciated and autoclastic flow tops are diagnostic but rare. A subtle fracture pattern is locally apparent that manifests in a distinctive angular fragment texture that is primary. It appears to be the result of autobrecciation and brecciation during flowage. This is well displayed by flows at the "Candy Stripe" outcrop. In other cases, because of residual heat in the flows, small (pegmatite-like amphibole–chlorite sweats) veinlets formed in the flows, perpendicular to flow contacts. These may contain gold. The felsic magma had low viscosity and turned more viscous and fragmented as devolatilization during decarbonation (i.e., exsolution of water and CO_2) attending flowage. Some diorite textures show probable rhyolite–basalt unmixing textures.

At the North and South breccias in the approximate center of the Côté Gold deposit, felsic rocks flowed into a small sea-floor depression or basin and incorporated the sulfide- and gold-impregnated chlorite–carbonate mud, forming the interstitial (green) matrix. There is minimal reaction of felsic magma with pelagic mud, and therefore limited rimming of fragments in a process similar to formation of peperites. The interstitial green chloritic material is a mixture of both sedimentary and hydrothermal origin. Its distinctive chemistry with high REE content may reflect higher temperatures and proximity to a volcanic vent.

6.3.3 Mineralization

Mineralization consists of rare native gold, gold in association with or within pyrite and chalcopyrite, sphalerite, galena, and gold contained in tellurides and sulfosalts. Gold occurs in either chlorite–carbonate–(sulfide) interflow sediments (either quartz "veins" [QU in Chapter 2] if silica-rich or basaltic chlorite–carbonate GCE units) and also disseminated or in fractures within felsic flows. The bimodal nature of volcanism may account for the association of chalcopyrite–gold with basaltic flows and gold and silica with the felsic flows. The gold itself originates from silica-rich hydrothermal fluids that exsolved from the felsic magma or accompanied extrusion of the flows. These either migrated up "vents," the expression of which today would be cross-cutting veins, or most likely in much of the Chester Complex, in a diffuse manner, from sea-floor seeps.

6.3.4 Alteration and Silica Flooding

It is not clear, from work to date, the extent to which postdeposition or alteration in the traditional sense has taken place. The authors think that most chemical and textural features of the felsic and basaltic volcanic can be considered due to differentiation and silica flooding. The North and South breccias may reflect local alteration and proximity to a heat source or possibly felsic agglomerates. The geochemistry of the breccia matrix and its elevated REE content and propylitic alteration are discussed by Kontak et al. (2013). Silica flooding or alteration and gold are closely related. The Blue Lagoon in Iceland shows how silica can be leached by chlorine-enriched, low-temperature water from basalt and precipitated as a colloid (i.e., not crystalline). This process may be invoked in part to explain the white, massive, uniform, quartzite-like unit that crops out in the middle of the Côté Lake deposit. In local terminology it was termed alaskite, but on the basis of a thin-section description, it is a plagioclase–quartz unit. Massive effusion of colloidal silica probably also underlies the large amount of silica-rich rocks: most probably there was copious effusion of colloidal silica from vents or seeps on the sea-floor into which the felsic flows were being deposited.

6.3.5 Structure and Metamorphism

A south-facing monolithic sequence overturned by thrusting from the north is interpreted by all mapping and drilling. Dips further steepen toward the south (from 53° at Clam Lake to 60–75° on the Chester Road). The strata may be overturned and thrust into a nappe-like structure farther to the south. Considerations of regional fold symmetry suggest that felsic–sericitic pyroclastics of the Marion Group, bordering the Chester volcanics to the north, are older than the Chester volcanics, but there is insufficiently detailed age dating. Proterozoic faults that offset Matachewan diabase dikes (2400 Ma) have also dismembered and re-oriented the Chester Complex stratigraphy. There is no evidence that low-grade greenschist facies metamorphism or deformation have had a significant effect on textures other than that refraction of the D1 axial plane foliation common to interflow GCE. The lamprophyre at the "Two Lamps" outcrop was intruded after deformation and isoclinal folding, and dating it may further constrain the timing of isoclinal folding and period of mineralization.

6.3.6 Volcanic Environment

Volcanism is bi-modal, and flows emanate from the same vent or vents in close proximity. The basalts are tholeiitic. The felsic lithologies are calc-alkalic dacite–rhyolite and also plot in the field of Archean TTG. Blue quartz eyes are common in both felsic and gabbroic rocks and suggest a cogenetic origin. Initial felsic volcanism was followed by an upward increasing proportion of mafic volcanism. The virtual absence of explosive pyroclastics in the sequence is consistent with deep water and an intermediate to distal volcanic environment. The abundance of interflow sediments (GCE units) suggests a relatively flat or low-slope sea-floor environment and relatively quiescent conditions. The environment seems to have been one of repeated episodic effusion of either exsolving basalts or felsic flows, followed by quiescence and formation of interflow chloritc tuffaceous sediments with localized and low-volume hydrothermal activity. On the other hand, elevated REE, and tourmaline in peperite matrix suggests higher formation temperatures and a proximal environment. As well, some quartz veins show well-terminated quartz crystals and miarolitic cavities that are indicative of shallow formation and boiling conditions on the sea-floor or in a shallow subsurface environment.

6.4 DISCUSSION OF CHESTER TWP VOLCANISM AND COMPARISON WITH THE NORANDA CAMP

6.4.1 Chester Felsic Volcanism

The common occurrence of GCE in felsic lithologies requires renewed focus on felsic host rocks. GCEs are time-stratigraphic markers of volcano-sedimentary origin that subdivide flows and flow sequences. Their presence in felsic-hosted gold deposits implies—indeed requires—felsic submarine extrusive volcanism rather than plutonic magmatism. Felsic intrusives have been proposed as a source of Au-bearing fluids by Burrows et al. (1986), Jemelieta and Spooner (1989), Spooner (1991), and Beakhouse (2007b) based on close spatial and temporal relationships with significant gold mineralization. Kerrich (1989) discusses geochemical aspects, which center on whether the stable isotope signature of fluids is magmatic or metamorphogenic. Whalen et al. (2002), Whalen et al. (2004), and Condie (2014) discuss TTG genesis.

In the Chester Complex and the Côté Gold deposit, textural and compositional variability can be attributed to a combination of unmixing of rhyolite–dacite from basalt–andesite and autobrecciation. Volcanism is bimodal, and flows probably emanated from the same vent or from vents in close proximity. Blue quartz eye phenocrysts are common in both felsic and mafic rocks, suggesting a common magma chamber and a cogenetic relationship (also suggested by Berger, 2011). Individual quartz-eye dacite, rhyolite, and gabbro-diorite flows vary from 30 cm to 20 m in thickness and are characterized by uniform grain size, absence of chilled margins, and a variable percentage of mafic, pyroclastic, and autobreccia fragments. Felsic units are short, stubby, and fragmented, and show autobrecciated flows that are laterally discontinuous, the result of increasing viscosity during devolatilization, de-carbonation, and flowage. Minor lithologies are pyroclastic units with lapilli to bomb-sized heterolithic fragments, andesite flows, and tuffs.

Gold in vein-type mineralization occurs primarily in GCE interflow sediments and chloritic tuffs, whereas disseminated gold occurs in fractures and sulfides in felsic flows and, to lesser extent, in thin cross-cutting veins that appear to represent feeders for flowing solutions, e.g., the Candy Stripe outcrop and in a similar area on the Chester Road.

Both the North and South Côté Lake breccia horizons, considered characteristic by Kontak et al., 2013, but nonrepresentational of the deposit from Kretschmar's mapping, are thin (<10 m thick), clearly stratigraphically controlled and interbedded with thin basalt flows and pelagic-pelitic interflow sediment with minor hydrothermal chlorite. The North and South breccias (Kontak et al., 2013) are different: the South breccia is heterolithic, the other monolithic. Both breccias have characteristics of peperite, a rock formed essentially in situ by disintegration of magma intruding and mingling with unconsolidated or poorly consolidated, typically wet sediments (Branney and Southren, 1988; White and Skilling, 2000; Skilling et al., 2002; David and Bell, 2007). Similar breccia textures have been described from the Olympic Dam deposit by Johnson and McCulloch (1995). Alteration of felsic fragments is minimal at the Côté Lake North and South breccias. This may be related to the heat capacity of the felsic fragments and extrusion rate as they enter the semi-consolidated substrate. Figure 6.6 shows rimmed fragments from an area that is higher in the volcanic stratigraphy than the North and South breccias. Their elongated outcrop pattern is a primary deposition feature. Kontak et al. (2013) describe its dimension as being 1.5 km × <300 m but provide no drill-hole data. Although an intrusive core to the Chester Complex is not excluded by the proposed synvolcanic model, there does not, at present, seem to be sufficient evidence for it. The Candy Stripe outcrop (Figure 6.12; also Figure 8D and 8E of Kontak et al., 2013) is stratigraphically below the main Côté Lake mineralization, and displays quartz–carbonate–pyrite veining and alteration that may represent possible streaming of hydrothermal solutions through a sequence of felsic flows or flowage of felsic lava into a sulfide-rich brine pool.

There are some similarities to Archean subaqueous flows described from Rouyn-Noranda, by De Rosen-Spence et al. (1980), although the Chester flows are thinner. In Noranda, felsic flows alternate with pillow basalts and with massive basalts bearing pillow breccia at their flow top. Subaqueous rhyolite flows are dome-shaped or tabular masses, up to 10 km long and up to 400 m thick, rising with a steep flow front. Larger flows show a distinct facies change from the feeding fissure to the flow front. The thickness of individual flows ranges from 30 m to more than 400 m, and the average is about 200 m.

The rheology of rhyolite lava is addressed by Yamagishi and Dimroth (1985) and references therein, from which the following is summarized. Apparently rhyolite lava behaves as a Newtonian viscous fluid at temperatures above the liquidus and has neither shear strength nor tensile strength. Its viscosity decreases with increasing temperature and water content. Below the liquidus temperature, rhyolite lava acquires a more complicated rheological behavior, in response to the growth of crystallites and vesicles, and also because the melt polymerizes. The rheologic behavior of rhyolite at temperatures below the liquidus can be approximated by a Bingham fluid, which does not flow unless the shear stress exceeds a critical value, called the shear strength; furthermore, the lava fails by brittle failure if the tensile stress exceeds a maximum value, called the tensile strength. The shear strength and apparent viscosities increase very rapidly with decreasing melt temperature and water content, whereas the tensile strength of the melt decreases simultaneously.

The morphology and petrography of submarine rhyolite flows are correlated, and the morphology of submarine rhyolite flows depends on the temperature of the lava. At a temperature of 700 °C, rhyolite lava has a high shear and tensile strength and consequently tends to form short and thick flows of block lava or lava domes mantled by breccia, morphologies that are characteristic of very viscous magmas.

The quenched phase of rhyolite hyaloclastites described by Yamagishi and Dimroth (1985) contains less than 10% crystals, attributed to extrusion at a temperature not much lower than the liquidus. Some similar flows described by De Rosen-Spence et al. (1980) from Noranda have aspect ratios of 1:40.

By analogy, it appears the viscosity of the Chester Twp rhyolite lavas was low, and it extruded at a temperature that was approaching the liquidus. Hydrothermal fluid in solution in the lava is, in part, retained if the magma extrudes below the sea and further reduces viscosity or slows devolatilization.

Textures and flow morphologies displayed at the "Candy Stripe" can be attributed to quenching of the flow surface and formation of perpendicular contraction fractures (e.g., Figure 11B of Yamaguchi and Dimroth, 1985) and minor subsequent deformation.

In Chester Twp, pyroclastic textures are rare, suggesting a deep-water environment according to criteria discussed by Downey and Lentz (2006). The common occurrence of interflow sediments and GCE suggests a relatively flat or intermediate to distal volcanic and sea-floor environment. The thickness of GCE units will be roughly proportional to the time that it took for them to accumulate. Archean sedimentation rates in volcanic settings are on the order of 0.1–2.5 mm/1000 years (Thurston et al., 2008); therefore, a 1-m-thick GCE may take 10,000 to 25,000 years to accumulate. For VMS deposits, independent modeling of hydrothermal fluid flow in the Panorama field shows that individual deposits could have formed in some 5000 years, whereas district scale hydrothermal activity may have lasted for >200,000 years (Schardt and Large, 2009). In a given stratigraphic section, total GCE thickness will be a measure of accumulation time and will probably be related to distance from the volcanic center. Studies of GCE should significantly enhance deposit-scale understanding. Kretschmar's alternative interpretation of the Côté Gold deposit origin emphasizes the importance of continually reviewing the genetic model for a deposit during exploration and exploitation, as more detailed information on deposit architecture becomes available.

For the present work, it is relevant that Re-Os dating of Mo intergrowth with Au gave ages of 2741 ± 1 and 2737 ± 1 Ma from lode veins in the Côté Gold deposit (Kontak et al., 2013), whereas Pb on zircon from host trondhjemite gave ages of 2741 ±" same age in the Chester Complex. This is independent permissive evidence for contemporaneous veining in the Côté Gold deposit. Also, the predominantly magmatic isotopic signature (Kontak et al., 2013) found for Côté Gold, do not preclude authors interpretation that the felsic units are flows rather than a felsic intrusion. Further research is continuing by Katz et al. (2014), based on the premise that the Côté Gold deposit is an Archean Cu-Au porphyry. This seems unlikely, because the demonstrated mineralization does not fit the model of fracturing and vein injection and local wall rock alteration seen in porphyry deposit (McBride, 1972).

CHAPTER 7

Understanding Hydrothermal Systems

CONTENTS

7.1 INTRODUCTION

Any genetic model for lode vein formation is complete only if the P-T-x formation processes are understood. These parameters define the conditions under which the gold mineralization was formed and give the character of the fluids. Commonly, these data are interpreted on the basis of a "geological model." Field work must establish the stratigraphy, position of and mode of mineralization, and structural influence on the gold-bearing system. In this chapter, experimental and observational data from various studies, including those of recent mineralizing systems on the ocean floor, are investigated and summarized to assist interpretation of ancient systems for which they are a parallel.

Important clues are provided by formation conditions of gold, commonly associated sulfides, carbonates, graphite (carbon), and most importantly quartz. To what extent do the extensive stable isotopes, fluid inclusion, and vein dating data in the literature support a syngenetic formation mode, and do we see lode veins forming on the modern seafloor? These subjects are each comprehensively treated in the literature, but it is necessary to examine them briefly in the context of a syngenetic environment of lode vein formation. Various disciplines of mineral deposit formation are considered in the following discussion to determine their contribution to understanding gold deposit generation. The first consideration is the information that can be gleaned from a typical "vein," which the authors will call the "Rosetta Vein."

7.2 THE "ROSETTA VEIN"

In Chapter 2 (Figure 7.1), a polished slab of quartz vein float from the Beaver Dam gold deposit in Nova Scotia was considered to be a "Rosetta Vein" by analogy with the famous Rosetta Stone, which permitted deciphering of the Egyptian hieroglyphic alphabet. While other interpretations of the textures

The Metallogeny of Lode Gold Deposits. http://dx.doi.org/10.1016/B978-0-12-803222-0.00007-1

are possible, from a syngenetic perspective the "vein" is clearly a silica layer on a pelitic substrate: the upper E division of Bouma turbidite.

At least four distinct "domains" are recognizable, which clearly represent different episodes of quartz deposition. The bottom 7-cm-thick bed consists of massive white quartz with a few apparent shale wisps on a shale base. The top of this layer is marked by a thin shale parting, which was deposited on an indented surface. The hollows in this surface are filled with thicker (up to 3 or 4 mm) accumulations of black shale or graphite. A reasonable interpretation is that the shale fills in and contours onto an indented surface. This therefore reflects younging upwards, and constitutes a geopetal feature. The next two quartz layers are 0.5 to 1 cm thick and consist of darker and more recrystallized quartz. They are separated by a thin, poorly developed bedding surface with small amounts of clayey material.

Separating the lower three quartz "domains" from the overlying fourth is a 3-mm thick pyrite–galena cap. The upper quartz "domain" consists of 7.5 cm of coarsely crystalline quartz with irregularly shaped inclusions, generally elongate perpendicular to bedding. The fragments are delicately laminated, and individual laminae consist of a combination of graphite or sericite and chlorite. The quartz filling interstices between these structures is whiter and less crystalline than the lower quartz layers. The lowermost layer displays fractures perpendicular to bedding. These terminate at layer Q2 in the figure.

FIGURE 7.1

Polished slab of quartz unit (QU) or quartz vein, the "Rosetta Vein." Beaver Dam gold deposit, Cambro-Ordovician Meguma turbidite sequence, Nova Scotia. Illustrated is the sedimentary nature of deposition. Distinct pulses of hydrothermal quartz and sulfide (gold) were deposited with ongoing pelagic–pelitic sedimentation during a hiatus in turbidite deposition (same as Figure 2.3). *Photo by U. Kretschmar.*

The features displayed by this specimen can be studied further and described in greater detail; and whatever genetic model is used, it must be able to explain them. Textures similar to this have been discussed by Mawer (1986) for Meguma veins, as was discussed in Chapter 2. The obvious asymmetry has apparently not been previously addressed in the literature.

If one considers the "vein" a sedimentary unit, the following is apparent, and an explanation must be accommodated in an empirical model of fluid evolution:

1. quartz deposition was episodic and differed in nature between the upper and lower domain;
2. quartz deposition episodes were interrupted by deposition of pelagic material (black slate) and an interval of chemical precipitation of sulfides and gold;
3. the uppermost layer represents the deposition of graphite (carbon), chlorite, and sericite, which are discussed later in Section 7.7.1 on Strelly Pool Cherts.

The different nature of quartz in the upper lower domains is the result of variable turbidity (silica content in solution), which reflect different salinities and/or formation temperatures and deposition rates. Silica appears to precipitate episodically as gel or quartz when concentration and temperature of brines fluctuate in response to cooling, sudden drop in pressure, seawater dilution, bottom currents, boiling and primary changes in the fluid composition. The remainder of this chapter is devoted to understanding the P-T-x conditions under which this may take place.

Of significance to a syngenetic model for vein formation-the main theme of this book-are the sedimentation features attesting to episodic interplay of chemical and pelagic-pelltic deposition. The asymmetry is also a geopetal feature which appears to reflect secular changes in solution chemistry and deposition conditions during sedimentation.

This "vein" is a quartz precipitate and one must determine the conditions under which it must have formed. Part of the explanation can be found in modern hydrothermal seafloor vent fields.

7.3 BRIEF INTRODUCTION TO SEAFLOOR HYDROTHERMAL SYSTEMS

An excellent summary of the key findings are presented by Scott (1997), Hannington et al. (2005) and Pirajno (2009). About 71% of the Earth's surface is water covered by seawater and roughly 90% of this environment consists of marine basins >1500 m deep. In these depths, circulation of seawater through spreading ridges in oceans and in back arcs accounts for at least 25% of the heat flux. This heated seawater extracts large quantities of metals and other elements from the crust. Abundant heat generated by near-surface igneous intrusions, combined with essentially unlimited seawater, results in widespread alteration of ocean crust and numerous hydrothermal deposits on and within the seafloor. The authors are particularly interested in the behavior of silica and gold.

Scott (1997) states "Knowledge of seafloor hydrothermal processes comes largely from mid-ocean ridges, plus back-arc spreading zones of island arcs. Venting was first discovered in the Atlantis II Deep of the Red Sea, where dense warm brines overly unconsolidated metalliferous muds" (Degens and Ross, 1969). In 1976, buoyant but relatively low-temperature fluids (<13 °C) were discovered venting the Galagagos spreading center (Corliss et al., 1978). In 1978–1979, sulphides and black smokers were discovered at 21° N Lat E on the East Pacific Rise (Francheteau et al., 1979; Spiess et al., 1980).

More than 100 sites spewing mineral-laden fluids at temperatures as high as 405 °C are now known from a variety of tectonic settings (Rona and Scott, 1993). The larger sites have built hydrothermal structures up to several 100 meters in diameter and tens of meters high within vent fields that can be traced for kilometers. As of 2015, there are about 600 known and inferred sites.

Low-temperature diffuse venting has been estimated to account for up to 90% of the total oceanic convective heat flux and a significant amount of the material flux from the ocean crust to the seafloor (Morton and Sleep, 1985), resulting in widespread deposits of sulfates, iron oxide–silica–manganese oxide and cobaltiferous manganese oxide crusts. Quartz and silica hydrothermal chimneys are known from the Galapagos spreading center (Herzig et al., 1988; Vanko et al., 1991). These are described in detail in Section 7.4.2.2.

Silica deposition is also reported by Hein et al. (2008) for the Blanco Fracture Zone within the East Blanco Depression (EBD), where a unique opal–CT (cristobalite–tridymite)–hematite mound was discovered. Silica-hematite mounds have not previously been found on the deep ocean floor. The light-weight rock of the porous mound consists predominantly of opal–CT and hematite filaments, rods, and strands, and averages 77.8% SiO_2 and 11.8% Fe_2O_3. The hematite and opal–CT precipitated from a low-temperature (>115 °C), strongly oxidized, silica- and iron-rich, sulfur-poor hydrothermal fluid; a bacterial mat provided the framework for precipitation. Samples collected from a volcanoclastic rock outcrop consist primarily of quartz with lesser plagioclase, smectite, pyroxene, and sulfides; SiO_2 content averages 72.5%. The formation of these quartz-rich samples is best explained by cooling of silica-rich hydrothermal fluids by upward flow in a low-permeability system (Hein et al., 2008). Opal–A, opal–CT, and quartz mineralization found in different places within the EBD hydrothermal field likely reflects decreasing silica saturation and increasing temperature of the mineralizing fluid with increasing silica crystallinity. As well, total rare-earth element concentrations are low. Ce and Eu anomalies reflect high- and low-temperature hydrothermal components and detrital phases. These investigators state that a remarkable variety of types of mineralization occur within the EBD field, yet a consistent suite of elements is enriched (relative to basalt and nonmineralized cores) in all samples analyzed: Ag, Au, S, Mo, Hg, As, Sb, Sr, and U; most samples are also enriched in Cu, Pb, Cd, and Zn. On the basis of these element enrichments, the EBD hydrothermal field might best be described as a base- and precious-metal—bearing, silica—Fe-oxide deposit.

If lode quartz veins are indeed hydrothermal seafloor crusts, where and how do we see them forming today? Reviews by Rona (1988) and Rona and Scott (1993) and references therein provide excellent summaries of early findings. The TAG hydrothermal field and Mid Atlantic Ridge (MAR) crests at show modern-day silica and quartz formation. A more current review is provided by Hannington et al. (2005). To date, it appears that modern hydrothermal vents at mid-ocean ridges that precipitate quartz and silica in major amounts are not common.

However, one can gain clear insights into P-T-x parameters during the formation of lode veins using the model for seafloor hydrothermal processes proposed by Scott (1997) or an examination of the chemistry and physics of vent fluids as they relate in particular to silica, sulfides, gold, and (carbon) within the context of high-temperature reaction zones at mid-ocean spreading ridges or seamounts (Figure 7.2). A brief review of sulfide phase chemistry is also warranted in order to understand how certain sulfides can help to understand formation conditions.

7.3.1 Temperature and Pressure

The temperature—depth relations in the high temperature reaction zone may be estimated from the silica geothermobarometer from the end-product (Fournier, 1985; Von Damm and Bishoff, 1987). Its basis is that the fluid has equilibrated with quartz at the pressure and temperature of the high-temperature reaction zone and has risen directly to the surface without the precipitation of silica (Von Damm et al., 1985). These assumptions seem valid. Quartz, not amorphous silica is the phase encountered in the source rocks by the hot fluid, and the high discharge velocity of the end-member fluids is consistent with their rapid ascent from the reaction zone only with adiabatic cooling. A diagram for quartz solubility in distilled water and 0.5 M NaCl (approximate concentration for seawater) in the P-T region of interest for seafloor hydrothermal systems is given in Figure 7.3 (Scott's Figure 16.15, p. 844). The pressure is hydrostatic, consistent with a hydrothermal system operating in

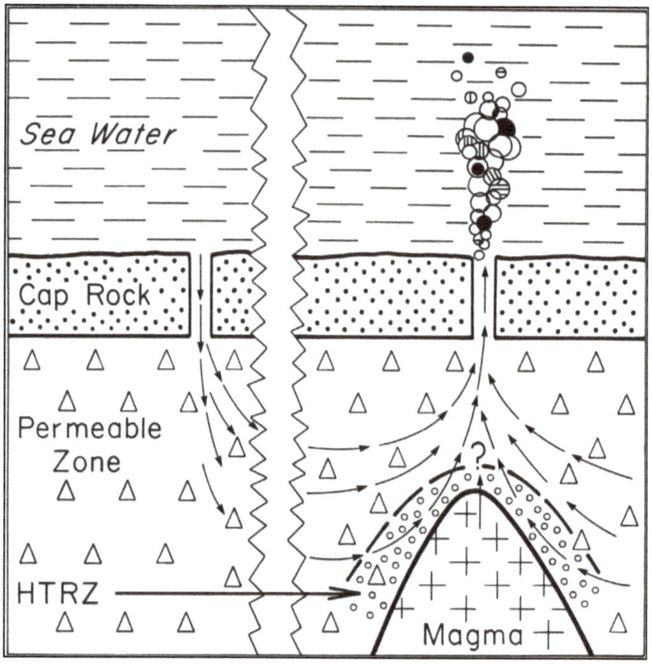

FIGURE 7.2

Model for hydrothermal circulation at a spreading ridge or seamount *(from Scott (1997).)*. HTRZ is the high-temperature reaction zone where deeply penetrating seawater reacts with hot rock and is converted to a hydrothermal fluid. The magma may contribute to the fluid.

a circulation loop and the known high permeabilities of ocean crust in axial regions and seamount. The shape of the silica isopleths being quite steep (i.e., relatively independent of pressure), they tend to provide better information on temperature than on pressures over most of the diagram, although, at higher temperatures, the rollover in the isopleths makes them more sensitive to pressure than to temperature.

7.3.2 Fluxes and Mass Heat Flow

According to Scott (1997) the thermal and mass fluxes are widely variable. The mass flow rate of seawater through the ocean ridges is estimated to be about the same as the discharge into the oceans of the Amazon River (Garrels and Mackenzie, 1971). According to these figures, the Earth's entire mass of ocean water is cycling through hydrothermal centers on spreading ridges every 5–11 million years" (Woolery and Sleep, 1976), and this may have been a factor of 100 faster in the Archean (pre-2500 Ma) when thermal gradients in ocean crust were thought to have been much higher (de Wit et al., 1992). The entire oceans must have been recycled through the ridges many hundreds of times.

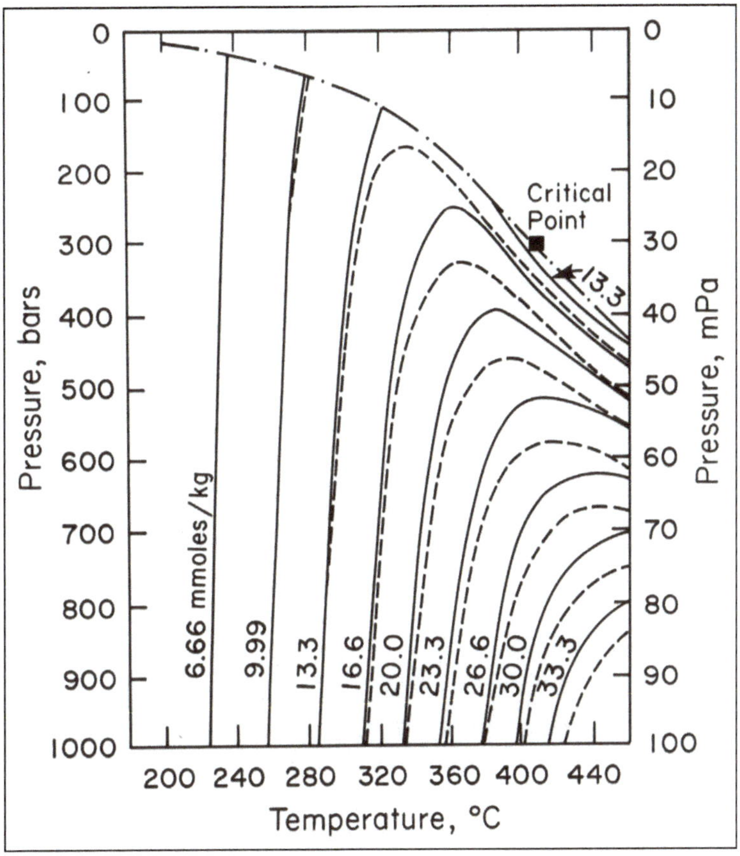

FIGURE 7.3

The silica geothermobarometer: silica isopleths as a function of pressure and temperature in 0.5 M NaCl and in distilled water (dashed lines) (Von Damm and Bishoff, 1987; Scott, 1997).

The rate of generation of new oceanic crust is about 5×10^{13} kg/a (Rona, 1984) or about 2–20 times small than the seawater flux. Therefore the mass of newly formed ocean crust is very small compared to the mass of circulating seawater (the water/rock ratio is very large), which results in extensive basalt–seawater interaction.

7.3.3 Salinity

Salinity (S‰) is defined as grams of dissolved solids per kg of seawater and chlorinity (C‰) is grams of Cl + Br + I per kg of seawater. S of normal seawater is 35‰. End member vent fluids range from 40% lower to 70% higher than seawater.

Variability, as well as very high salinities, are seen in fluid inclusions from QV in the modern ocean crust, the Cretaceous-age Troodoos ophiolites and

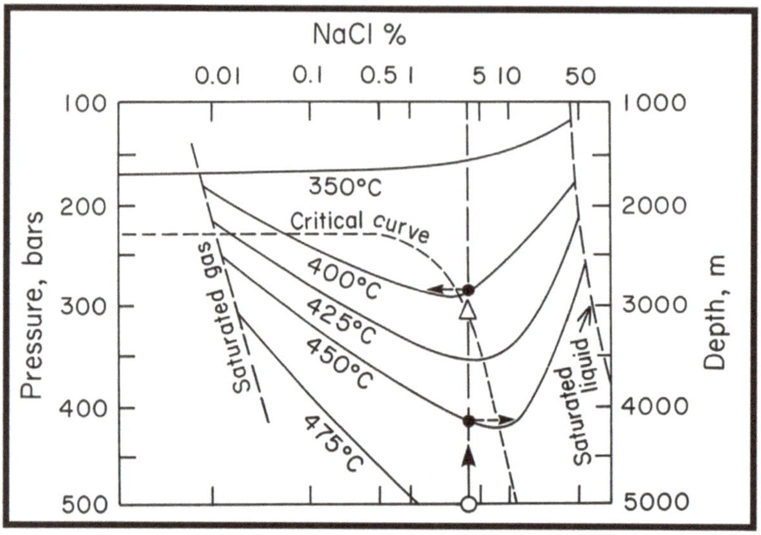

FIGURE 7.4

Phase diagram for the system NaCl-H_2O. Examples are given for 450 °C and 400 °C fluids of seawater salinity decompressing as they rise toward the seafloor and separating into a conjugate pair, one of high salinity and the other of low salinity, where they pierce their respective solvus. *Figure 16.16 of Scott (1997).*

Archean veins. Among the several mechanisms discussed by Scott (1997, p. 854) that can produce variable salinities, phase separation in the supercritical region for seawater or boiling at lower pressures is an effective means of producing fluids with widely varying salinities. The phase diagram for the binary system NaCl-H_2O at pressures and temperatures of sub-seafloor hydrothermal conditions, has an extensive solvus separating one fluid ("liquid") with a very high salinity, (as much as 50%) NaCl, from another ("gas") with a very low salinity (0.01% NaCl). At pressures above the solvus dome, the NaCl solution exists as a single-phase supercritical fluid. Although end-member vent fluids contain many species, they are overwhelmingly NaCl solutions.

Seawater has the same P-V-T properties as a 3.2% wt% NaCl solution and a critical point in the range 403–406 °C, 285–302 bars. The P-T trace of seawater along the top of the solvus places an effective upper limit on its maximum temperature of circulation in the crust (Figure 7.4).

7.3.4 Density of Vent Fluids (Summary from Scott, 1997)

Calculated and sampled densities for selected modern and ancient vent fluids all have a density less that of ambient bottom water (1.03 g/cm^3 at 2 °C, 2000 m depth) although some approach this value. Black smokers have the highest density contrast (about 0.2–0.3 g/cm^3) and produce buoyant hydrothermal plumes that rise well above the seafloor. The fluids of Fe-Si-Mn oxide deposits have much less density contrast (about 0.01 g/cm^3) and flow at velocities of only a few cm per second (Binns et al., 1993a).

Geologists have presumed from the study of some ancient syngenetic ore deposits, particularly those in fine clastic sediments ("shale hosted"), that some ore-forming fluids must be denser than seawater, despite their apparent high temperatures, and should therefore be bottom-seeking. These deposits were originally conformable, blanket-like and are commonly well banded. None of the sampled vent fluids discussed by Scott (1997, Table 16.11) are bottom-seeking when they reach the seafloor.

Lonsdale et al. (1982) describe an unanalyzed 10–15 °C fluid that sank and dispersed downhill upon being emitted from an iron oxide deposit in a submarine caldera. However, this fluid is not producing a large metal sulfide deposit, which would require a higher temperature fluid capable of carrying abundant metals.

Bottom-seeking fluids can be generated as evidenced by the unusual situation (i.e., dissolution of evaporites) in the Atlantis II and other deeps of the Red Sea and also by the fluid inclusion data from the ancient Corbet, Silvermines, and Sullivan deposits. However, Silvermines and Sullivan had ore-forming fluids that barely exceeded the density of seawater.

Meguma vein textures almost certainly reflect coincident precipitation of a silica gel and pelagic–pelitic turbidite sedimentation.

7.4 HOW SULPHIDATION EQUILIBRIA, GEOTHERMOMETERS, OR GEOBAROMETERS HELP TO UNDERSTAND THE GOLD-MINERALIZING ENVIRONMENT

There is a relatively complete understanding of ore formation processes, but can this help us find more ore? Some answers arise from an understanding of boundary conditions for solution parameters such as pH, fO_2, T, and P. These will allow us to narrow down the specific deposition environment and in combination with what geologists find in outcrop or drill core; together these can provide additional exploration tools.

A significant milestone that provided a thermochemical framework toward understanding ore deposit formation evolved from the electrum tarnish method of Barton and Toulmin (1964). This work was published while Kretschmar was an undergraduate student at McMaster University and inspired him by bringing simplicity and elegance to the study of sulfide phase chemistry to better understand ore formation. The original work used the degree of tarnishing by sulfur vapor in evacuated glass tubes of electrum (gold-silver alloys) of different compositions. Numerous studies followed this breakthrough; one of the more important applications was the temperature and pressure dependence of iron content of sphalerite (Barton and Toulmin III, 1966) and its subsequent expansion into the concept of ore forming environment (Barton, 1970). There was a major flourishing of studies, and the complex field of sulfide mineral chemistry is addressed in excellent reviews by Barton and Skinner (1979) and Hannington and Scott (1989).

7.4.1 The Ore-Forming Environment

The ore-forming environment consists of a grid, based on laboratory experiments and theoretical calculations on which are plotted the stability fields of common ore forming sulfides in temperature and sulfur fugacity space. The upper limit is the condensation curve of sulfur vapour, which marks approximately the upper limit of geologic interest. The lower limit is marked by the equilibrium between metallic iron and troilite. Practically all sulfide deposits must have formed at sulfur fugacities above this equilibrium, because native iron is very rare and iron sulfides are very common. Most of the region of geologic interest lies between these two equilibrium curves, and the pyrite–pyrrhotite curve lies in the middle of the field. Gold is commonly associated with copper minerals, which suggests that it follows mineral precipitation in the intermediate to high sulphidation states (Figure 7.5). This diagram was designed for porphyry copper deposits, but the same remains true for all gold deposits with copper accessory minerals. A good example is the McIntyre Mine in Timmins, Ontario, which was mined for gold; later the central area was mined for low-grade copper. This copper zone was so close to the gold workings that the same mine infrastructure was used (B. Buffam per communication with McBride, 1975). The same mineralizing system is inferred.

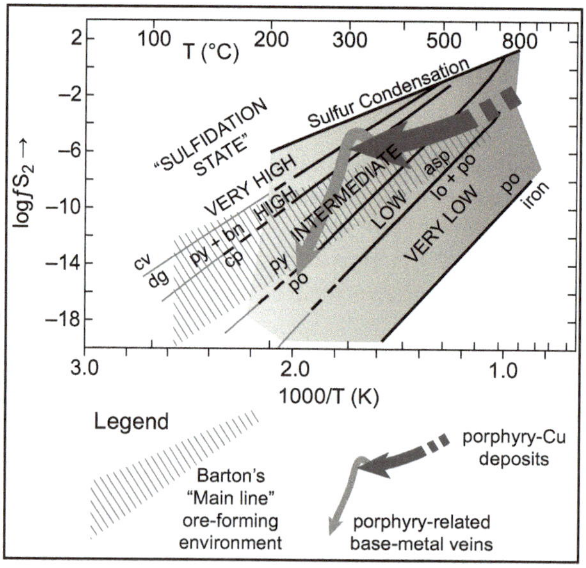

FIGURE 7.5

Sulphidation state and the "ore-forming" environment based on *(Barton (1970) and redrawn from Einaudi et al. (2003).)*. Shown are sulphidation equilibrium curves for common ore-forming minerals.

Of direct interest is what sulphide minerals in lode veins tell us about formation conditions. Arsenopyrite and, to a lesser extent, sphalerite are common minerals associated with gold deposits. It is important to understand the advantages and the limitation of using geothermometers, geobarometers, and other indirect and analytical techniques. Their main utility is to supplement and refine our understanding of formation conditions; common sense must prevail (ie, what are the rocks telling you?) in applying chemical analyses, experimental results, and theory.

7.4.2 Arsenopyrite Geothermometry

An excellent illustration is provided by the temperature determinations using arsenopyrite geothermometry. This geothermometer is widely applied, and Kretschmar has some familiarity with it (Kretschmar and Scott, 1976). In Meguma quartz veins at Harrigan Cove (Figure 2.14), arsenopyrite occurs as.

1. euhedral to irregular crystals in beds of black shale host, and
2. as irregular masses in the quartz veins on internal beds or occasionally cross-cutting the contact of quartz–slate contact (Figure 7.6).

Growth and sector zoning, S/As variation, and the variables responsible were studied by Vesselinov and Kerestedjian (1995). They describe a simple quartz–arsenopyrite vein assemblage which formed at 280–200 °C in the Madan lead–zinc ore district, Bulgaria. Zoned arsenopyrite with crystal sectors showing variable S/As reflect processes taking place very close to the growing crystal surface, which are superimposed on those produced by larger-scale variations in the hydrothermal environment.

If the S/As variation is assumed to reflect temperature variations, an improbable fluid temperature fluctuation of about 200 °C is required to explain the observed S as follows from the phase diagram of Kretschmar and Scott (1976). Other mechanisms such as pressure shock or compositional changes of the fluid can be ruled out, because there is no evidence for them either in the hand specimen or in the geological environment. According to crystal growth theory, resistance to growth and kinetic factors rather than variable super saturation can account for formation of oscillatory zoning. The processes take place very close to the growing surface and involving a continuous interplay between face growth rates, in this case, the amount of sulphur absorbed on the surface, and transport processes inside a boundary layer.

The point is that S/As in arsenopyrite reflects growth conditions rather than temperature complexity of mineral chemistry and may affect age dating (Figure 7.7).

7.4.3 Sphalerite Composition

The abstract of Keith et al. (2014) is quoted as follows: "Experimental studies have shown that temperature, pressure, sulfur fugacity (fS_2), and oxygen fugacity

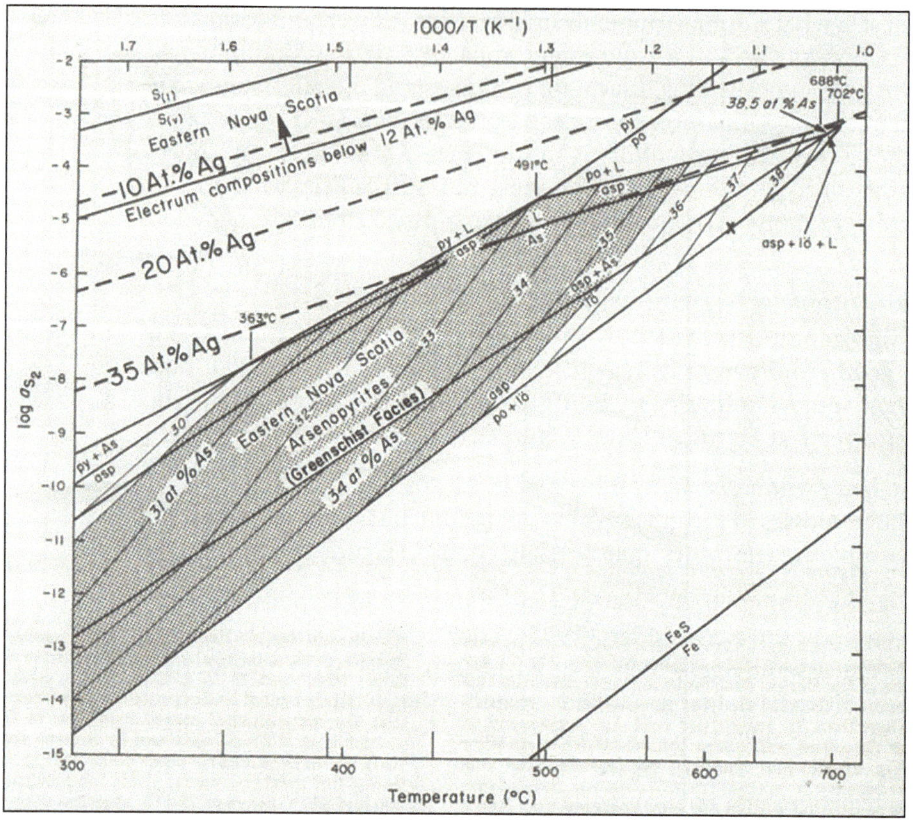

FIGURE 7.6

Meguma arsenopyrite and electrum compositions plotted on log fS2-T grid. *Figure 7 of Haynes (1986). Stability field and As/S contours of arsenopyrite after Kretschmar and Scott (1976). Electrum composition contours from Barton and Toulmin (1966).*

(fO_2) influence the Fe content of sphalerite. We present compositional in situ data on sphalerite from submarine volcanic massive sulfide (VMS) ores of hydrothermal vents from different plate tectonic settings and with variable host-rock compositions. Sphalerite from sediment-hosted vents has systematically higher S contents and Fe/Zn ratios than those of the sediment-starved vents, reflecting an influence of fS_2 and fO_2 on Fe partitioning between fluid and sphalerite. The Fe/Zn ratios of sphalerite from sediment-starved vent systems apparently increase systematically with the fluid temperatures of the corresponding vents. The conclusion is that the composition of sphalerite can be used to:

1. distinguish between sediment-hosted and sediment-starved hydrothermal processes, and
2. estimate minimum fluid temperatures of sphalerite precipitation from inactive sediment-starved hydrothermal vent sites and fossil VMS deposits."

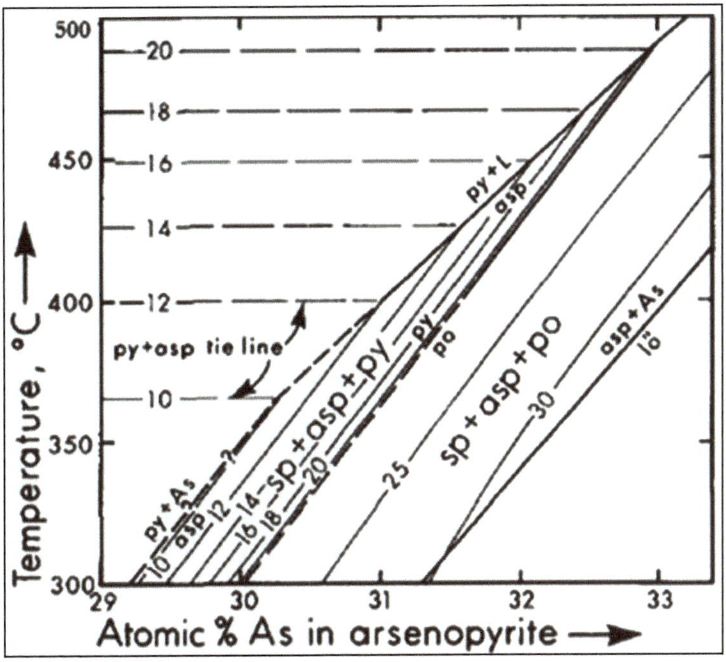

FIGURE 7.7

Compositions of sphalerite and arsenopyrite co-existing with pyrite and pyrrhotite at 1 kbar confining pressure. Contours are mole % FeS in sphalerite. *From Scott (1983).*

7.4.4 Seawater Boiling Curve

Combined with the boiling point curve and maximum vent water temperatures from Hannington et al. (2005), Figure 3, p. 115, are the effects of adding NaCl and CO_2. The critical point of normal seawater (3.5 wt% NaCl) is 405 °C at 3000 bars. The curve provides a useful context within which to the consider formation of lode veins and the seafloor environment in which they form (Figure 7.8).

Further important insights into Archean seafloor physical–chemical conditions can be also be obtained from the 3.45 Ga Strelly Pool cherts in the Pilbara craton Lindsay et al. (2005) and DeGregorio (2006), as is shown in Section 7.7.1. First, however, an examination of the solubility and hydrothermal chemistry of quartz throughout both the high and low temperature ranges of the silica–water phase diagram is warranted. There are two common sources of silica–gold–rich hydrothermal solutions. One is the result of hot seawater circulating through crustal lithologies, which quickly become quartz saturated at high temperature and precipitate silica in some form when cooled. In a related process, silica-saturated solutions may also exsolve from felsic and

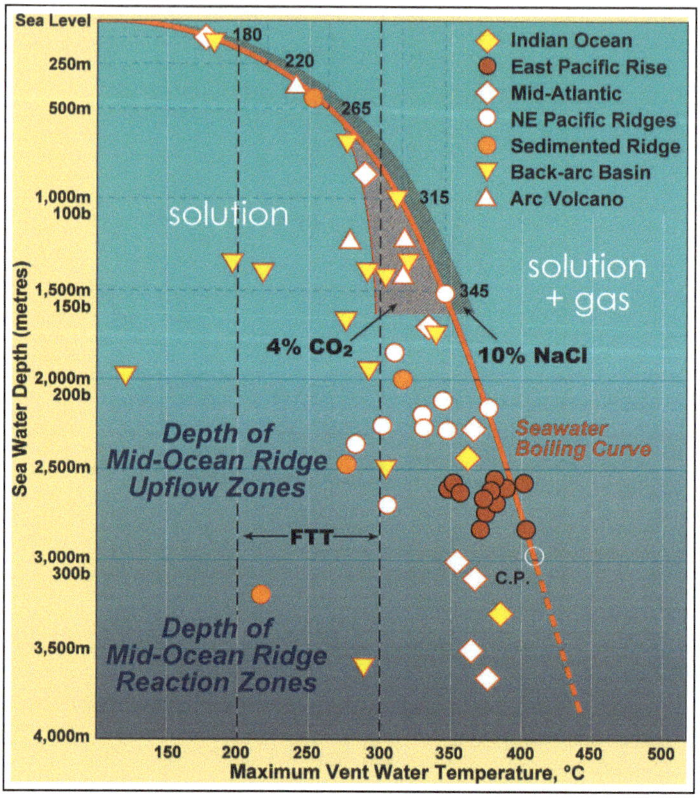

FIGURE 7.8

Seawater boiling curve and depth of mid-ocean ridge reaction and upflow zones *(redrawn from Hannington et al. (2005).)*. The effect of increasing NaCl concentration from normal seawater (3.5 wt.% NaCl) to 10% NaCl is to raise the boiling point at any given pressure. Increasing the CO_2 content will decrease the boiling point at any given water depth.

mafic volcanics (discussed in the context of the Cote Gold deposit, Chapter 6). In Meguma turbidites, nonmineralized quartz veins are ubiquitous, as we have seen in Chapter 2. At Softwood Island (near Mink Island, Figure 2.2 and 2.16), for instance, well distant from gold districts, quartz veins constitute up to 21% of a stratigraphic column (Kretschmar, 1983). High percentages of quartz were also described from Bendigo by Thomas et al. (2011). Often, textures support deposition of quartz as a gel. (Large, 2007 at Sukhoi Log). SiO_2, which is usually found as quartz or "chert" in ancient ores, occurs on the seafloor primarily as amorphous silica, in some cases with identifiable opal—A or —CT structures (Scott, 1997).

7.5 QUARTZ

7.5.1 High Temperature Quartz—Water System

An excellent overview of the quartz—water system is provided by Holland and Malinin (1979). Subsequent experimental and theoretical work is extended

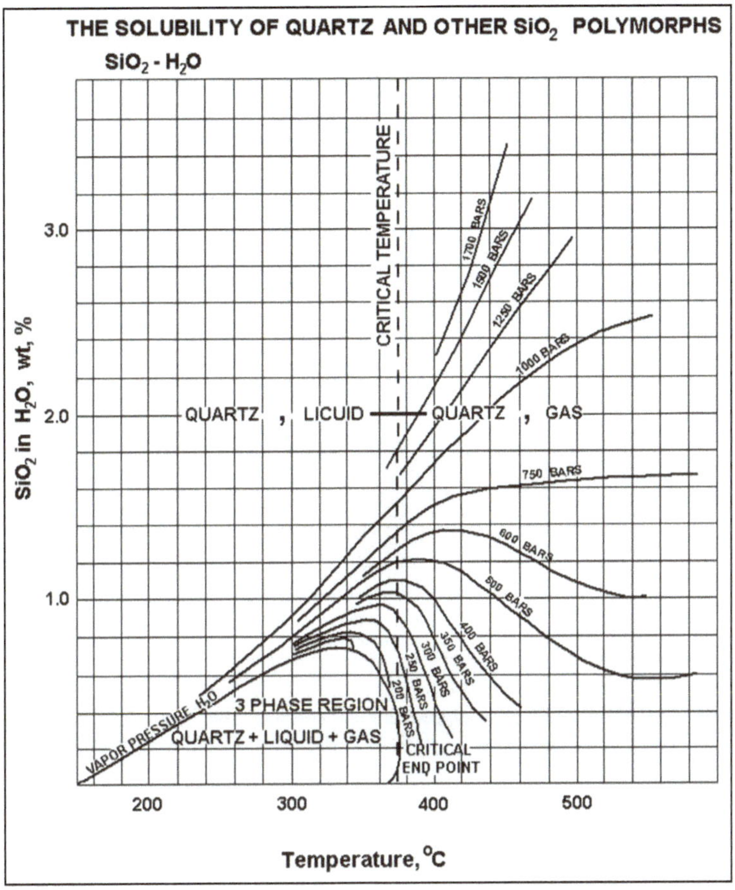

FIGURE 7.9

Solubility of quartz in water at temperatures up to 500 °C and pressures up to 1750 bars. *Figure 9.6 of Holland and Malinin (1979).*

and summarized by Fournier (1985), Von Damm et al. (1991), Dove and Rimstid (1994), Fournier (1999), and Hofstra and Cline (2000). The classical study of the solubility of quartz in water over a large portion of the temperature and pressures of interest to ore-forming hydrothermal system is shown as Figure 7.9.

At a given pressure, the solubility of quartz increases with temperature except in the region near the critical point of water, where the density of the solutions is quite small. At a given temperature, the solubility of quartz increases with pressure, except as shown in Figure 7.9.

In seawater or water with dissolved NaCl, the solubility of quartz is higher than in pure water. The combined effect of increased pressure and variable NaCl content is shown in Figure 7.9.

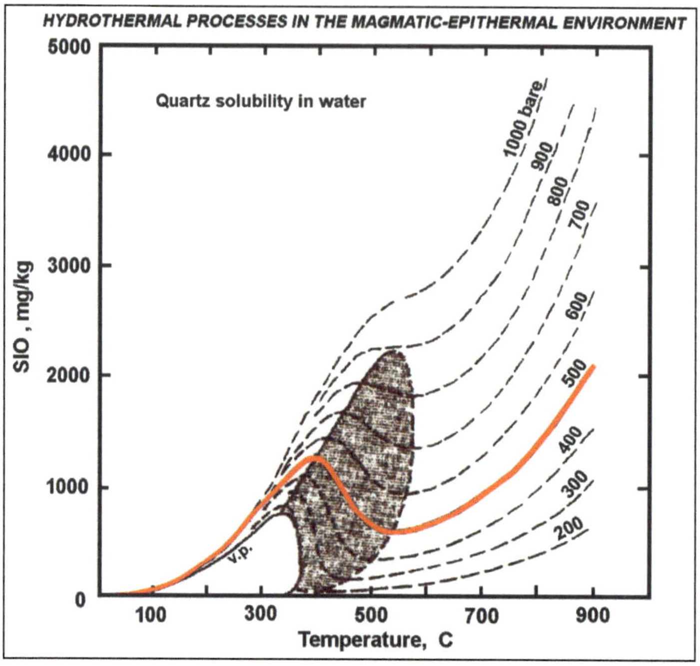

FIGURE 7.10

Calculated solubilities of quartz in water as a function of temperature and pressure. The 500-bar isotherm is shown in red. The stippled pattern shows a region of retrograde solubility in which the solubility of quartz decreases with increasing temperature. *From Fournier (1999), Figure 3.*

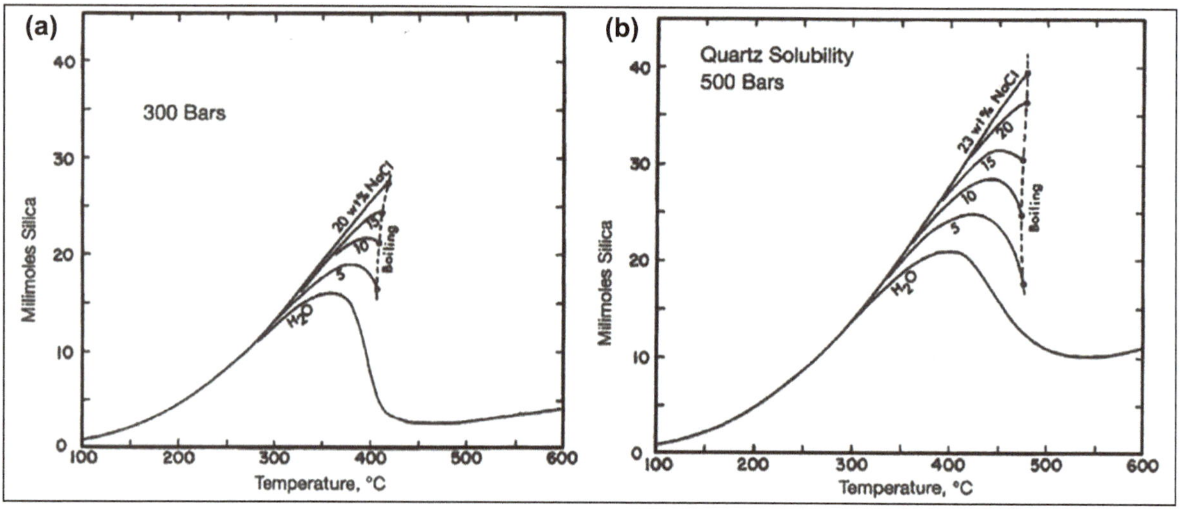

FIGURE 7.11

Calculated solubilities of quartz at 300 and 500 bars and variable NaCl content *(from Fournier (1999), Figure 4.)*. The critical (boiling) point at 500 bars is about 470 °C, and 1 L of hydrothermal solution with 7.5% NaCl will hold approximately 1.3 g of dissolved silica.

Using these figures shows that, at 500 bar pressure with 7.5 wt% NaCl, at the critical (boiling) point (approx. 470 °C [that is what the diagram shows, but it is too high] rather than 403 °C at 3000 bars with 3.5 wt% NaCl), the hydrothermal solution will contain about 23 mmol silica, equivalent to about 1300 ppm silica per 1 L of solution. It is clear that hydrothermal solutions saturated with silica at these pressures and temperatures have the capacity to precipitate a large amount of silica as it debouches onto the seafloor. A typical disc-shaped quartz vein, say 250 m in diameter, 3 m thick in the center, thinning to 0 cm at the edge, will contain about 5.0×10^6 kg silica, and in theory may form from about 4.0×10^6 million L of solution or 1.1×10^4 L/day over 1 year. Actual discharge parameters from hydrothermal seafloor vents are worth examining first, after which the low-temperature portion of the silica-water phase diagram and seafloor quartz formation can be examined to obtain a rough estimate of silica behavior.

7.5.2 Vent Fluids, Silica and the Cauliflower Garden

7.5.2.1 Silica Content of Vent Fluids

High concentrations of silica are typical for hydrothermal solutions discharging at oceanic spreading centers, based on experimental seawater/basalt reactions at elevated temperature, as summarized by Mottl (1983) and Rosenbauer and Bischoff (1983) as well as direct analyses of hydrothermal fluids at the East Pacific Rise (EPR), 21°N. Amorphous silica is also a common constituent of submarine polymetallic sulfide deposits, e.g. at 13°N 23, 21°N. EPR, the Endeavour Segment of the Juan de Fuca Ridge, and the Atlantis II Deep, Red Sea.

The actual measured composition of vent fluids and a different perspective on the behavior of silica are presented by Scott (1997), who states that among the major components in end-member vent fluids, silica concentrations show comparatively little variability within the three major tectonic settings, which are:

1. sediment starved mid-ocean ridges,
2. sedimentary sites, and
3. back-arc spreading zones of island arcs. The SiO_2 contents of end-member fluids at sediment starved open ridges, with the exception of southern Explorer Ridge, lie in the narrow range of 17.5–22.9 mmol/kg n (calc ppm 0). Most likely, vent fluids equilibrate with quartz in the high-temperature reaction zone more or less all at the same temperature. This is the basis of the silica geothermometer/geobarometer discussed below.

SiO_2 contents of vent fluids from back-arc spreading zones of island arcs, above, are about two-thirds of those from sediment starved mid-ocean ridges, despite being associated with more siliceous lava. The SiO_2 contents of vent fluids from sedimentary sites are about one-half, despite the sediment being a source of excess silica. This is the result of the fluids being saturated with respect to quartz in all three cases, but at lower temperatures for back arcs and sedimentary ridges than for sediment-starved ridges.

Quartz formation details from two modern seafloor hydrothermal vent locations provide insights into modern silica crust ("vein") formation. Hydrothermal material dredged from the summit of a young volcanic seamount near the East Pacific Rise at 14°09′ N, 104°22′W is part of a massive sulfide seafloor hot-spring deposit with an unusual abundance of primary quartz, some of which has fluid inclusions. Details summarized from Vanko et al. (1991) are as follows:

The seamount is underlain by crust about 150,000 years old 15 km west of the EPR. Lavas and hydrothermal deposits seem to be fresh and to have formed recently. The volcano is roughly 4 km wide and 7 km long at its base, and is 800 m high above the surrounding 3000-m deep seafloor. It has a broad, flat cross-section, similar to other young East Pacific volcanoes. It has no caldera, similar to roughly 20–60% of young deep-sea volcanoes examined. The summit contains several high points separated by broad saddles.

Dredge samples of the summit recovered basaltic lava and several tens of kilograms of hydrothermal material, which consists mainly of massive sulfide, earthy, low-density reddish orange crusts, and poorly consolidated sediments as well as several denser and harder pieces of massive sulfide, including rare crusts coated by euhedral crystals of atacamite ($Cu_2Cl(OH)_3$). The massive sulfide is gray, fine-grained, and porous, with 5–20% porosity resulting from vugs and irregular tubes, about 1 mm across and several millimeters long. Some of these tubes are lined by druzy, (subhedral crystalline) quartz, and others are filled completely with quartz. The sulfides are mainly quartz and pyrite, with variable proportions of sphalerite up to 25%. Accessory minerals include chalcopyrite <30% and traces of covellite, digenite, and galena. The sulfides are a mosaic of fine-grained (0.1 mm and less) anhedral quartz containing blebs and stringers of colliform pyrite, which commonly align into curved stringers or domal features up to several centimeters long. Pyrite occurs with this coarser quartz as cubes measuring about 0.5 mm. Transparent isotropic sphalerite occurs as interstitial grains up to 0.25 mm across, in stringers and thin colloform layers, and rarely in euhedral grains 2 mm in diameter. Chalcopyrite forms small interstitial patches and irregular replacements within sphalerite; it is rimmed and in places completely replaced by covellite and digenite. Galena(?) forms tiny inclusions in sphalerite.

Trace element content of selected trace and heavy elements in massive sulfides were recalculated to a silica-free basis and were shown to contain 0.2–2.3 ppm Au and 50–150 ppm Ag, Cu, and Zn values in the range of 1.6–3.8% and 2.3–7.0% respectively, and Pb is 0.02–0.03 wt %.

Most of the dredged material is porous, low-density, earthy, reddish orange rock and crusts. The earthy parts contain optically isotropic red translucent grains and abundant red translucent filaments, typically 2–4 mm wide and up to tens of micrometers long. They exhibit curved and convoluted branching or intertwined forms and probably represent iron-encrusted filamentous bacteria or the molds of such bacteria.

Quartz in vugs is coarser and clearer in pockets and small veinlets that measure up to about 2 mm. Almost all quartz grains exhibit undulose extinction. In some filled vugs, concentric layering and plumose growth-habits are common, indicating that the quartz here may be primary. Some mosaic quartz encloses faint spherulitic outlines, or represents quartz replacement of primary opaline silica. Therefore some early silica was precipitated as opaline or amorphous silica. Such material later recrystallized, and vugs and veins were subsequently filled by primary quartz.

Quartz grains commonly contain fluid inclusions, particularly the primary vug-filling. These fluid inclusions that consist of liquid and a vapor bubble. The inclusions are equant but have irregular walls. They are isolated and are interpreted to represent primary trapping of fluids from which the quartz was precipitated or which induced recrystallization from pre-existing amorphous silica. Temperatures of homogenization of fluid inclusions into the liquid phase (Th L-V (L), designated Th) vary from about 160° to 350 °C, with a strong mode around 216 °C. At 2600 m water depth (260 bar ambient pressure), the pressure correction to Th is about 25 °C $\pm$ 5 °C, resulting in a mean trapping temperature of the fluid of about 240 °C $\pm$ 35 °C (1 σ), which is a good approximation of the temperature of quartz growth.

The massive sulfides described by Vanko et al. (1991) fall in the middle of the silica spectrum and are similar to sulfides from other East Pacific Rise seamounts described by Alt et al. (1987) and Hekinian and Fouquet (1985) that contain abundant amorphous silica (opal), as well as quartz that has both replaced opal and has crystallized directly in vugs and fractures. The silica–sulfide–sulfate spires from Axial Seamount (Hannington and Scott, 1988) similarly contain abundant amorphous silica. Alt et al. (1987) estimated temperatures of quartz precipitation on the basis of oxygen isotopic values. Assuming seawater isotopic composition ($\delta^{18}O = 0‰$), the quartz grew between 230° and 250 °C. If the fluid was shifted to δ18O = +2.0‰ then the corresponding temperature range was 280° to 320 °C, in excellent agreement with fluid-inclusion temperatures of 240°$\pm$ 35 °C obtained by Vanko et al. (1991).

These authors conclude that moderate-temperature hot springs (in the mid-200 °C range) may form substantial silica°sulfide deposits either within seafloor mounds or beneath the seafloor. Black smoker fluids are generally supersaturated with respect to quartz as well as amorphous silica at seafloor pressures but quartz and, to a lesser extent, amorphous silica are rare within most-black smoker chimneys. Such fluids apparently rise nearly adiabatically from a deep reaction-zone where the fluids are in equilibrium with quartz at high temperature and pressure.

Most importantly, these authors propose that during rapid upwelling, quartz nucleation is suppressed, and dissolved silica ends up being flushed out into the ocean. Amorphous silica or quartz with sulfides in mounds or chimneys attests to a process that allows nucleation of silica to occur, and one possibility is that fluid encounters a regime of slower and more diffuse flow, perhaps combined with substantial conductive heat loss.

There may be a complete natural spectrum between the silica-rich, sulfide-free chimneys from the Galapagos spreading center (Herzig et al., 1988) and sulfide-rich chimneys. One that best describes modern seafloor precipitation of silica is the "Cauliflower Garden," a description of which follows below.

7.5.2.2 The Cauliflower Garden Silica Chimney Field, Galapagos Spreading Center

Within the project area GARIMAS ("Galapagos Rift Massive Sulfides") during 1984 and 1985, the Galapagos Rift between the Inca Fracture Zone and the De Steiguer Deep was mapped at various scales, and several hydrothermal sites were found (Figure 7.12).

These include sulfide deposits, mounds composed of silicates and Mn oxyhydroxides, and chimney fields almost entirely formed of amorphous silica (Figure 7.13).

Two such fields (locations C and G, C = 0°46.05′–0°46.15′ N, 5°53.39′–85°53.55′ W) were named "Cauliflower Garden" by Herzig et al. (1988), who describe the occurrence as follows. The Cauliflower Garden hydrothermal field occupies an area of about 250 × 250 m of the rift valley floor, extending from the axis to the base of the northern talus slope. The bedrock is mainly blocky sheet lava, partially covered by sediments and hydrothermal products. Prominent are very irregular silica heaps and towers 1.0–2.0 m in height and 0.5–1.0 m in diameter that resemble gigantic cauliflowers. The sediments at their base frequently show black Mn hydroxide staining, centering around small, slightly active conduits. The silica stacks are characterized by a gray to brownish shelly surface structure, except when some of the shells are broken off, showing bright yellow to ochre colors. In the central part of the specimens, white to light yellowish colors dominate. There are no worm tubes, as known from sulfide chimneys, but numerous subconcentric tube-like structures occur. These may represent small vent conduits. Locally, thin Mn oxyhydroxide layers and crusts were found, either within the silica host or covering the stacks.

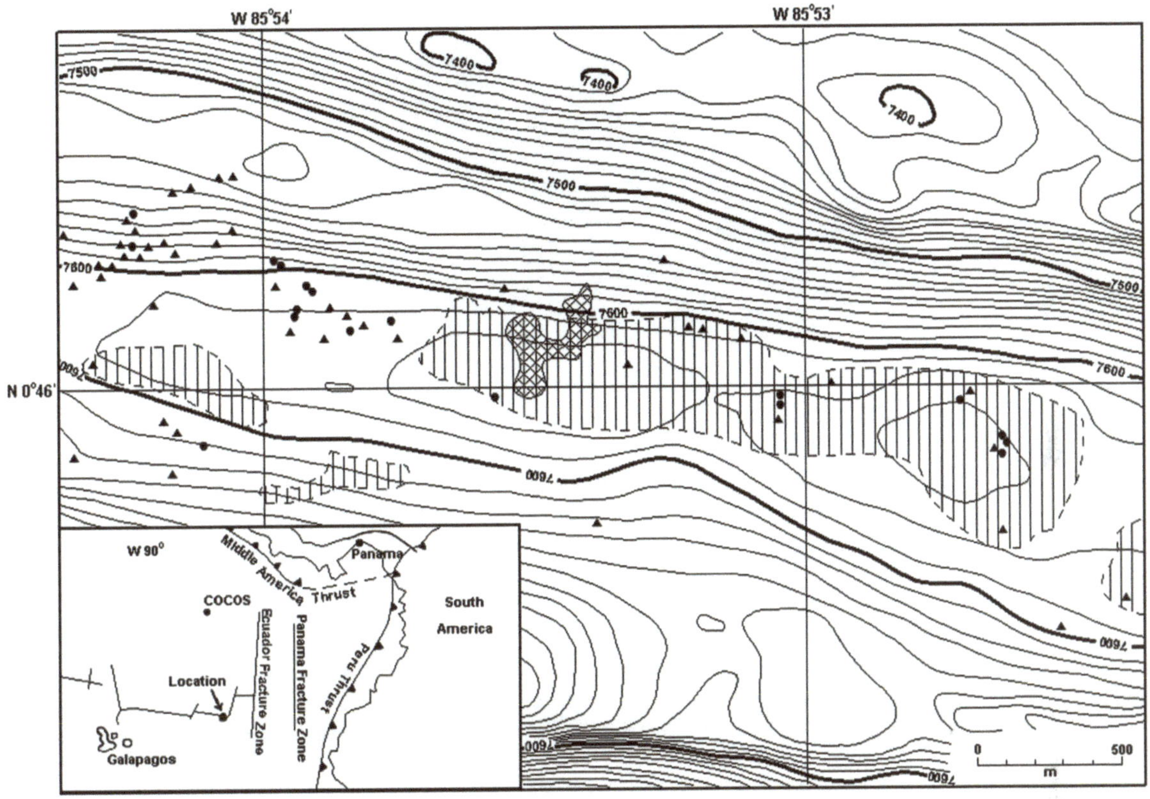

FIGURE 7.12

Location of "Cauliflower Garden" hydrothermal silica chimney vent field (Herzig et al., 1988). Cross-hatched area delineates the hydrothermal field of location C, the Cauliflower Garden; filled triangles indicate other hydrthemal products outside the this field, and filled dots indicate where faunal elements related to hydrothermal vents were observed. The vertically hatched areas are covered by ponded lavas.

FIGURE 7.13

Seafloor photograph of Cauliflower Garden silica chimneys at location C, covering about 8 m^2. The hydrothermal products consist mainly of silica with minor Fe-Mn crusts. The light colored sediments are locally stained by dark Mn oxyhydroxide precipitates (Herzig et al., 1988).

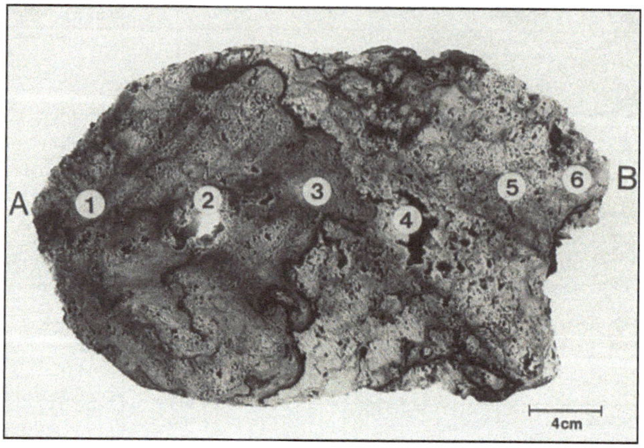

FIGURE 7.14

Photograph of slab of silica chimney, location of mini drill core samples and central fluid conduits 2 and 4. *Figure 3 of Herzig et al. (1988). Note similarity of this texture to barren Meguma quartz vein Figure 2.11 from Goose Island, Nova Scotia.*

Mineralogical and chemical studies were carried out on 2-cm minicores from a silica chimney about 1 m in height, sampled at location C (Figure 7.14).

The results were compared to silica material associated with Fe– and Zn– sulfide stacks analyzed from location B (0° 46.00′-0 °46.02′ N, 85° 54.64′-85° 54.71′ W) about 2.5 km west of the Cauliflower Garden.

At location C, the samples consist of mainly porous and cell-like, soft silica with spongy or honeycombed appearance and abundant cavities and vesicles. Locally a few more compact layers occur, providing stability to the whole structure. The friable and partly solid material characteristically has a very low density. Subordinate, reddish-brown to ocherous Fe hydroxide layers, as well as black, umber-like layers consisting of Mn oxyhydroxides, occur within the yellowish-grey to partly greenish silica.

A sample traverse across specimen I (Figure 7.14) shows two fluid conduits, interpreted as separate generations. Sample sil-2 is probably the centre of an early silica stack, marked by an outer irregular to subconcentric layer of Mn oxyhydroxide. This vent is filled by solid, whitish-gray silica. Sample sil-4 is a probable later stage filling in which silica precipitated onto and partly around the older stack, probably due to plugging of the first vent and opening of a new fluid outlet. The center of this younger vent consists of extremely soft, powdery silica.

X-ray powder diffraction shows that all sulfide-free silica samples are monophase, with a broad, diffuse peak centred at about 4.0 A, which is typical for amorphous silica (opal-A, $SiO_2 \cdot n\ H_2O$) according to the nomenclature of Jones and Segnit (1971). The mineralogy was the same for silica, varying in

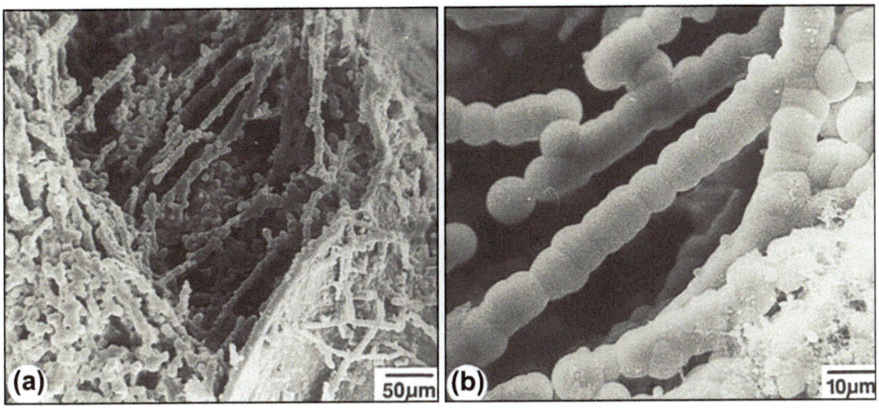

FIGURE 7.15

(a and b) Scanning electron photomicrograph of silica precipitate from Cauliflower Garden silica chimney *Figure 5 of Herzig et al. (1988).*

color or consistency. Mn encrustations are a mixture of birnessite and todorokite, common constituents of seafloor manganese nodules. In addition there are traces of δ-MnO_2 and gypsum.

In thin section, silica occurs as void-filling, cryptocrystalline, chalcedony-like aggregates showing spherolitic extinction. Disseminated and globule-like reddish-brown Fe-hydroxide is abundant. Scanning electron microscopy revealed a highly porous three-dimensional network consisting of chains build up by individual silica globules and spheres (Figure 7.15(a)), averaging about 10 μm in diameter (Figure 7.15(b)). Partly tabular aggregates with relatively smooth surface structure occur (Figure 7.15(a)).

Chemistry: Silica is remarkably uniform in composition, containing SiO_2 91.95 to 96.34 wt%, Fe_2O_3 0.37 to 3.11 wt%, and extremely low contents of all other major elements including SO_3. LOI varies between 3.04 and 5.09 wt%, which is apparently low for opal—A, and more typical for opal—C.

Maximum trace element values for all samples are 34 ppm Zn, <20 ppm Cu, Co, and Ni (ppm), as well as for other trace elements. One sample showed 134 ppm Ba and 96 ppm Sr and another 120 ppm Cu, 190 ppm Zn, 1400 ppm Sr, and 2.5% Ba. The Au content determined in two silica samples was found to be extremely low, showing a maximum of only 5 ppb.

Oxygen isotope ratios for calculation of formation temperatures, obtained with a precision of ±0.16‰ vary between δ O = +27.8‰ (sil-2) and +29.9‰ (sil-7) for pure silica samples. For sulfide-bearing samples the values were +δ18O = 18.8% (sil-12) and +δ18O = 18.6% (sil-13), respectively. Equilibration temperatures, calculated for pure silica, using the quartz-water oxygen isotope fractionation equation of Knauth and Epstein (1975) and

Kita et al. (1985), range from 32.0 °C to 42.2 °C, assuming equilibrium with normal seawater (+δ18O = 0%). The highest temperatures were determined for the central vents. The surrounding material formed at slightly lower temperatures. The two sulfide-bearing samples yielded distinctly higher formation temperatures of 100.8 °C (sil-12) and 102.3 °C (sil-13), similar to 60–172 °C obtained by Skirrow and Coleman (1982) from the Galapagos Rift.

One silica sample gave an upper age limit of 1440 ± 300 years, using a 230Th/234U method outlined by Ku (1976).

The silica chimney field "Cauliflower Garden" is the first known occurrence of a large hydrothermal site consisting entirely of pure amorphous silica. Thus it appears to be a new variety of seafloor hydrothermal deposit.

7.5.3 Low Temperature Quartz, Silica Gel, and Seafloor Conditions

Vein textures (Chapter 2) and also GCE and QEA asymmetry are governed by several physical principles that underlie formation of quartz. The first of these is illustrated in Figure 7.10, which shows that at lower temperatures, the solubility of the metastable polymorphs of SiO_2 (cristobalite and amorphous silica) is higher than that of quartz, and the difference decreases toward higher temperatures, as shown in the figure (Figure 7.16).

From this diagram, in practical terms, we can envisage a saturated solution with, say, log −1.0 g silica ponded in a hollow (assuming fluid density >1.0; see below) with temperatures fluctuating between 50 and 100 °C due to convection, diffusion, or addition of fresh brine. Silica will precipitate as quartz, cristobalite, or amorphous silica, which may form corresponding layers on the sea floor. Amorphous silica gel will later crystallize to quartz.

In greater detail, Hofstra and Cline (2000) and Fournier (1985) show an inflection in the quartz solubility curve (Figure 7.17) from which it is apparent that a superaturated solution may result from the mixing of two saturated solutions. At temperatures <180 °C to 200 °C, simple cooling of a quartz-saturated solution leads to supersaturation. This is particularly significant below 150 °C and underpins quartz and lode gold vein formation (Fournier, 1985; Rimstidt, 1997).

At temperatures below about 250–300 °C the quartz solubility curve is concave upward, so that mixing of hot and cool quartz saturated fluids result in precipitation of quartz, opaline silica, or amorphous silica, depending on the degree of supersaturation.

The quartz solubility diagram and hypothetical mixing lines provide a framework within which to think about this precipitation process, but the inflected curve in Figure 7.17 is the "lower 3 phase" line, i.e., the solubility of quartz in the presence of a vapor, at equilibrium (G.M. Anderson, personal information).

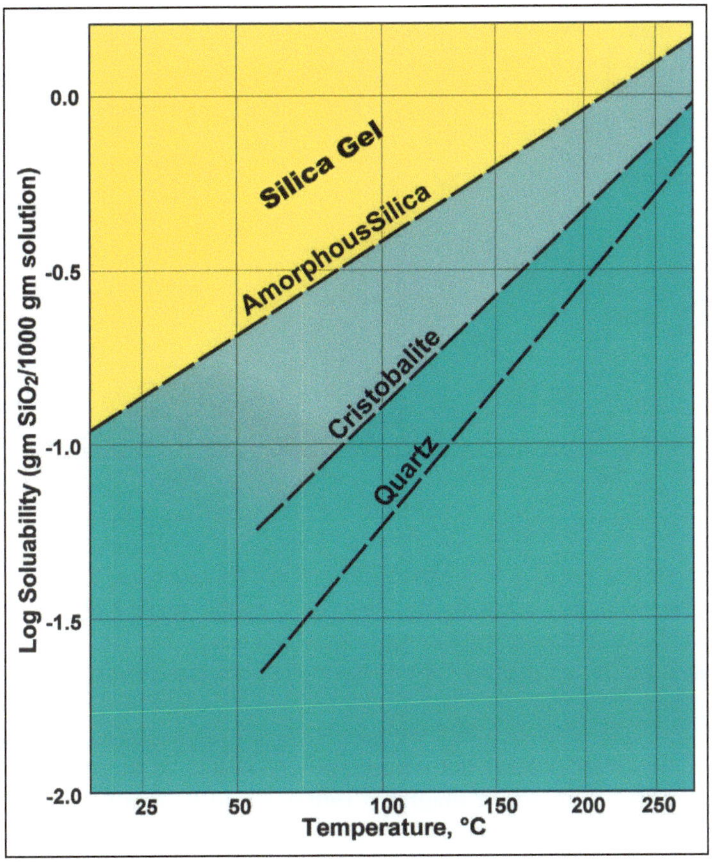

FIGURE 7.16

Solubility of quartz, cristobalite and amorphous silica in water between 25 and 250 °C. *Figure 9.3 of Holland and Malinin (1979).*

As we saw in Figure 7.9–7.11, the effect of P is to increase the solubility of quartz in water, but if the physical situation is as discussed below, this does not matter much, and neither do the mixing lines on the silica solubility figure.

In trying to understand the physical process, it invariably will involve diffusion, which is very slow unless there is turbulence. An analogous situation for Mississippi Valley–type ores was discussed by Anderson (2008), and many of the same questions can be posed regarding the formation of quartz veins in an open system under essentially hydrostatic pressure.

Where fluids exit on or close to the sea floor from relatively shallow depths in a volcanic environment (high T), there is no time for diffusive mixing unless they are ponded in a depression. Rapid cooling, essentially quenching, depositing silica gel, or amorphous silica, or cryptocrystalline silica of some kind, but it

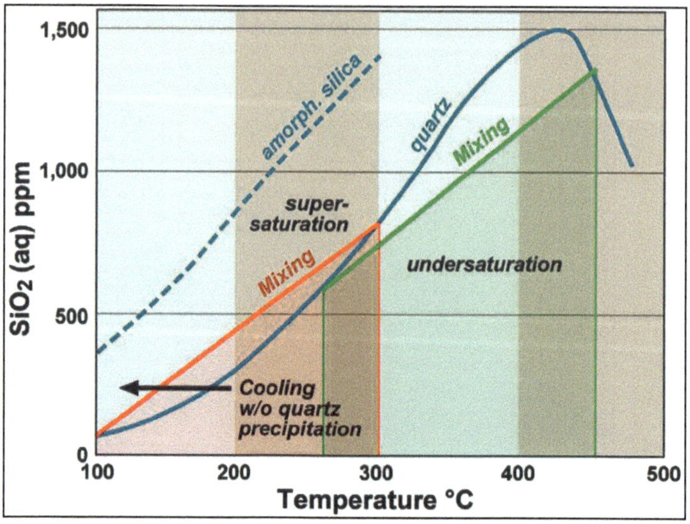

FIGURE 7.17

Solubility of quartz and amorphous silica as a function of temperature in a 5 wt% NaCl solution at 500 bars *(adapted from Fournier (1985).).* The thin lines show that mixing quartz-saturated 450 °C and 250 °C solutions results in undersaturation, whereas mixing quartz-saturated 300 °C and 100 °C solutions results in supersaturation. At temperatures less than 180–200 °C, quartz may not precipitate during cooling, resulting in supersaturation (arrow). Two saturated solutions mixing can result in supersaturation. *Figure 18 of Hofstra and Cline (2000).*

will later (soon) dehydrate and crystallize as quartz. It does not matter what part of the phase diagrams (what T and P) this solution starts from. It is not likely on the three-phase curve, but slightly above it at some pressure which prevents a vapor phase from forming, although boiling and vapor phase formation may well happen near the surface. The point is that wherever it comes from, it quickly reaches sea-bottom T and P, and it quenches. Therefore, the 250–300 °C interval or any other interval is essentially not important, if the precipitation process is fast. This is how you get a gel (G.M. Anderson, personal communication, 2013).

Further details regarding the mechanism of silica precipitation at lower temperatures can be gained if diffusion and conductive cooling are considered. The important principle regarding solubility and precipitation of quartz and amorphous silica as a function of temperature, diffusion, and conductance is summarized by Scott (1997, Figures 7.13 and 7.14). These in turn are developed from the experimental and theoretical work of Walther and Helgeson, (1977), Holland and Malinin (1979), Rimstid and Barnes (1980), Janecky and Seyfried (1984), and Herzig et al. (1988). The principles demonstrated by existence of the "Cauliflower Garden" silica chimneys by Herzig et al. (1988) are particularly important for the theme of this book, because they prove that quartz can form on the seafloor in an open system.

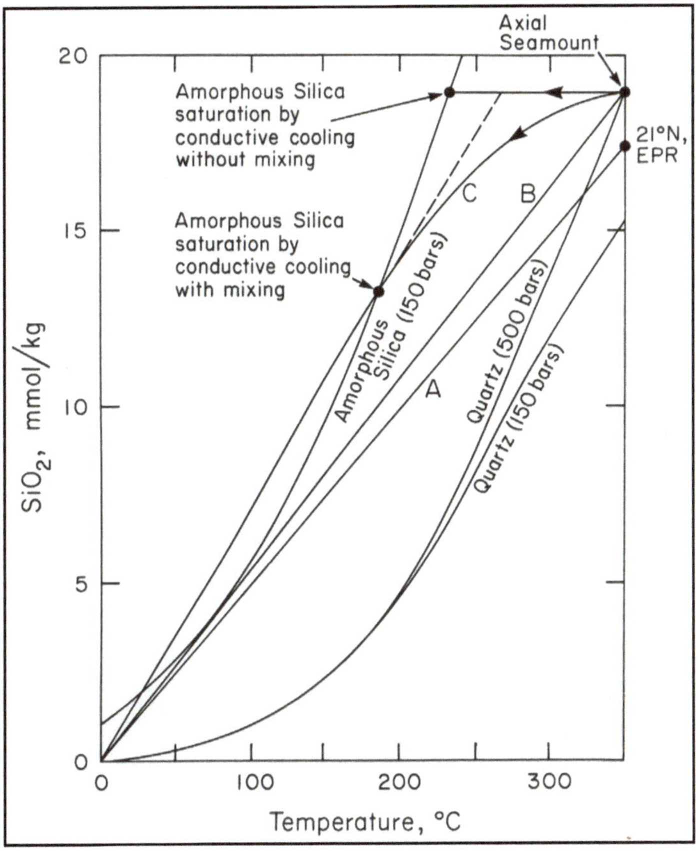

FIGURE 7.18

Silica precipitation in cooling vent fluids using data from CASM site of Axial Seamount at 21 °N East Pacific Rise (EPR) as example. Saturation curves are shown for amorphous silica and quartz (Holland and Malinin, 1979). Both the Axial Seamount (154 bars) and 21 °N EPR (260 bars) fluids are supersaturated with quartz at seafloor depths, but quartz doe not precipitate for kinetic reason, nor does simple mixing with ambient seawater result in precipitation of amorphous silica (lines A and B). Silica saturation occurs only where there has been conductive cooling or a combination of conductive cooling and mixing, as shown for Axial Seamount. *From Scott (1997).*

In greater detail, vent fluids become supersaturated with quartz as soon as they cool below 350 °C, but, for kinetic reasons, quartz is not precipitated. Considerable conductive cooling is necessary before the saturation curve for amorphous silica is reached.

Figure 7.18 shows that the simple mixing of vent fluid with cold ambient seawater (lines A and B) does not result in precipitation of amorphous silica at any temperature. A combination of conductive cooling and mixing (e.g., line C for Axial Seamount) can result in silica saturation at various temperatures below that due to conductive cooling alone.

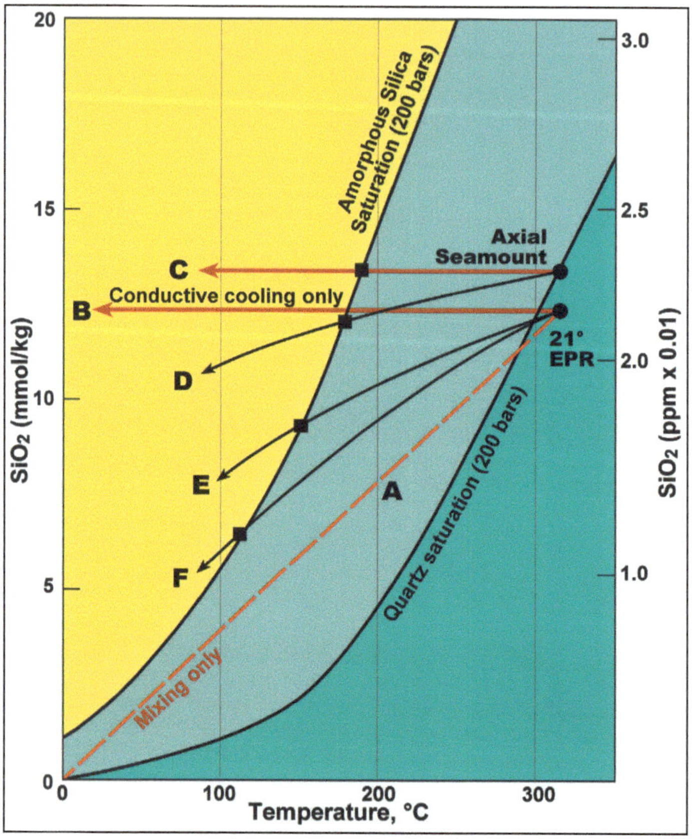

FIGURE 7.19

Use of fluid inclusions to constrain amount of conductive cooling combined with mixing to cause the precipitation of amorphous silica in chimneys from two different fluids at Guaymas Basin. Neither quartz nor amorphous silica precipitate for kinetic reasons by simple mixing of the vent fluids with ambient seawater (line A). However, amorphous silica does precipitate as a consequence of conductive cooling and mixing. The relative amounts of mixing may be determined accurately from the intercepts of the fluid inclusion homogenization temperatures with the silica saturation curve (lines B–F) *Original Figure 16.9, Scott (1997).*

Figure 7.18 illustrates why fluid inclusions in silica on seafloor sulfide deposits (or quartz veins) can never record the end-member temperature, because some conductive cooling, perhaps accompanied by mixing with ambient seawater, must take place for silica to precipitate.

Fluid inclusion data however, can be used to estimate the amount of mixing that has occurred. For example, Figure 7.17 shows the mixing behavior for two end-member fluids from Guaymas Basin. Line A represents simple mixing and B conductive cooling. Lines C–F represent cooling paths for increasing amounts of mixing, as determined by the intercept of the silica solubility curve at temperatures indicated by fluid inclusions (Figures 7.19 and 7.20).

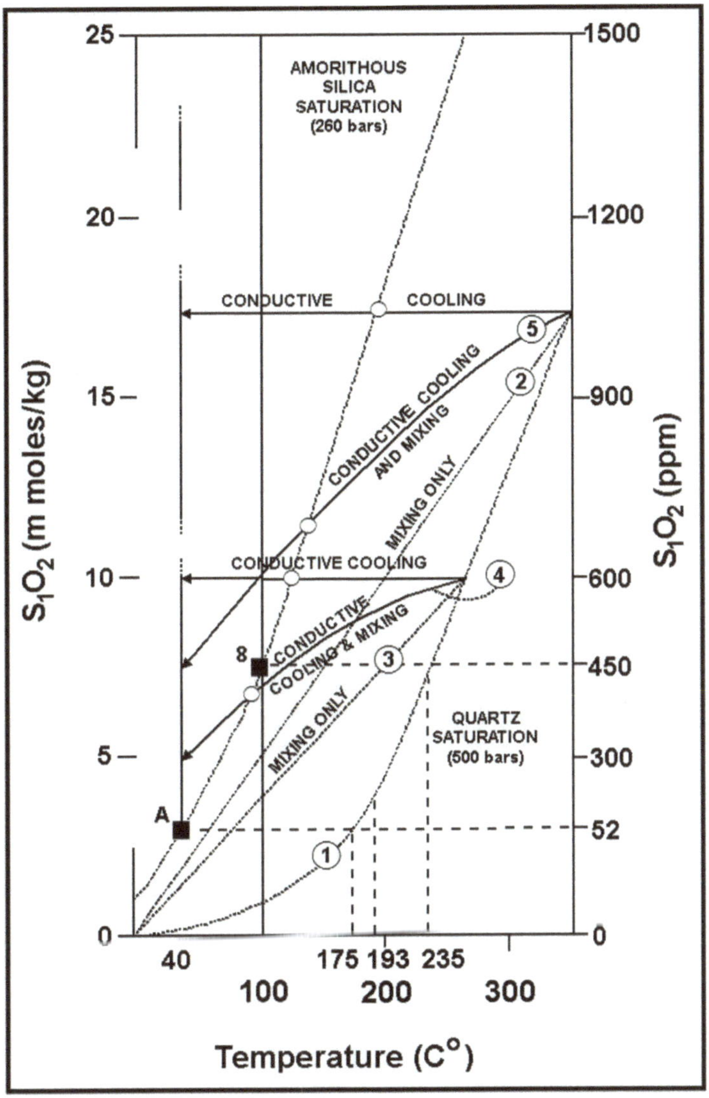

FIGURE 7.20

Demonstration of how quartz and amorphous silica form, specifically in the Cauliflower Garden hydrothermal vent field. The curves in Figure 7.15 show the SiO_2 solubility—temperature relations for amorphous silica at 260 bars (Walter and Helgeson, 1977) and quartz at 500 bars (After Herzig, et al., 1988).

Point 1 is at the minimum initial temperature and silica concentration of a hydrothermal fluid in equilibrium with quartz at 500 bars and saturated with respect to amorphous silica at 40 °C and 260 bars. Lines 2 and 3 are possible mixing lines between normal seawater (2 °C, 10 ppm SiO_2) and hydrothermal fluids in equilibrium with quartz. Mixing Line 2 represents mixing of seawater and an EPR 21° N-type fluid (350 °C, 1055 ppm SiO_2) according to Janecky and Seyfried (1984). Reaction paths 4 and 5 are for a combination of conductive cooling and mixing of seawater with hydrothermal fluids allowing for supersaturation and deposition of amorphous silica at any temperature between 40 °C and <193 °C. Solid horizontal lines represent possible reaction paths for pure conductive cooling. Points A and B represent the saturation with amorphous silica for the pure silica chimneys from the Cauliflower Garden (40 °C formation temperature).

The dissolution of silica in the source region of the high-temperature reaction zone is controlled by quartz, but its precipitation at, or just beneath, the seafloor (Figure 7.9) is controlled by the higher solubility of amorphous silica or opal.

Why is quartz not more common on the modern seafloor? Bowers et al. (1985) think that the scarcity of quartz and predominance of amorphous silica in rapidly formed seafloor vent deposits can be explained by the slow nucleation kinetics of quartz, which forms the basis of the work of Rimstidt and Barnes (1980). In addition, as Vanko et al. (1991) stated, hydrothermal fluids rise nearly adiabatically from the high-temperature reaction zone, and during rapid upwelling, quartz nucleation is suppressed and dissolved silica ends is simply flushed out into the ocean. If fluid flow is slower and more diffuse with heat loss due to conduction, say, in a ponded hollow, nucleation of silica will be favored and laminated quartz crusts may form. These are our "lode veins."

7.5.4 Fluid Inclusions, Quartz Decrepitation, and CO_2 in Hydrothermal Fluids

Fluid inclusion data are widely used to provide insight into the chemical and physical conditions of ore forming fluids. Fluid inclusions and baro-acoustic decrepitation principles are summarized on the Website of Kingsley Burlinson (www.appliedminex.com) and "What do fluid inclusions tell us?" http://www.appliedminex.com/viewpoint/porphyryfi.htm as well as his several publications (Burlinson, 1991, 2007, 2013).

Of considerable exploration interest is his method to determine the approximate total gas content of quartz, since this frequently closely correlates with mineralization potential. CO_2-rich fluid inclusions develop high internal pressures at low temperature resulting in decrepitation long before the inclusion reaches homogenization temperature. The decrepitation gives rise to distinctive low-temperature peaks on an acoustic decrepigram, which can be used to recognize the presence of CO_2-rich fluids. This led to the recognition that CO_2 measurements by acoustic decrepitation could be used as an exploration tool for gold deposits, since these are often associated with CO_2-rich hydrothermal fluids.

7.6 GOLD CHEMISTRY

7.6.1 Gold in Hydrothermal Vents

The literature on the source and origin of the gold is vast, and the chemistry and thermodynamics of the system are well enough understood (e.g. Large, 1977; Kerrich, 1983; Hannington and Scott, 1989, 1997) and will be only briefly summarized. Of interest are the boundary conditions for ore solutions, that is:

1. where does the gold come from?
2. what is its behavior in hydrothermal solutions and sulfide minerals?
3. at what depth is the fluid generated, and how does the combination of water depth, salinity, and CO_2 control silica precipitation when the hydrothermal solution emerges on the seafloor?

7.6.1.1 *Gold Chemistry in Vent Fluids*

The expected concentration and behavior of gold in hydrothermal solutions has been calculated from thermodynamic data (Scott, 1997, p. 841).

These calculations suggest that seafloor hydrothermal vent fluids contain as much as 0.5–10 nmol/kg Au. Gold is delivered to the seafloor in relatively low-temperature fluids (300 °C) as a gold bisulphide complex, Au $(HS)_{2-}$, and is precipitated during late stages of formation of the deposit when this fluid is thoroughly mixed with cold ambient seawater. Evidence for this consists of.

1. the high sulphidation state of associated mineral assemblages (sulfur-rich sulfide minerals, low iron content of zinc sulphide);
2. chemistry of analyzed vent fluids that is favorable for gold bisulphide complexing (high H_2S content, near-neutral pH, near-seawater salinity (Hannington and Scott, 1989a);
3. the absence of gold (less than −0.2 ppm Au) in high-temperature copper–iron sulphide veins beneath the seafloor; and
4. typically associated high abundance of barite for which the sulphate is demonstrably of ambient seawater origin.

7.6.2 Fluid Inclusions and the Chemistry of Lode Vein Forming Solutions

Increasingly detailed and sophisticated understanding of fluid and gold chemistry, metal transport, and deposition is apparent from reviews by Thomas et al. (2011), Robert et al. (2007), Dubé and Gosselin, (2007), Goldfarb et al. (2005), Groves et al. (2003), Hageman and Cassidy (2000), Bierlein and Crowe (2000), Poulsen et al. (2000), Kerrich et al. (2000), Kerrich (1983, 1993), Hodgson (1990), Colvine et al. (1988), and others. An excellent and comprehensive study that has stood the test of time is by Kerrich (1983). More current summaries are by Ridley and Diamond (2000).

From these studies, we can summarize that hydrothermal ore-forming fluids are dilute, aqueous, carbonic fluids, with salinities generally less than or equal to 6 wt % NaCl equivalent, and $CO_2 \pm CH_4 \pm N_2 \pm H_2S$. They possess low (approx. 60 ppm) Cl, and relatively high (approx. 1000 ppm) S.

Primary fluid inclusions are:

1. H_2O-CO_2,
2. CO_2-rich with variable CH_4 and small amounts of H_2O, and
3. two-phase H_2O (liquid–vapour) inclusions.

Inclusion types 2 and 3 are often interpreted to represent liquid immiscibility of the original ore fluid. Immiscibility can be triggered by fluid pressure drop during coseismic events, and possibly shock nucleation, leading to highly variable homogenization temperatures in an isothermal system.

7.6.3 What Causes Precipitation of Gold?

Regardless of the actual concentration of Au in the hydrothermal ore-forming fluids, to form an economic ore body, it is necessary to induce a change in the fluid chemistry that will efficiently remove gold from solution by destabilizing aqueous Au–S complexes. Based on available data, destabilization could be accomplished by a number of processes:

1. cooling of the fluid,
2. oxidation of the fluid,
3. reduction of the fluid,
4. decreasing of pH of the fluid, or
5. lowering of S in the fluid.

This modification of the fluid chemistry can be achieved by any of the following:

1. large-scale pressure and temperature gradients along the fluid plumbing system;
2. reaction of the fluid with the wallrock surrounding the fluid conduit;
3. transient pressure fluctuations inducing phase immiscibility in the fluid;
4. fluid mixing; or
5. chemisorption.

According to a summary by Robb (2004), gold preferentially complexes with bisulfide ion and to a minor extent with chloride. Gold typically occurs in a monovalent state (Au^+) in aqueous solutions and will form the complex AuHS at low pH and $Au(HS)_2^-$ under weakly acidic to basic conditions. The bisulfide complexes predominate in most hydrothermal fluids and are stable over a wide temperature range. Under highly oxidizing, saline, and acidic conditions, it is possible to form both Au +−chloride and Au 3+−chloride complexes in the form of $Au(Cl)_{2-}$ and $Au(Cl)_{4-}$, respectively. The former complex might prevail in certain high-temperature environments such as porphyry coppers, whereas the latter is likely to be important only at low temperatures in near-surface settings.

7.6.4 Gold Solubility

Experiments carried out by Loucks and Mavrogenes (1999) on dissolved gold contents in chloride brines that were experimentally saturated with gold, magnetite, iron sulfides, orthoclase, and muscovite at 550° to 725 °C and 100−400 MPa show that the solubility of sulfide species in supercritical brines is far higher than previously supposed. Dissolved gases such as molecular hydrogen sulfide that are ineffective metal-complexing agents at low pressures evidently become highly effective at high pressures.

7.6.5 Separation of Au from Base Metals

Kerrich and Hodder (1982) studied the distribution of metals in gold and base deposits in Canadian Archean greenstone belts. They found that gold deposits were enriched in gold, silver, arsenic, mercury, boron, tungsten, selenium, tellurium, and bismuth, but with low or negative enrichment of copper, lead, and zinc, even though these metals are more mobile. In contrast in the VMS deposits, gold, silver, copper, zinc, and, in some cases, lead were enriched significantly, but gold did not seem to be preference for these base metal sulphides. It may be that this differentiation has more to do with the source rocks; gold deposits usually occur at the top of mafic volcanic sequences, whereas the base metal deposits are associated with intermediate rocks. Gold was enriched in a pyritic ore pile at the past producing Betts Cove Mine (1874−1886), a VMS deposit. The miners had mined this material with the copper ore and differentiated it for future processing. A composite grab sample from the top of the pile averaged 15 g/T. In the Heath Steele Mines B Zone, a gold-rich pyritic zone lies west and near the source in comparison to the zoned copper−lead−zinc B Zone; there was recoverable gold and silver associated with the copper−lead−zinc. It appears that two zoning patterns are present: the gold precipitates with the initial pyrite, but some is transported and precipitated with the base metals.

For the assumptions used to derive this diagram, Figure 7.21, the reader is referred to the original paper of Hannington and Scott (1989), which uses data from Seward (1973). Contours of gold solubility (bold dashed lines: ppb Au(HS)−) calculated assuming an activity coefficient of 1.

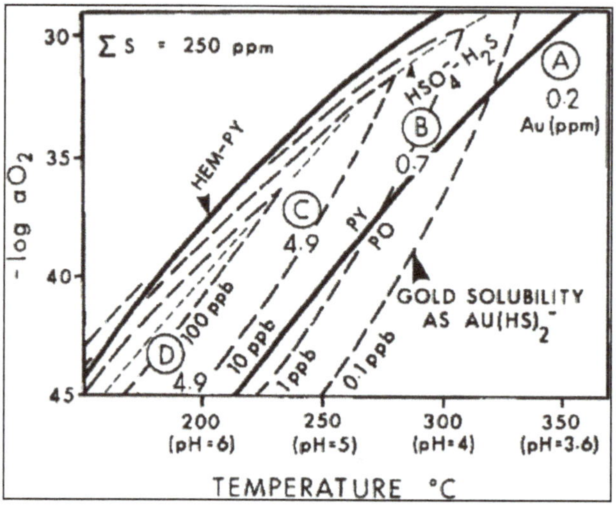

FIGURE 7.21

Calculated solubility of gold as Au(HS) in hydrothermal seawater as function of temperature, pH, and aO_2. *Diagram from Hannington and Scott (1989).*

A constant H_2S concentration (250 ppm H_2S) equivalent to the end-member value for 350 °C vent fluids at 21° N, N, East Pacific Rise (Von Damm et al., 1985a) is used to reflect the high average H_2S content of known vent fluids.

Characteristics of fluid chemistry are plotted at the midpoints of estimated conditions of or (A) 21° N E at Pacific Rise(Haymon and Kastner, 1981; Styrt et al., 1981; Von Damm, 1983; Janecky and Seyfried, 1984; Zierenberg et al., 1984), (B) southern Explorer Ridge (Tunnicliffe et al., 1986; Hannington and Scott, 1988; Scott et al., 1990), and (C and D) two different stages of mineralization at Axial Seamount (Hannington and Scott, 1988). Data for 21° N (A) are from 350 °C vent fluids with a measured pH of 3.6, and log O values of −30 to −31 based on measured pH. Data for Explorer Ridge (B) are from 276° to 306 °C vents with measured pH of 4–5 and calculated log O values of −32 to −36 based on FeS contents in sphalerite. Conditions of mineralization at Axial Seamount are based on fluid inclusion trapping temperatures and silica geothermometry estimated pHs, and calculated log O′ values of −34 to −38 at 235 °C (C) and −44 at 185 °C (D). The fields of principal aqueous sulfur species (H_2S/HSO boundary: dashed line) and stability fields for pyrite (PY), pyrrhotite (PO), and hematite (HEM) (bold lines) are calculated from Helgeson (1969), Helgeson et al. (1978), Robie et al. (1975), and Bowers et al. (1985).

Average gold content of sulphides at each deposit are 0.2, 0.7, and 4.9 ppm Au, respectively. Temperature and chemical changes are a consequence of mixing

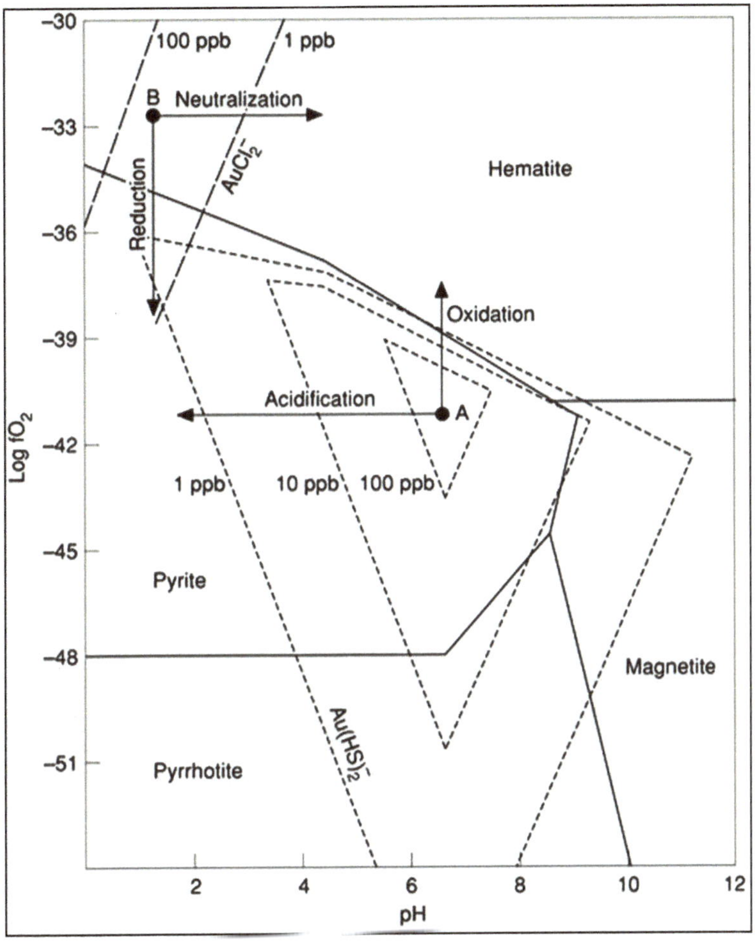

FIGURE 7.22
Log f02-pH plots showing controls on gold solubility. *Figure 03–14, Robb (2004).*

between high temperature end-member fluids and seawater; estimated pHs correspond approximately to the mixing model for a 21° N end member fluid from Janecky and Seyfried (1984) (Figure 7.22).

7.6.6 Gold Abundance

According to Robb (2004), the average crustal abundance for selected metals and typical concentration factors that need to be achieved in order to produce a viable ore deposit is as follows (Table 7.1):

Table 7.1 Average Crustal Abundance for Selected Ore Metals and Typical Concentration Factors Reflected in an Economically Viable Ore Deposit According to Robb (2004)

	Average Crustal Abundance	Typical Exploitable Grade	Approximate Concentration Factor
Fe	5.6%	50%	9
Cu	55 ppm	1%	180
Ni	75 ppm	1%	130
Zn	70 ppm	5%	700
Sn	2 ppm	0.5%	2500
Au*	4 ppb	5 g/T	1250
Pt	5 ppb	5 g/T	1000

Note: Au. The average grade of gold deposits in production today is about 1.0 g/T.*

These or similar abundances can be used to calculate how much of a given rock must be leached to provide gold for a given-sized gold deposit. Phillips et al. (1987) indicated that all the significant gangue and metal components in the giant Golden Mile ore body could have came from a block of greenstone 8 × 8 × 5 km. They use a metamorphic-replacement model for gold mineralization in which all ore components derive from devolatilization of greenstones at amphibolite facies or above which is above the metamorphic grade of most "greenstone lode gold deposits."

7.6.7 Source of the Gold

Pyrite in marine sedimentary rocks and in greenstones is likely a leachable source mineral with high concentrations of gold. Similarly, radiogenic isotope studies at Muruntau suggest that elements in scheelite and, by association, the gold itself were derived from the Palaeozoic metasedimentary host-rock sequences (Kempe et al., 2001).

Gold solute concentrations in the ore fluids have been estimated at approximately 20 ppb, requiring commensurately large quantities of hydrothermal fluids, or about 6000 cubic kilometers for the Abitibi terrain deposits. Gold abundance through time is discussed by Large et al. (2015). They say that gold content in sea water can be correlated with the gold contents in gold-rich komatiitic and magnesium-rich basalts and atmospheric oxygen over time.

7.6.8 Fluid Source: Evidence from Stable Isotopes

Phillips et al. (1987) suggest that the rather uniform nature of Archean Au-bearing fluids implies a uniform fluid-generating process and that devolatilization of deeper greenstone belts offers such a process, particularly if premetamorphic alteration zones are involved.

Currently unexplained in models of lode gold vein genesis, according to Bierlein and Crowe (Question 4, Gold Deposit Models) and Kerrich is that "wall-rock alteration mineralogy, geochemistry, and isotopic halos all indicate that the hydrothermal assemblages are not in equilibrium with the surrounding host rocks," nor should they be if they are deposited on the sea floor from a vent and subsequently buried by rocks from a distal source.

A simple explanation emerges from our syngenetic mode of vein formation. If we assume that the hydrothermal silica–gold–rich fluids are in equilibrium with the reservoir rocks from which they were generated (i.e., the underlying turbidites, volcaniclastics, or black shales), and as it emerges and silica is precipitated as a gel, its immediate environment is the pelitic/pelagic substrate on the sea floor, which has a different isotopic signature (i.e., Q is not in isotopic equilibrium with E).

Far from being evidence against syngenesis (Bierlein and Crowe, 2000, p. 136), vein-host rock stable isotopic disequilibrium constitutes further robust proof of syngenesis, because fluids with homogenized isotopic composition of the leached source rock will in general be out of equilibrium with seawater and host pelagic–pelitic sediments where they are deposited. The extent of this nonequilibrium is related to the origin of the hydrothermal fluid and the deposition environment, as is briefly discussed below.

With regard to what extent radiogenic isotope systematics are useful in pinpointing the source of gold, Bierlein and Crowe answer: "Pb, Sr, and Nd systematics can provide insights into the nature and composition of potential source areas, but:

1. gold is mono-isotropic and cannot be directly traced;
2. the source of the Pb-Sr, or Nd-bearing minerals is not necessarily coeval with the gold;
3. input of common lead from a mantle-like source is likely to get swamped by radiogenic crustal lead and
4. large scale homogenization processes commonly obscure the likely heterogeneous signature of various source rocks (Figure 7.23).

Fluid inclusion research has become increasingly sophisticated, as shown in a study by Gaboury (2013). Fluid inclusions from selected deposits were analyzed by solid-probe mass spectrometry and were found to be aqueous–carbonic, with variable amounts of N_2, CH_4, C_2H_6, Ar, H_2S, H_2, and He. For fluids with CH_4 and C_2H_6, their ratios (C_1/C_2) range from 2.6 to 25.5. Variable proportions of carbon-bearing species and hydrogen (CO_2, CH_4, C_2H_6, and H_2) can be attributed to a hydrothermal reaction in which C_2H_6 is degraded

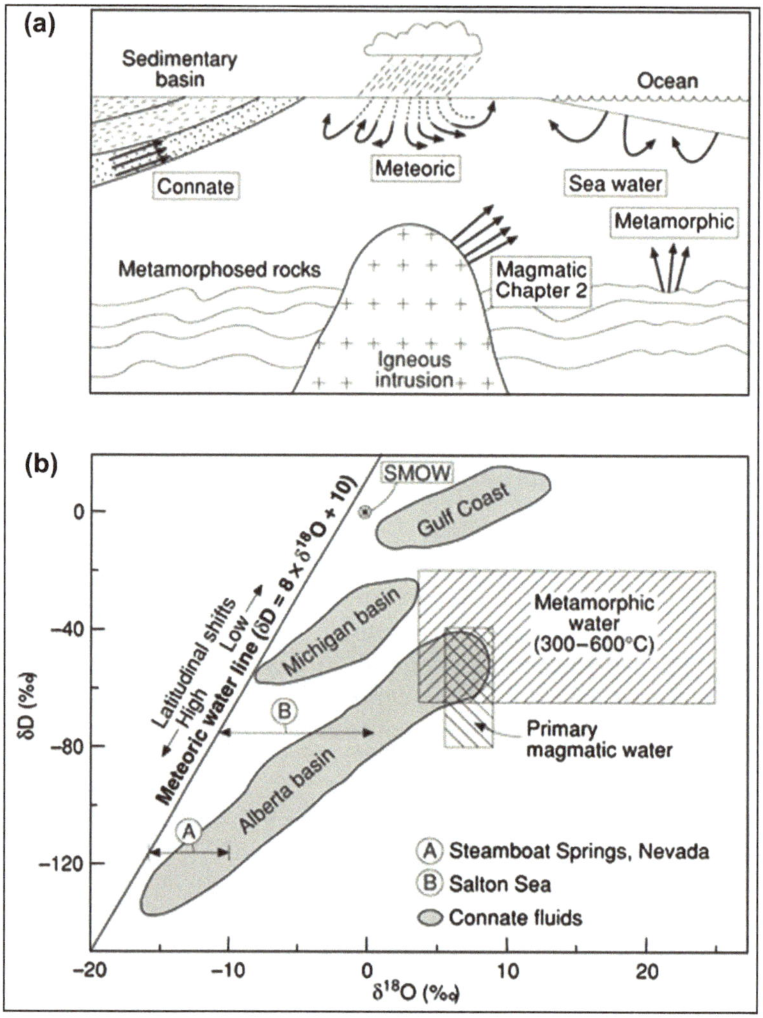

FIGURE 7.23

Oxygenisotopic composition showing the fields of primary magmatic and metamorphic water. *From Robb (2004).*

to CO_2 by water consumption. This may account for CO_2-rich, H_2O-poor fluids associated in some gold districts. Fluid samples from Mesoarchean to Cretaceous gold deposits indicate that C_2H_6 are sourced from thermally degraded organic matter. Conditions needed for C_2H_6 degradation are the same as those fundamentally required for forming gold deposits, namely HS−-enriched fluids for carrying gold and local weakly oxidizing conditions for promoting gold precipitation. The study provides further support for a general model in which fluids and gold are sourced from deeply buried, carbon-rich and pyrite–gold–bearing sedimentary and volcanic rocks.

7.7 CARBON (GRAPHITE) AND WHAT IT SAYS ABOUT THE DEPOSITIONAL ENVIRONMENT OF GOLD DEPOSITS

The widespread and well-known association of carbon or graphite and gold is seen by its association with gold deposits. The present is the key to the past, and the authors have learned that the geological system repeats itself. Therefore, the relationships in more recent rocks can be applied to ancient rocks (i.e., Paleozoic to Archean). This section explores the environment for carbon deposition the volcanic and sedimentary environments. Gold is common in stratabound beds in these rocks; this discussion will expand to include evidence that the presence of carbon in stratabound rocks is conclusive evidence for syngenetic deposition of associated gold deposits.

Geologists will not question the fact that carbon or hydrocarbons are of biogenic origin in modern rocks. The problem arises when ancient rocks are discussed. Experienced exploration geologists were likely exposed to the testing of electromagnetic conductors in the search for volcanic hosted massive sulphide deposits in the 1960s and 1970s. Many were located and drilled, only to reveal graphitic schists with varying amounts of sulphides, mainly pyrite and pyrrhotite and minor amounts of chalcopyrite and sphalerite; results of these programs are in the government assessment files. McBride drilled some in the Middle Ordovician Tetagouche Group of the Bathurst Camp of New Brunswick in the late 1970s and the younger Ordovician rocks of the Meductic Area in the southwest part of the province. All electromagnetic anomalies proved to be carbon or graphite in sedimentary rocks with or without associated volcanic rocks. The origin of these carbonaceous rocks was considered to be biogenic (Berry and Wilde, 1978).

In the older rocks from the Grenville supergroup dated at 1.2 to 1.3 By, graphitic intervolcanic sedimentary rocks have been drill tested (Tulonen and King, 1997). Numerous similar zones were identified in the Archean, including some mapped by McBride. One was studied as part of his B.Sc. thesis at Queen's University.

7.7.1 The Presence of Graphite or Carbon in Volcanic and Sedimentary Rocks

In the summer of 1967, the author McBride was employed as a geologist mapping properties staked on the basis of an airborne electromagnetic survey. Each conductor had to be assessed as a potential drill target. At that time, very little was known about these conductors, but it was soon learned that most of them resulted from graphite and sulphides in intervolcanic sedimentary bands. In the 1970s, similar exploration projects in the Cambro-Ordovician of New Brunswick, Canada, identified the same carbon–sulphide–bearing sedimentary environment for conductors in the Tetagouche Group volanco-sedimenatary rocks that host the famous volcanogenic massive sulphide deposits of Brunswick and Heath Steele Mines and younger sedimentary-volcanic sequences it the southwestern part of the province.

McBride obtained drill core from a conductor that he had mapped. This conductor was caused by a black schistose clastic sedimentary unit sandwiched between felsic and mafic volcanic rocks. The carbon or graphite is easy to recognize because of its color and schistose nature; it is the only black, nonmetallic, opaque mineral found in sedimentary rocks. During the author's study, Dr R. W. Boyle of the Geological Survey of Canada, recognized for his work on gold deposits, discussed the origin of the ancient graphitic conductors with him; Boyle had tried using sulphur isotopes, but could not distinguish a sedimentary or volcanic source. It was later discovered that organisms were present around the vents, which led to the conflicting results. To McBride's knowledge, his work was never published. Dr L. G. Berry, of Queen's University, provided a sample from McIntryre Mines that had been sent to him for identification because it was interfering with gold recovery. The material turned out to be amorphous carbon. It was interpreted that the carbon in these clastic sedimentary units was biogenic carbon. Boyle (1979) made reference to carbon and its propensity in gold deposits.

Regional metamorphism does not reduce limestone to graphite; instead it recrystallizes it to marble, or crystalline limestone which may have calcite crystals to 15 cm across with bedding still preserved, as observed by the author McBride at the Long Lake Lead-Zinc Mine in the Grenville Geological Province in Eastern Ontario. Any original carbon is preserved as minute black graphite crystals.

As part of the Northern Empire project regional mapping around the mine property in 1981, many airborne electromagnetic conductors were identified that proved to be graphitic or cherty pyrrhotitic graphite conformable bands in the Archean volcanic stratigraphy or along its contacts with the overlying Temiskaming clastic sedimentary rocks. Similar graphitic sediments host the Ordovician gold mineralization at Nugget Pond and the Cambro-Ordovician of the Meguma gold fields of Nova Scotia. The Nugget Pond deposit occurs in moderately dipping red, green, and black mudstones sandwiched between two volcanic sequences. According to Swinden et al. (1990), the Nugget Pond mineralization is stratabound and lacks significant metamorphism, metasomatism, hydrothermal alteration, and shearing, and is only tilted. This lode gold deposit is discussed in detail in Chapter 9.

The association of gold with graphitic bearing rocks is more the rule than the exception in volcanic and sedimentary rocks of all ages from the Archean to the Cretaceous of Haiti and Recent rocks in Indonesia. McBride has worked in gold exploration in all these areas, but his work is found only in company reports. The author stressed the significance of this association again in 2009 (McBride, 2009). Goldcorp's Red Lake Mine has high grade mineralization in black mudstones (Figure 7.24).

Springer (1985) describes two types of graphite occurrences in gold mines in the Abitibi belt from the following mines: Kerr Addison, Dome, Kidd Creek, and Pamour Porcupine. Sedimentary black argillites with carbon contents of up to 5% C contain up to 0.5 ppm gold and graphitic schists, interpreted as schistose black argillites, have gold values up to 15 ppm that reflect inherited concentration in the sedimentary precursors or later hydrothermal introduction of gold and its preferential deposition in calcareous carbonaceous rocks by reduction at vein margins. He does not explain how or when the gold was introduced, but points out that the conditions deduced for formation of gold deposits according to Colvine et al. (1984) imply that deposition took place from subcritical, dilute chloride solutions highly charged with CO_2 at temperatures of 200–500 °C, with pressures of 1–3 kbar (100–300 MPa). He notes that these conditions are not unlike those used for commercial preparation of active carbon. Springer (1985) therefore suggests that the reactivity of the carbon has been enhanced by hot hydrothermal fluids flowing up shear zone conduits, but the conditions are well within the range at vents on the ocean floor.

Recent studies on carbon have been summarized in the December 2014 issue of the journal Elements. Buseck and Beyssac (2014) and Rumble (2014) explore the diversity of graphitic carbon in the natural environment and present a simple illustration to show its universality in the geological cycle (Figure 7.25). Windley (2015) says that the release of oxygen in the Precambrian by cyanobacteria, a class of prokaryotes that are unicell organisms with rudimentary internal organization, used the process of organic photosynthesis, which produced oxygen as a byproduct and was responsible for the build-up of oxygen in the Archean atmosphere.

FIGURE 7.24
Gold in black graphitic material, Goldcorp Mine, Red Lake,Ontario.

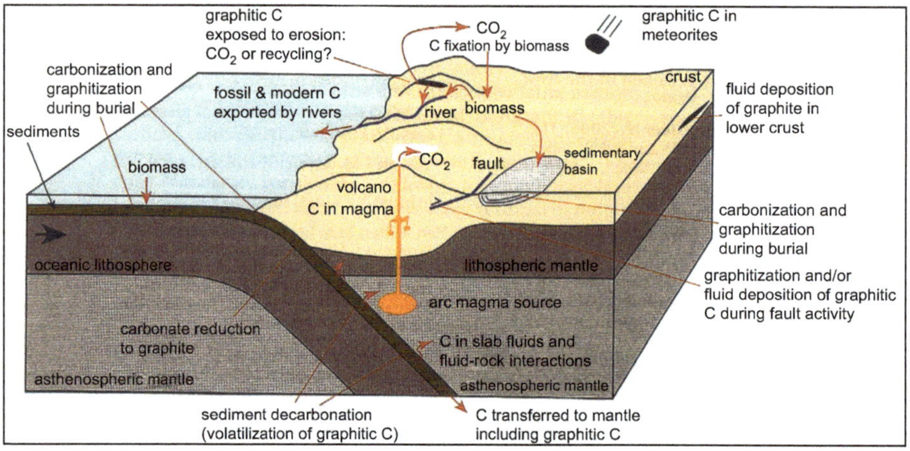

FIGURE 7.25

The geological cycle of graphitic carbon. The diagram illustrates the geological processes and carbon fluxes involving carbon; the examples are not exhaustive. *Figure 5, Buseck and Beyssac (2014) and Rumble (2014).*

In a later paper, Rumble (2014) investigates hydrothermal graphitic carbon and states that carbon may be remobilized from carbonate and organic-rich sediments during metamorphism. This graphitic carbon records the presence of carbon-bearing aqueous fluids along grain boundaries, in fluid inclusions, and in fissures. Most of the reactions cited to support this mobility take place at temperatures above 400 °C. In the same issue, Buseck and Beyssac (2014) demonstrate the progression from organic molecules to graphite (Figure 7.26). They state that in the carbonization stage, which takes place during diagenesis, the organic materials go through the keogen transformation that produces smaller molecules, like oil and gas, and leave a residual carbonaceous residue, coal, which is enriched in carbon. This residue remains in the host environment and is subject to the regional metamorphism in orogenic belts or older sedimentary rocks. Determining the biogenic or abiogenic precursor source of the carbon is difficult and subject to intense controversy. Bernard and Papineau (2014) discuss the problems distinguishing ancient biogenic carbon from abiotic carbon. They present a list for assigning a biogenic origin for graphitic carbons Table 7.2.

Much of the controversy hinges on the understanding of carbon in the metamorphic environment. In many greenstone belt deposits, the carbon is not crystallized to graphite. Supporting evidence can be found by comparing the geological settings of more recent carbon-rich sediments. Few geologists would question that the black muds in recent sediments or older ones such as the Halifax and Goldenville Formations of the Meguma Group are biogenic, but many question the concentration of carbon in the older Precambrian rocks, even though the setting is

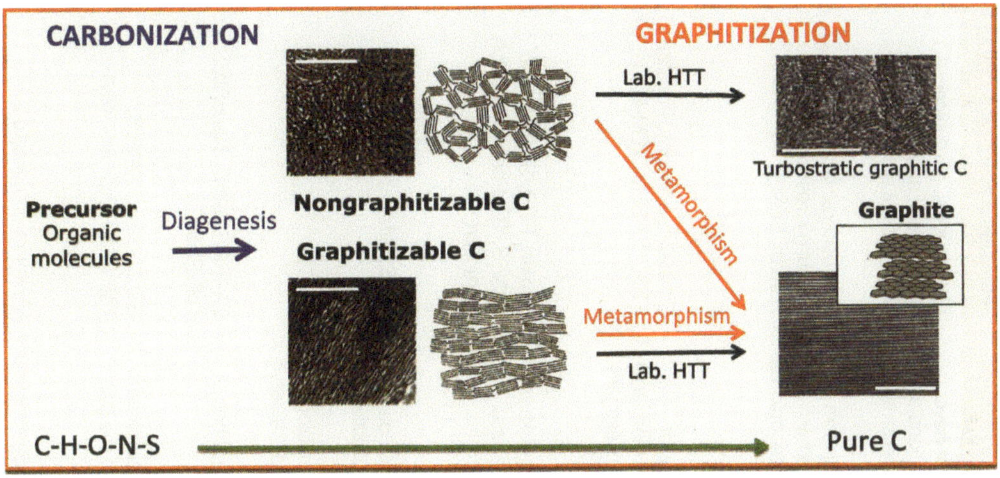

FIGURE 7.26

Schematic representation of the transformation of organic compounds into graphite through the carbonization and graphitization stages. Nongraphitizable carbon retains its poorly ordered stacking after laboratory heat treatment, whereas it transforms into graphite under high-grade metamorphism (T > 700 °C). Scale bar for TE mages is 10 nm. Lab. HTT = high-temperature treatment in the laboratory (up to 3000 °C). Structural models for graphitizable and nongraphitizable carbon are from Franklin (1951) *Figure 1 of Buceck and Beysssac (2014).*

similar. These zones are identified commonly as graphitic schists because of the shiny black color observed in the field. The widespread nature of "graphitic" conductors in sedimentary rocks is strong evidence for the source of carbon being biogenic, especially in low- or moderate-grade metamorphic rocks.

The association of carbon in intervolcanic sedimentary bands and gold is discussed in Chapters 8 and 9, but there is not any reason to think that the source of carbon associated with gold deposits is any different from that in any similar intervolcanic sedimentary and exhalite horizons. Recent work on black, sulphide-bearing sediments at the Golden Mile Deposit, Kalgoorlie, Western Australia, had shown that the nodules of pyrite contain gold that shows no evidence of being introduced after deposition or diagenesis (Steadman et al., 2015). Their photographs (Figure 4) show the controlling sedimentary textures. In fact, in this sedimentary or any similar tuffaceous environment, none would be expected for the simple reason that fine-grained rocks deposited in this environment lack any porosity and thus the permeability to pemit fluid flow without obvious fracturing and/or veining. Introduction of any metal after diagenesis is not possible without the presence of cross-cutting veinlets coming from well outside the host rocks.

Recent studies by Harris et al. (2009), Lindsay et al. (2005), and De Gregorio (2006) on the 3.45 by old Strelly Pool cherts in the Pilbara Craton of Western Australia present a wealth of information directly applicable to lode vein

Table 7.2 List of Suggested Criteria for Assigning a Biogenic Origin to Graphitic Carbons.

	Required Criteria	Cautionary Notes	Techniques Useful for Assessment*
Geological context	Sedimentary nature of the protolith	Chemical sediments might have formed in multiple generations and under the seafloor or in the crust.	Field studies, geological, petrological, and/or geochemical studies
Petrographic context	Occurences of graphitic carbons should be numerous within a single thin section and demonstrably indigenous. Morphological and textural structures may be important clues but may be ambiguous.	Possible artifacts resulting from sample preparation and other sources of contamination should be identified and avoided.	Macroscopic observations, optical microscopy, CLSM, Raman, SEM, EA
Mineralogical context	Graphitic carbon occurrences should be included within the crystalline mineral matrix and not in pore spaces. Associations with diagenetic minerals can be useful.	Occurrences associated with neoformed minerals, alteration, and skarn phases should be investigated with special care.	Optical microscopy, Raman, FTIR, FIB-SEM, TEM-EDS
Structural and microstructural characteristics	Proofs of syngeneity: The crystalline structure of the graphitic carbon is consistent with the peak metamorphic conditions experienced by the host rock.	Care is needed for hosts rocks that have experienced a multiphase geological history.	Raman, FTIR, STXM-based XANES, TEM-based electron diffraction
Isotopic composition	^{13}C-depleted carbon isotope compositions consistent with a biological origin, taking into account possible metamorphic fractionation and reequilibration of inorganic carbon phases.	Adequate standards should be used, and artefacts should be identified and avoided. Care is also needed as nonbiogenic carbons may exhibit similar values.	IRMS, SIMS, NanoSIMS
Elemental composition	Graphitic carbons of biogenic origin might still contain heteroatoms (O, N, P, S) as traces.	Contamination by recent carbon, if any, should be identified.	EA-IRMS, SIMS, NanoSIMS, ToF-SIMS
Molecular structure	Trace levels of molecular functional groups might remain present and be consistent with heteroatom content.	Their amount should be consistent with the thermal history of the host rock.	STMX-based XANES, ToF-SIMS, GC–MS, NMR
Reproducible observations	Reproducible observations in other samples from other localities appear necessary.	Absence of evidence is not evidence of absence.	Minimum of triplicate observations

Table 1 of Bernard and Papineau (2014).

FIGURE 7.27
Comparing vein textures in hand specimen. Upper Left, Beaver Dam "Rosetta Vein," Meguma turbidite-hosted gold vein. Upper Right, gold-bearing vein from Lamaque mine, Quebec. Bottom Right, Detail of Strelly Pool vein. *Photo credits: Lindsay et al. (2005), "Rosetta Vein": U. Kretchmar; Lamaque Vein, Ann Rept. cover.*

formation, including the suggestion that stromatolite-like structures and stylolites in veins (e.g., upper Q3 domain of the Beaver Dam, the "Rosetta Vein," (Figure 7.1)) and numerous gold-bearing veins from producing mines contain carbon and carbonaceous compounds. In Strelly Pool black cherts, the compounds could have been generated abiotically by a process similar to Fischer-Tropsch synthesis (200–300 °C and 500 bars) at a palaeodepth of 2 km within the hydrothermal system, according to Lindsay et al. (2005), or in the ocean at similar depths (Figure 7.27).

The Strelly Pool veins provide key evidence that permit interpretation of features common to lode veins worldwide. The evidence indicates that the carbon, associated with lode gold deposits, was formed as part of the deposition on the sea floor, similar to the "graphitic schist" intervolcanic sedimentary strata of all ages.

7.8 SUMMARY OF P-T-x CONDITIONS FOR VEIN FORMATION

Given the parameters discussed above, we can make the following specific statements about the P-T-x conditions of quartz–gold–forming fluids:

1. they were dilute, metal poor;
2. they barely exceeded the density of seawater and were bottom-seeking or sat in indentation of the seafloor, such as hollows or ripple marks?

Precipitation of silica took place by slow diffusion and conduction from lower silica concentration solutions, rather than dilution of fluids or rapid quenching of silica rich brines;

3. temperatures were in the range 150–430 °C;
4. pressures were in the range of 25–350 bars
5. in many deposits, they were formed in the presence of carbon from biogenic activity around the vent systems.

The combination of data from hydrothermal systems on the modern ocean floor, laboratory analyses and experiments, and field studies leads to new insights into formation of lode gold quartz veins.

The seafloor studies are providing quantitative data on the composition of vent fluids and their role in the ore-forming process. Laboratory analysis of the coexisting solid phases being precipitated from the vent fluids, for which temperatures, water depths, flow rates, and other variables can be measured, is aiding the interpretation of mineral chemical, fluid inclusion, and isotopic data from ancient ores.

For VMS deposits, field studies on land have indicated the geological setting in which seafloor sulfide and oxide deposits are likely to be found and are directing current research to the back arcs, which is a favored tectonic setting for ancient economic ores. Although we have learned much of relevance to ore-forming processes over the past three decades of seafloor research, there are many things we do not understand. It is clear, for example, that the primary mineral compositions and textures that are produced on the seafloor are changed through time by early diagenesis and subsequent low-grade metamorphism and other processes attending burial and preservation of an ore body. These changes must be kept in mind when relying on conventional geochemical methods to interpret the ore-forming process in ancient hydrothermal systems.

A recent discovery of supercritical water discharged from a vent that holds a mechanism for increasing the leaching rate of silica, gold, and minor constituents. The Archean crust was thicker and hotter, and there was much less land. Ocean water temperatures were likely higher and the oceans shallower. The recent discovery of a Blue smoker in Okinawa Trench adds an additional parameter that remains to be evaluated.

7.9 QUARTZ FORMING DURING DEFORMATION

A description of quartz forming during deformation conditions of chevron folding, and en echelon vein formation on bedding-plane slip zones in the Hartland Quay area (Davison et al., 2004). McBride (1978) describes the geology of the 180-square-kilometer Ecum Secum map area. All of the mineralized veins were bedding-parallel and folded. A few linear veins were found perpendicular to the regional trend and nonmineralized. A complete review of the structural geology and its importance to understanding and dating gold mineralizing events is the subject of Chapter 8.

7.10 THE SIGNIFICANCE OF YOUNGER VEIN AGE DATES

Some issues associated with interpreting the significance of vein age dates have been summarized by Kerrich and Cassidy (1994). The importance of understanding the geological context of the sample is clear. On occasion it is not apparent exactly what is being dated, and therefore the significance of the dates obtained cannot be clear either. Lamprophyre dating was discussed in Chapter 5, and Rock's statement that wherever detailed field and/or geochronological evidence is available, lamprophyres overlap gold mineralization in both time and space satisfies our syngenetic criteria. Similarly, for the Cote Gold deposit, Re-Os vein ages on molybdenum coincide with ages obtained from the host Chester Complex (Kontak et al., 2013; see Chapter 4).

Both K/Ar and Re-Os techniques were used for Meguma veins, and ages are considerably younger than the host Meguma sediments (Morelli et al., 2005; Kontak and Horne, 2011). In the Meguma, Waldron et al. (2009) show that zircons from eroded older granitic terrain occur higher in the stratigraphic sequence, the result of basin subsidence and sediment sourced from higher and older basement. This granitic detritus may have imparted a basement component to the stable isotope signature of the Meguma, which was pointed out by Kontak and Kerrich (1997).

The Goldenville Formation is interpreted to represent an abyssal plain fan deposit, formed by sedimentation related to turbidity currents, whereas the overlying Halifax Formation represents turbiditic deposition in the lower part of a continental rise prism that prograded northwestward over the Goldenville fan deposit.

Based on palaeontological constraints: Tremadocian fossils (the Tremadocian lasted from 485.4 ± 1.9 to 477.7 ± 1.4 million years ago www.wikipedia.com) in the Halifax formation overlay Goldenville turbidite. See also: member. http://earthsciences.dal.ca/people/schenk/Meguma/Meguma_Field_Guide.html

Deposition of Type I sills, in the study area, were emplaced during deposition of turbiditic sequences over a span of time that extends from the lower (latest Neoproterozoic) part of the Goldenville Formation through to the upper (earliest Ordovician) part of the Halifax Formation, a time span of about 70 million years (White and Barr, 2003).

7.11 DISCUSSION

The physical and chemical systems have been examined; they provide insight into the likely P-T-x "window" for lode vein formation. These are briefly:

1. Brines are generated at various depths due to elevated temperatures in a reaction zone in the crust. The crust may consist of any major rock type, and those with suitable concentrations of gold or base metals will represent a source for the silica and metals.
2. The precipitation of gel from hydrothermal solutions requires that dense brines are ponded in hollows and depressions on the seafloor, and that they cool primarily by diffusion. This in turn requires the presence of quiescent seafloor conditions that are either intermediate to distal from the vent, or proximal to overlying seeps. Turbulent mixing of hydrothermal fluids will result in dispersal as colloidal silica. Very silica-rich solutions will rapidly precipitate and quench silica as gel. Common metals associated with these fluids are precipitated with the silica including native gold.

CHAPTER 8

The Role of Structural Geology and Remobilization

CONTENTS

8.1 INTRODUCTION

The influence of deformation has to be deciphered before any study of gold deposits can be fully understood. Terms such as shear zones and remobilization are mentioned in almost all papers on gold deposits, but are rarely described and documented. It will be demonstrated in this section that all observed structures associated with gold mineralization postdate the gold mineralization itself and have no effect on the gold mineralization or their host rocks beyond what would be expected during the deformation of any other "noneconomic" rock unit in the deformed sequence. Numerous papers refer to the remobilization of fluids along shear zones that gives rise to gold deposits. It should be noted that, prior to the work of researchers in the 1960s (McAllister, 1960; Hutchinson, 1965; Constaninou and Govett, 1972; Stanton, 1959), the same explanation was given for volcanogenic massive sulfide deposits (VMS). The work of McBride (1976b) and Luff (1973) demonstrated that the Heath Steele Mines B and Brunswick No. 12 VMS deposits were deposited as a stratum in the volcanic–sedimentary sequence and deformed by the first and subsequent deformations; thus they are syngenetic and not epigenetic, confirming the work of the early researchers McAllister and Stanton.

Massive sulfide deposits share many common characteristic with gold deposits, in that they are found in similar rocks and share similar histories of deformation in many orogenic or mobile belts. The authors wonder why many geologists ignore this commonality and prefer to classify gold deposits as shear zone replacement deposits. One Canadian exploration geologist mused about the "shear zone hosted gold model being an economic and academic failure" (Bottrill, 1990). He argued that, firstly, "as an exploration model it was not successful and secondly as an academic model, it failed to provide parameters to be used as exploration guides."

Before discussing existing mineral deposits in deformed rocks, it is important to understand what really happens when mineral deposits in mobile belts are

The Metallogeny of Lode Gold Deposits. http://dx.doi.org/10.1016/B978-0-12-803222-0.00008-3

deformed. Deformation is a process of physically compressing rock sequences to cause them to slide up and over adjacent cratons (overthrusting) or to exert a compressing force that reduces the width of a mobile belt, producing steeply dipping, folded sequences by flattening and elongating the rocks sequence perpendicular to the stress. In addition to deforming the rocks, an axial plane cleavage is formed with an intensity that is directly influenced by the amount of phyllosilicates minerals in the precursor rocks and the amount of strain superimposed (Stanton, 1988). This fabric is the first recrystallization; sericite and chlorite are commonly the first minerals formed, but they are extremely fine-grained and not useful for dating.

Greenstone belts around the world host many of the gold deposits that fall into the class of lode gold deposits. These belts show ductile fold patterns that generally produce major folds and many magnitudes of parasitic folds down to microscopic. The following discussion, Section 8.2, describes this style of folding and the resulting effects on all strata in any given sequence. McBride has spent the last 45 years studying ore deposits and the effects of structural deformation on them and has used the results as a successful exploration tool.

8.2 FOLDS AND CLEAVAGE

Regional deformation of mobile volcanic belts transforms subhorizontal unconsolidated rocks to steeply dipping metamorphosed ones. Usually the major folds have flat to moderately plunging hinges, whereas the parasitic folds can have flat to steeply plunging folds, especially in the major synclinal folds. As the rocks are compressed, considerable dewatering and recrystallization of the phyllosilicates takes place to form new minerals with a lower molecular water content. The planar fabric produced has been termed a solution cleavage, but the terms cleavage, slaty cleavage, schistosity, fracture cleavage, or gneissosity based on megascopic or microscopic appearance are commonly used. It is axial planar to folds of all pre-existing planar fabrics including bedding and any previous cleavage.

It is important to realize that in any fold system, the folded units had to be present to be folded. The axial planar cleavage defines the axes to these folds. Together, with associated lineations they, define a deformational event. S0 is used to refer to the bedding or similar layering; D1 refers to the first observed deformational event with its folds F1, cleavage S1, lineations L1. The second or subsequent deformations use the same terminology with D2, D3, etc., replacing D1.

D1 folds may be open to isoclinal. Around the Heath Steele Mines, some of the first folds have boudinaged hinges indicating the high intensity of deformation. The first generation of folds are usually the most difficult to detect because the intense deformation forms, isoclinal folds, that may appear to be parallel or en

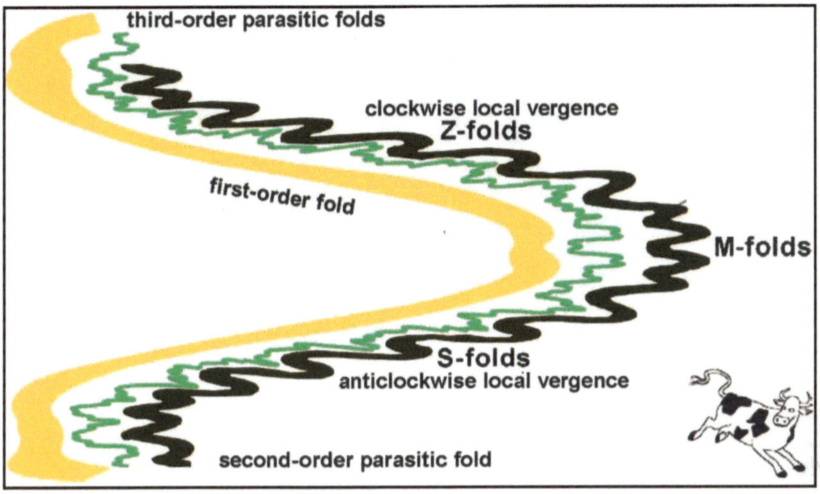

FIGURE 8.1

The relationship between major and parasitic folds, illustrating the "S," "M," and "Z" form of the parasitic folds and the relationship of the axial plane cleavage and bedding, showing the major fold position relative to the angle between the two and whether the bedding is left or right of the cleavage (www.files.ethz.ch/structuralgeology/jpb/files/english/8foldspdf, p. 220).

echelon lenses. Subsequent deformational events refold the older cleavage and bedding, and usually appear as crenulations or minor folds on the S1 cleavage plane; the later major folds are usually open to closed in the volcanic–sedimentary belts.

The relationship of parasitic folds to major folds is not well understood in the exploration community. In any fold system, both are present (Figure 8.1) and may be detected on all scales from major to microscopic. In the figure, the major fold and second- and third-order minor (parasitic) folds are represented. Their form defines their position in the next larger fold. In addition, the angle and direction of the folded surface relative to the axial plane cleavage can direct one to the major fold. For example, bedding to the right of the cleavage indicates that the major anticline is to the right and the syncline to the left. If the bedding is left of the cleavage, the fold form is reversed.

By convention, the shape of the parasitic fold is defined by looking down plunge at a fold in outcrop. Looking up, as in a mine, the shape is reversed. Consider the anticline in Figure 8.1: the minor fold at the top mimics the letter "S," that at the crest the letter "M," and that at the bottom the letter "Z." Thus the fold shape can assist in determining the location of the major fold. Their presence in the folded rocks shows that they were folded together as a series of layers. This rule applies regardless of the composition of the layer (see Section 8.2).

The depositional layering is commonly preserved throughout the deformation. In a normal heterogeneous sequence, the incompetent beds are compressed

FIGURE 8.2

Diagram showing the second- or third-order parasitic folds with harmonic and disharmonic parasitic folds and axial planes (www.files.ethz.ch/structuralgeology/jpb/files/english/8foldspdf, p. 218).

and adapt to the form of the competent beds. Depending on the competency contrast, the incompetent beds will form second- and third-order harmonic to disharmonic parasitic folds. The difference is that the harmonic preserve continuous axial planes, but the disharmonic do not. Nevertheless, all the resulting folds are the product of a single deformation (Figure 8.2). It is important to understand that the composition of the beds does not matter; they can be volcanic tuffs, mudstones, siltstones, quartzwackes, massive sulfides, iron formation, quartzites, or quartz veins. Any of them involved in the deformation will have the same shape superimposed on them. Conversely the shape says that they all were there, unfolded prior to the deformation as part of a layered sequence.

8.3 CLEAVAGE AND ITS RELATIONSHIP TO THE CHARACTER OF THE PRECURSOR ROCKS

The folding and cleavage in rock sequences are well established facts that can be demonstrated worldwide. It is never questioned that the mud and sand layers in the Meguma Group of Nova Scotia, or the volcanic–sedimentary rocks of Canadian greenstone belts, represent anything other than deformed depositional layers (Figures 8.3 and 8.4). The deformation may produce folded bedding or may boudinage and rotate the more competent beds, but the composition of the beds remains intact. The cleavage can vary from bed to bed, depending on the precursor clay mineral percentage, and may also refract between beds; but it will always remain axial planar to the fold formed during the same deformational event (Ramsay and Huber, 1983).

The presence of clasts, augens, fragments, crystals, or crystal aggregates in a deformed sequence can be dated relative to the deformation by their character.

FIGURE 8.3
Roadside outcrop showing an upright fold of quartz arenites cut by the vertical slaty cleavage, Ecum Secum Map Area. *McBride Photo, 1975.*

Those existing in the rock or formed prior to deformation have the cleavage deflected around them and lack an internal expression of the cleavage. Some minerals are formed during deformation; they grow over the developing cleavage in the rock and are in turn deformed by the deformation to produce "snowball" textures. Garnet is well known to do this. Finally, those formed after the deformation overgrow the cleavage, but show no deformation. Thus visual inspection provides the answer to the question of when a mineral is formed in the sequence. Spry (1969) provides an excellent summary of these textures, and the reader is referred to his Chapter 10 for a detailed discussion. He makes it clear that the growth of minerals during or after deformation does not push the cleavage aside but, rather, overprints it (Figure 8.5). Ramsay and Huber

FIGURE 8.4
Folded mafic tuff from Tudor Township, southern Ontario with axial plane schistosity in more schistose tuffs adjacent to the fold. Note pale buff layer along right edge of fold. *McBride Photo, 2011.*

(1987) provide excellent examples of the structural overprinting superimposed on the various rocks in the sequence.

These relationships are of paramount importance, since the Goldenville Formation of the Meguma Group contains minerals that can be shown to predate the deformation and are parallel to bedding. The first example is a folded andalusite sedimentary sequence near Canso, Nova Scotia. Typical fine clastic sedimentary beds outline a 10-m "M" fold. The minor folds are outlined by 1-cm sand layers and the larger fold by the bedding (Fletcher and Faribault, 1887). One bed in the fold contains 1- to 2-cm andalusite crystals that have been rotated so that their long axes are parallel to the cleavage and the cleavage bends around them. It is apparent that these crystals predate the deformation and were formed by syngenetic or diagenetic processes and then incorporated into the deformation. The second example is a 2-cm pyrite crystal the Ecum Secum Map Area of McBride (1978). Many of these crystals are found within the study area; they help to define a stratigraphic section of the Goldenville Group, which, together with disseminated iron carbonate, hosts all of the gold deposits. Their position in the geological evolution is defined by the relations documented by Spry (1969). The pyrite crystal has deflected the cleavage and therefore predates it and the deformation. Other than some minor bleaching in the pressure shadow area at the end of the crystal, alteration is totally lacking, indicating that it is syngenetic or diagenetic (Figure 8.6). The

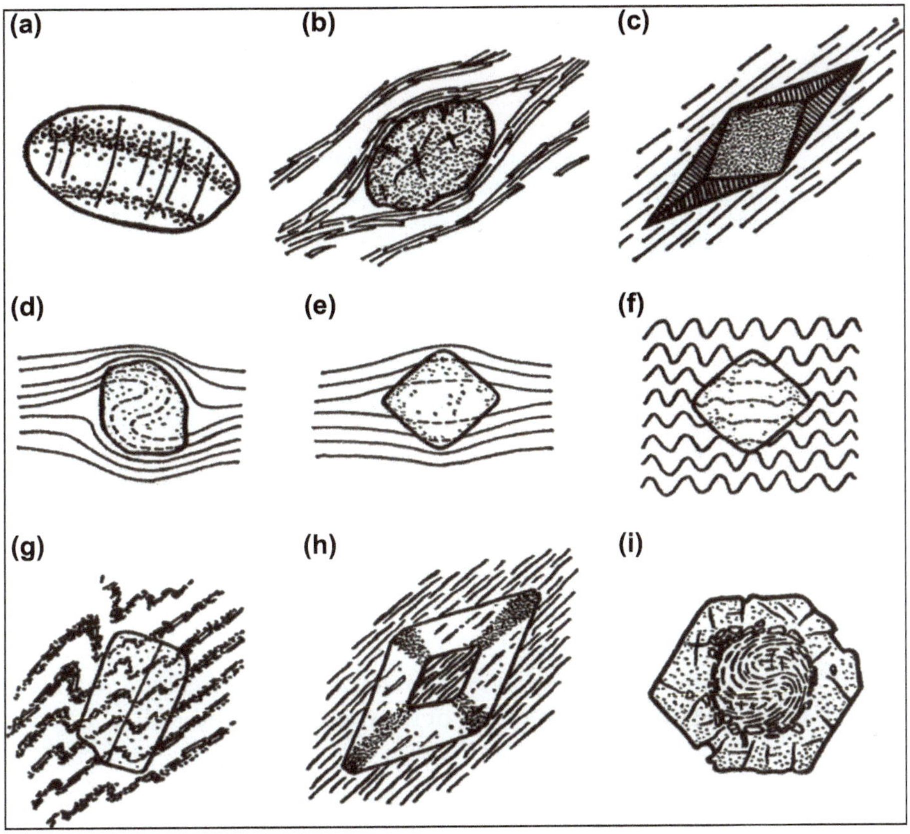

FIGURE 8.5

Mineral textures and their growth relative to the deforming event. Pretectonic crystals: (a) undulose extinction and deformation lamellae in quartz; (b) cracked garnet wrapped around by the foliation; (c) pressure fringes around pyrite. Syntectonic growth of crystals: (d) snowball garnet; (e) andalusite porphyroblast that grew during flattening of pre-existing foliation; (f) porphyroblast that grew during crumpling of pre-existing foliation. Characteristics of post-tectonic crystals: (g) helicitic structure in albite, Si (S surface within crystal) concordant with Se (S surface outside the crystal); (h) chisatolite with central helicitic and outer cross-structure and boundaries discordant with the foliation; (i) garnet with syntectonic rotational core and posttectonic idioblastic rim. *Spry (1969), from Figures 58, 59, and 61.*

importance and hence the timing of geological events of this pyrite and associated iron carbonate are discussed in the next chapter; here it suffices to say that they outline a preferred stratigraphic horizon.

The S1 cleavage documented in the Ecum Secum Map Area is a regional feature. In the Meguma, it is found throughout the group from Yarmouth to Canso, a distance of 440 km, and is pervasive to some degree in all rock types, as mapped by Faribault and summarized by Malcolm (1912). McBride (1978) presents the results of the detailed structural analysis of the Ecum Secum Map

FIGURE 8.6

A 4-cm pyrite crystal in fine-grained quartz arenites, showing later slaty cleavage being deflected around it (highway road cut, Ecum Secum) *McBride collection, 1975.*

Area, which clearly shows the stratigraphic and structural parameters that control the present distribution of the Goldenville and Halifax Series and the gold deposits within them (see Section 8.4).

Multiple periods of deformation were identified and differentiated at the Heath Steele Mines VMS B Zone deposit in the Bathurst mining Camp of New Brunswick (McBride, 1976b) and by Luff (1973) for the Brunswick 12 deposit. These studies demonstrate the types of structural investigations that must be carried out to determine the position of the mineralizing event in the geological evolution of a gold deposit in complicated structural settings.

The same structural rules apply to the Archean greenstone belts; a penetrative cleavage is always present and has been documented in numerous government and private reports (Pye, 1951; Milne, 1967; McBride, 1987). It can be found throughout the belts and is much more obvious in the finer-grained, clay mineral–rich precursor rocks; however it can be detected in the coarser units as well. It is the first fabric to be developed and is axial planar to the folds; thus it is important for dating the gold mineralizing event.

8.4 WHAT IS A SHEAR ZONE?

Most recent gold deposit studies refer to the "shear zone" as being the conduit for mineralizing solutions from some unknown source. This approach was known as the "source bed concept" and was used to explain volcanogenic hosted massive sulfide copper–zinc–(lead, silver, gold) deposits until the 1960s, when geologists discovered that they were in fact syngenetic. The shear zone model was developed over 100 years ago and saw resurgence only with the rise in gold prices in the 1970s.

But what is a shear zone? It is described as a structure. If it is a structure, it must have the basic structural criteria, mentioned above, to define it as a deformational event. A fault is defined as a fracture along which there has been movement. It is defined by a fault plane, possible strike-slip minor folds, and by the direction of movement (normal, reverse, or strike slip). A regional cleavage, on the other hand, is defined as a planar fabric that is pervasive throughout a rock sequence. Its presence defines a structural event. For example, if the deformation is the first structural event that affects the bedding, only the bedding is folded. Subsequent deformation events will deform the bedding and the any pre-existing deformation fabrics. Most volcanic–sedimentary belts have only one intense deformation and one or two minor crenulation cleavages that usually produce very open cross-folds (McBride, 1976b). The Bathurst Camp is more complicated, with five deformational events, each with its own cleavage; it seems to be an exception because it was involved in two major orogenic events, the Taconic and Acadian (McBride, 1976a).

A shear zone, in contrast, is a very important structural dislocation in the earth's crust and upper mantle. Shear zones form as a response to inhomogeneous deformation partitioning strain into planar or curviplanar high-strain zones (www.wikipedia.com). Dewey et al. (1998) calls it a zone of transpression or transtension producing strike slip on plate boundaries that postdates the folding within the plates. Burchfiel (2004) gives an excellent example of a moving plate with shear boundaries and the rate of movement. Schwerdtner et al. (2005, 2010) give examples of tangential shear to illustrate the repeated movements between plates. Within these shears, complex recrystallization produces snowball growths from pre-existing minerals in the high-strain and high-grade metamorphic environment. These researchers point out that shear zones do not represent the first period of deformation and mark the boundaries between previously deformed terrains. These shear zones have defined plate boundaries. None is reported to have economic mineralizing systems.

A shear zone, by its name, suggests a cleavage or planar fabric that is restricted to a zone or band of a limited width. Therefore one should be able to separate it and describe it as a unique structural event, and define its position and importance in the geological history. In the study of gold deposits, the above constraints seem lacking; shear zones are rarely described, they just exist because gold mineralization in schistose rocks is present (Goldfarb et al., 2005; Robert et al., 2005; Heiligmann et al., 2008; Helt et al., 2014; Lawrence et al., 2013). Nearby parallel nonmineralized schistose zones are ignored even though they may be only a short distance away, i.e., a few tens of meters. In fact, in many studies, the fact that these deposits are stratabound and lie in a specific part of the volcanic–sedimentary sequence is forgotten; all of the rocks are "altered"! Many studies do not show the host rocks to the veins that are being described. Field relationships are not mapped, or, if they are, the database is often incomplete. The present authors are frustrated to the point that they must do their own field structural analyses, as none exist around the mineral deposits in their study areas.

All "shear zones" associated with lode gold deposits are described as steeply dipping structures that are parallel to the mineralization and, more importantly, to the regional stratigraphy. This is very important, because it implies that the shear zone must have developed after the rocks and mineralized strata were folded into their present position. If they predated the D1 fold deformation, they would have had to be flat faults in unconsolidated sequences because the entire system has been rotated approximately 90° to the vertical. This is a highly improbable situation!

It has long been recognized that Canadian greenstone belts have been deformed, but at the time of mapping, the structural geology, especially folding, was not well understood. The authors have mapped many of the greenstone areas as part of property evaluations. In all situations, a schistosity or slaty cleavage is present that is pervasive throughout the greenstone belt. McBride (1987) demonstrated this in the Onaman Lake area and east of Beardmore, where a stereonet of over 600 schistosity measurements shows a single deformation fabric throughout the southern volcanic belt and adjacent sedimentary rocks (Figure 8.7). There is no evidence of refolding of this schistosity or of a separate shear fabric. Other maps show a shear zone along the northern boundary of the volcanic band. Mapping of the contact by McBride (1980a) shows a gradational contact over a few meters, but no evidence of a shear zone.

In the Porcupine Camp, only one fabric is present and is a regional schistosity. Ferguson (1968) called it a second cleavage and shows isoclinal first-period folding without a folded cleavage. This interpretation is not correct, because bedding and the cleavage are present. Any cleavage between these events would also be preserved. The schistosity axial planar to the folds is the first cleavage. In the authors' experience, all greenstone belts in Ontario and Quebec have a single major deformation in rocks varying from low greenschist to amphibolite-grade metamorphism. The Destor-Porcupine Fault postdates the folding of the mineralized sequence. It cuts off favorable Tisdale rocks and places them against the non-gold-bearing Deloro Group. It has an apparent sinistral horizontal displacement of several kilometers. An exposure of it on an old road south of Porcupine shows the regional schistosity to be folded on itself but only over a width of a few tens of meters that marks the Destor-Porcupine fault. Across the fault, the

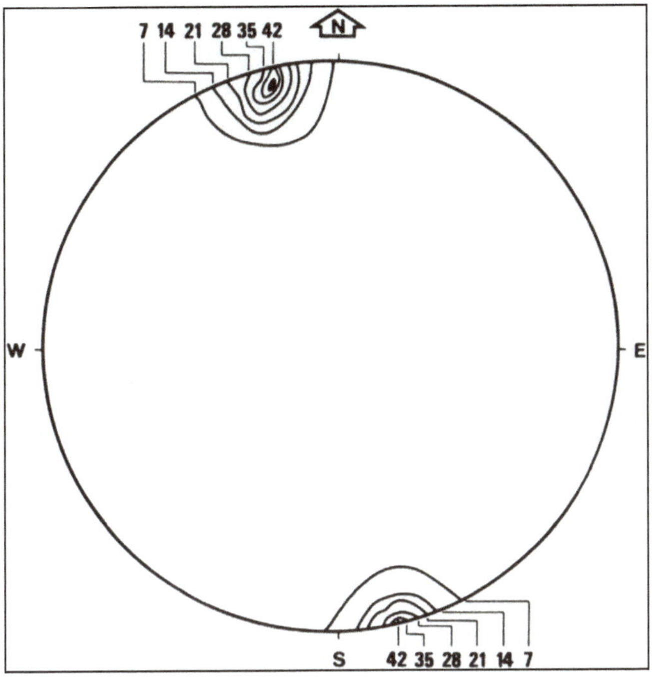

FIGURE 8.7

Plot of 600 measurements of the fine-grained schistosity showing the presence of a single, nondeformed, regional tectonic fabric. *Data are from surface mapping of the Pan Empire claim block (McBride, 1981).*

geology is entirely different. The 1948 map of Dunbar has been reinterpreted to incorporate the folding and Destor-Porcupine Fault (Figure 8.8).

In the Grenville Greenstone Belt of eastern Ontario, the same penetrative first cleavage is present throughout the volcanic–sedimentary sequence (McBride, 1992, 2009). Similarly work in the Bathurst Camp volcanic–sedimentary rocks shows that the schistosities and crenulation cleavages are pervasive throughout the camp (McBride, 1976b). A "shear zone" or strike-slip fault is documented south of Heath Steele Mines (see below).

The same can be said for the entire Meguma Group of Nova Scotia (Malcolm, 1912) and has been confirmed in the Ecum Secum Map Area (McBride, 1978). It is important to understand that in these volcanic–sedimentary belts, the regional schistosities are pervasive throughout the rock sequence and represent the first and only major deformation.

So where does a "shear zone" fit into this scenario? It has been documented above as a high-strain zone of finite width separating stable deformed blocks. In a mobile belt it can be the same; but then it has to be a second or later deformation zone. It is a plane or band with more strain, separate from the

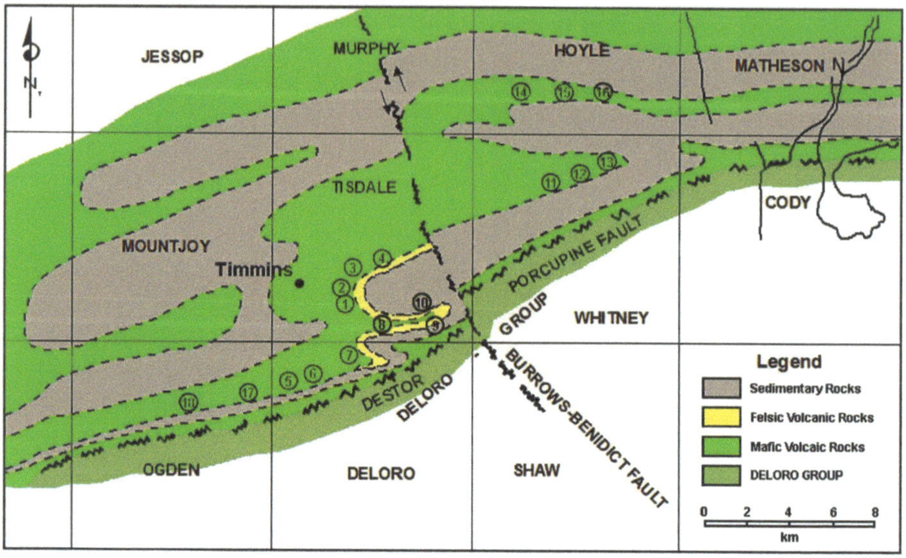

FIGURE 8.8

Map of the Timmins area showing the distribution of the folded mineralized strata and the Destor-Porcupine Fault with a regional structural interpretation; mines 1) Moneta, 2) Hollinger, 3) McIntyre, 4) Coniaurum, 5) Delnite, 6) Aunor, 7) Buffalo-Ankerite, 8) Paymaster, 9) Preston East Dome, 10) Dome, 11) Hallnor, 12) Broulin, 13) Pamour, 14) Bell Creek, 15) Owl Creek, 16) Hoyle Pond, 17) Naybob, 18) De Santis. *Revised from Dunbar (1948); Pyke (1982).*

regional schistosity, and must postdate both the bedding and schistosity. In the Timmins Camp, the deposits follow a preferred section at or near the top of the volcanic sequence around the Porcupine Syncline and other associated folds (Figure 8.8).

Regional distribution of deposits shows a large-scale control of mineralization in a preferred part of the stratigraphy. This association is seen on the local level in the description of individual mineral deposits. The Destor-Porcupine fault terminates the fold structures around the Porcupine syncline and associated folds. It separates Tisdale from Deloro rocks; the Tisdale rocks contain economic gold deposits, but the Deloro is lacking in significant gold deposits in the same area. This planar structure is a shear zone using the criteria of Schwerdtner et al. (2005, 2010). It postdates the mineralization and the regional folding. A similar structure was identified by McBride (1976b) just south of the Heath Steele Mine, where a narrow shear zone separates the mine sequence of sedimentary rocks, crystal, and fine-grained tuffs from a thick sequence of fine-grained ashfall tuffs. It is a strike-slip fault or shear zone with a lateral sinistral movement of the mine sequence for more than 8 km (Figure 8.9). This fault has cut off the D Zone and the earlier folds, and appears to be folded by later crenulation cleavages.

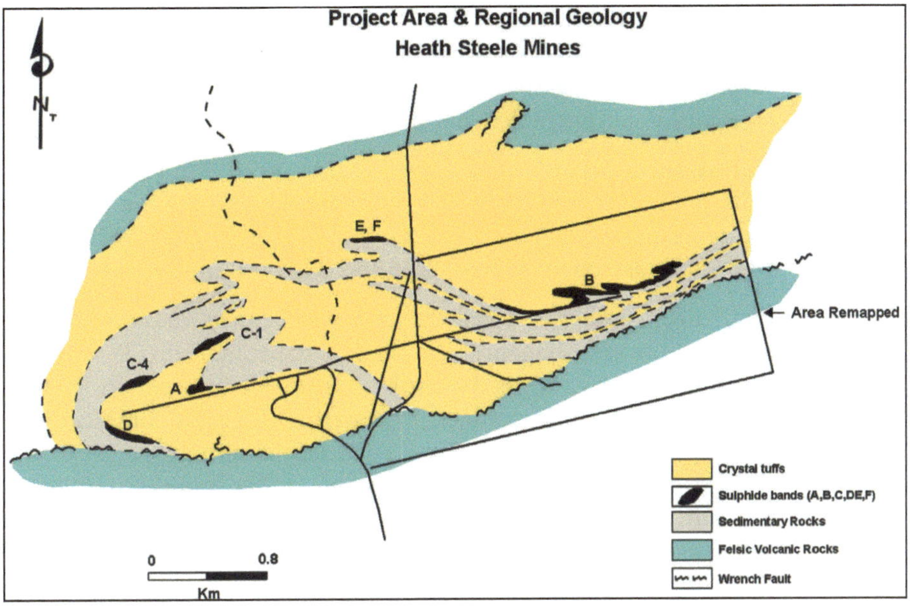

FIGURE 8.9

Geology of the Heath Steele Mine property showing the Folded mine sequence terminated by a major east—west strike-slip fault (shear zone).

In the Meguma Group, the folding is pervasive; the only shear zone that fits the structural criteria is the Cobiquid-Chedabucto Fault, which marks its northern boundary and is recognized as separating two cratons.

In conclusion, it can be shown that lode gold deposits that are found in volcanic—sedimentary and sedimentary belts are parallel to bedding and show a fold pattern of a D1 deformational event that has affected the entire host succession including the mineralization. Its overprinting on the mineralizing systems is discussed in the following section. Examples are taken from a variety of mineralized systems to demonstrate that what actually takes place is what is to be expected from any rock sequence, whether or not it contains elements or metals of present economic value. Any documented shear zones can be shown to postdate the folding events and hence the mineralization.

8.5 THE REALITIES OF REMOBILIZATION

Gold is often spoken of as being remobilized, but does that happen in the natural environment and, if so, to what degree? Any study on metallogeny shows a strong relationship between the type of mineralization and its host rock. The authors prefer to reference Bilibin (1968, an English translation of his lectures in 1951—1952 shortly before he died), as he is considered the father of modern metallogeny. His work in 15 mobile belts varying in age from Archean to

Cenozoic, in Russia in the 1930 and 1940s, established that host rocks were the most useful tool in the search for any specific type of economic deposit. Such a correlation suggests that the original economic concentration of a given metal is part of the syngenetic–diagenetic depositional system.

The important questions are the magnitude of remobilization, when it takes place, and what is remobilized. Electro-negativity plays a part in which metals are deposited first and are the least mobile in the geological environment; these will be the last to remobilize. Gold and other precious metals fall into this group.

Once deposited, any sequence of metal-bearing rocks is buried, leading to compaction and diagenesis as part of consolidation. Dewatering is a main feature of these changes. Latent heat from the volcanic rocks in the sequence may accelerate these changes. Dewatering channels commonly cut the rocks. They appear in sedimentary rocks as wispy channels (Figure 8.10), and in volcanic rocks as sealed microfractures and ladder veinlets that traverse the units. The authors have not observed any economic minerals in these veinlets in spite of having mapped many hundreds of square kilometers of these rocks.

Deformation is a physical process involving pressure and heat, and takes place when the rock sequence is compressed and folded as discussed above. At an early point, recrystallization takes place first as a slaty cleavage, and then as deformation progresses, it becomes a schistosity or solution cleavage. During this process, the porosity is greatly reduced as the rocks dewater. Eventually the porosity is reduced toward zero and the rocks become impermeable to fluid flow. The regional strain prevents large dilation zones from developing. Thus any movement of metals through the system must take place very early in the tectonic history of any mobile belt, but does this happen?

FIGURE 8.10
Dewatering channels in the Goldenville arenites from the Ecum Secum Map Area. *McBride Photo, 1975.*

8.5.1 Heath Steele Mine studies

Two studies in the early 1970s cast doubt that "remobilization" takes place in the sense used in the "shear zone hosted gold models." The first was a lithogeochemical study of the rocks around the B-Zone at Heath Steele Mines (Whitehead, 1973; Whitehead and Govett,1974). This study was undertaken to determine whether the rocks surrounding the 23-million-ton massive sulphide copper–lead–zinc ore body contained a geochemical halo that could be used to locate hidden deposits. Whole-rock samples were analyzed for multiple elements. It was determined that lead and zinc showed anomalous values of 40 ppm for lead and 60 ppm for zinc for 400 m above the massive sulfide zone. These data are important because the magnitude of movement is only of geochemical value and not of economic significance. These values represent original deposition, remobilization during dewatering, diagenesis during burial, and subsequent deformation to lower biotite-grade metamorphism.

McBride (1976b) demonstrated that the massive sulfide band of the Heath Steele Mine B zone had been folded into a series of sub-vertical, steeply plunging isoclinal folds and was subjected to a total of five periods of deformation. The B Zone has been traced for more than 1600 m along strike and 600 m down dip. Base metal zoning was identified in the B-1 Zone, within the most easterly and largest of the folds along the band. Figure 8.11 shows the normal copper footwall and lead–zinc hanging wall zones between the folds. The B-1 Zone is a complex "S"-shaped isoclinal fold; the base metal zoning was defined by 15 m spaced drill sections established as the grade control for mining; the maximal sample interval was 1.5 m. Mineralization was composed of a chalcopyrite–pyrrhotite lower unit and a bedded galena–sphalerite–pyrite upper unit. It is evident from Figure 8.12 that the base metal zoning was preserved throughout the fold and that there has not been any remobilization at the assay sample interval. Thus the physical action of deformation did not remobilize metals or minerals when rocks were deformed to the level of isoclinal folding up to lower biotite-grade metamorphism. These characteristics define the metamorphic limits superimposed on most lode gold deposits.

These studies show that there has not been movement of metals or minerals during burial, diagenesis, and regional deformation in the volcanic–sedimentary rocks hosting the volcanogenic hosted massive sulfide deposits of the Bathurst Mining Camp.

Do these rules hold for other mining camps? The metamorphic processes superimposed on the New Brunswick deposits are the same as or very similar to those in many mobile belts, so there is not any reason to expect entirely different processes elsewhere.

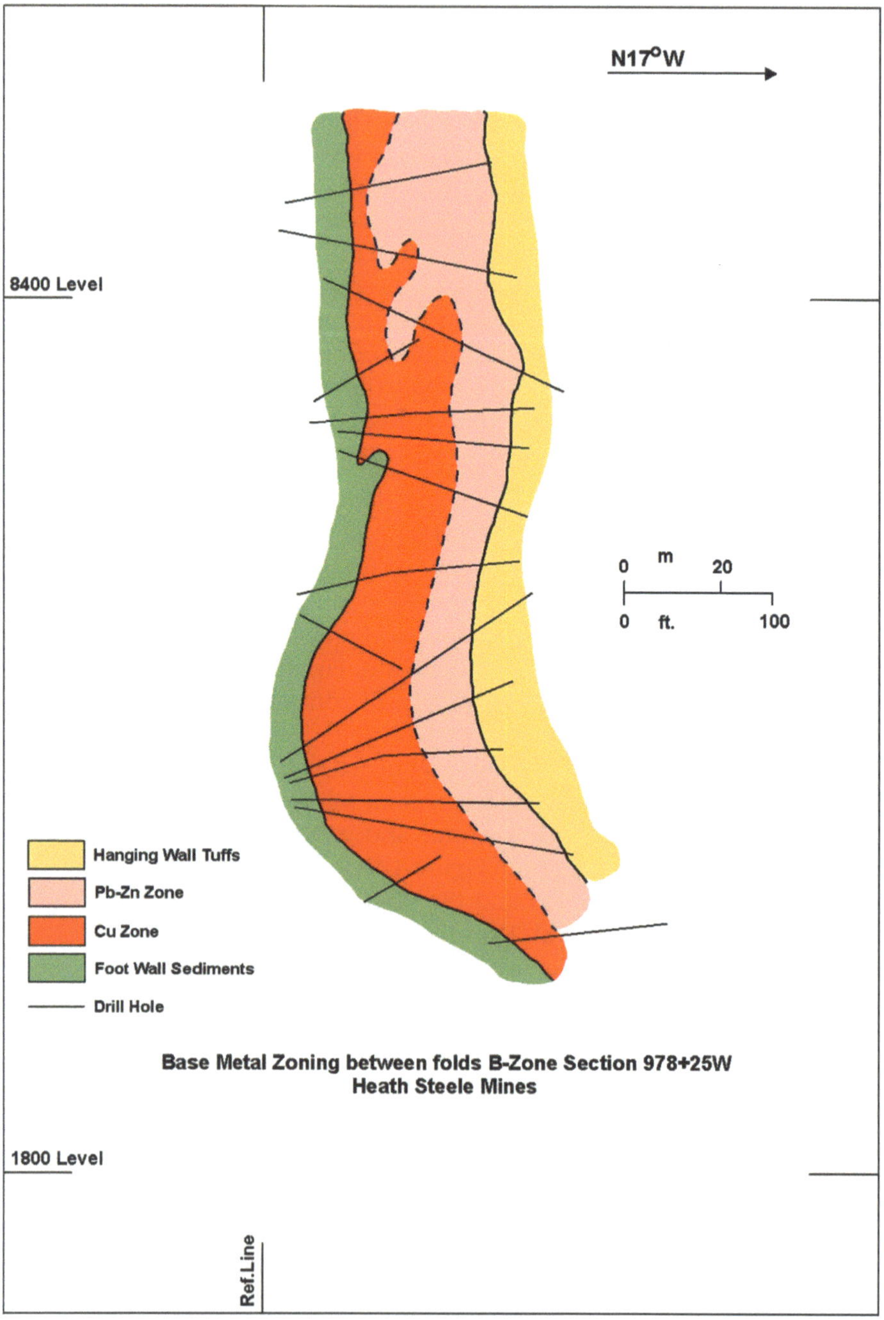

FIGURE 8.11

Base metal zoning in the Heath Steele Mines B-1 Zone showing the copper—pyrrhotite footwall and lead—zinc—pyrite hanging wall; tops are to the right.

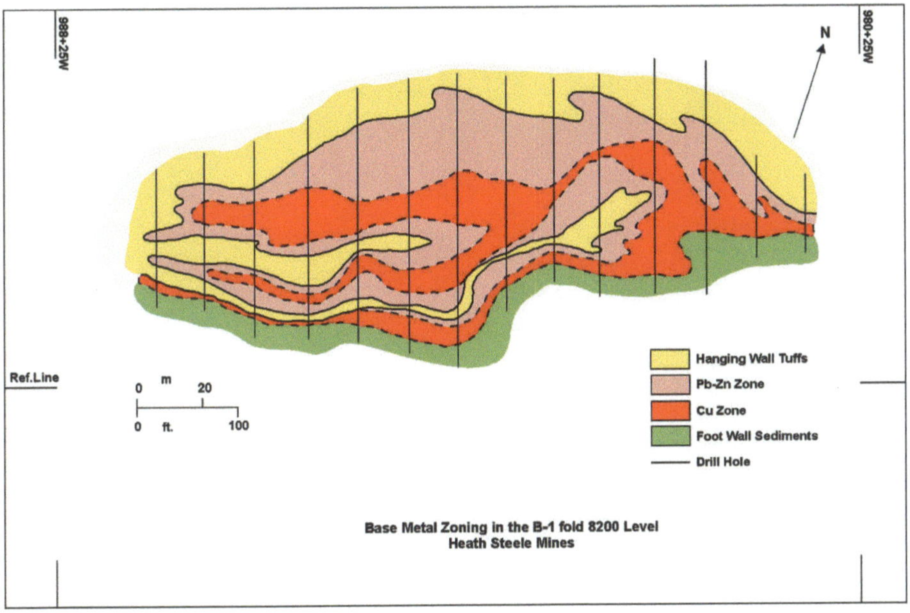

FIGURE 8.12

Base metal zoning preserved in the complicated isoclinal "S" fold that makes up the B-1 zone. There is no evidence of remobilization on the 1.5-m sample interval.

8.5.2 Deformation of Various Economic Minerals

The authors now extend this discussion to examine rocks of various compositions to illustrate that, during folding, there is no evidence that metals or minerals are remobilized relative to one another. The cohesiveness of the folded system preserves the depositional character. It is only when "economic metals" are found in the bands that the accepted rules of structural deformation are not accepted. Figure 8.4 shows a fold of a massive bed of mafic volcanic tuff within finer, more easily weathered tuffs in the sequence. This bed does not have any economic significance, and shows that silicate rocks define the fold form as a unit without losing their depositional character.

Folds in the Meguma Group of Nova Scotia clearly demonstrate that the quartz arenite bedding is preserved on the regional and outcrop scales, and that there are not any quartz veins penetrating along the schistosity. In Figure 8.13, the lake in the upper center is approximately 0.5 km long. These beds can be followed for many kilometers around the large fold structures without any evidence of mass movement of material between the beds. On a smaller scale, a roadside outcrop shows part of a fold approximately 1 m high (Figure 8.3). Again, the structures are preserved without disruption from remobilization. Similarly, folds on all scales have been well documented by the mapping of

FIGURE 8.13
Regional fold from the Ecum Secum Map Area of McBride (1978). This is a Google satellite image; the lake in the center of the photo is 0.5 km in length.

Faribault, as summarized in Malcolm (1912) and confirmed by many others since. Even the photograph of the cleavage in the crest of an anticline at Ecum Secum above the mined-out vein, barely visible in the lower part of the photograph, does not show evidence of remobilization along the schistosity (Figure 8.14), nor do complicated second order "M" folds of sandy beds at Sober Island, Nova Scotia (Figure 2.19). From all of these examples, the question arises as to why only economically significant metals and minerals remobilize when common ones do not seem to do so.

In the following discussion, "economic minerals" will be investigated using samples from mines and mineral deposits. Iron formation has well-known fold patterns (Jolliffe, 1966; Davies and Smith, 1966; Dubuc, 1966). The sample in Figure 8.15 is from the Bathurst Mines (Austin Brook) iron deposit, which is the continuation of the Brunswick Mines No. 6 deposit. The iron oxide beds form the upper part of a sulfide–oxide iron formation south of the mine. The folding in this deposit is well documented (Boyle and Davies, 1964). This sample shows the cherty bed tightly folded and the fine bedding of the iron oxide minerals following that fold pattern. Even the very fine bedding is preserved showing that, during the deformation, the fine bedding in the iron formation is the same as in any other bed. The schistosity is parallel to the long side of the sample and is outlined by minute hematite flakes.

McBride's work at Heath Steele Mines also documented folding of bedded sulphides on a small scale (Figure 8.16) and of hanging wall iron formation (Figure 8.17). The fine preserved detail in the samples show that the depositional features have not been modified during the deformation, except to fold them in the same pattern as any other bed in the mine sequence.

FIGURE 8.14
Photograph from Malcolm (1912) showing the horizontal bedding and vertical schistosity above a mined-out quartz vein (black strip at bottom of face); note that there is not any remobilization of the quartz vein along the schisitosity.

A sample from the Lyon Lake Mine in the Sturgeon Lake greenstone belt (Figure 8.18) shows that the fine bedding is perfectly preserved in the low-grade metamorphosed rocks. There is not any evidence of remobilization between the mono-mineralic pyrite and sphalerite beds. Similarly, a sample from the Henderson Mine in the Lac Dore Complex, Chibougamau, Quebec, shows the same folded monomineralic beds (Figure 8.19).

The grade of metamorphism does not destroy the depositional features; they have been recognized in deposits metamorphosed to the upper amphibolite facies. At the Geco Mine, Manitouwadge, Ontario, the volcanic hosted massive sulfide deposit has been metamorphosed to the upper amphibolite facies with sillimanite, cordierite, and some cross-cutting pegmatite dikes. The primary sulfide minerals have been recrystalized but remain intact. The presence of pyrite demonstrates the stability of the primary sulfide mineralogy (William et al., 1991).

FIGURE 8.15

Folded finely bedded cherty iron formation from Bathurst Mines site; scale is in centimeters. *McBride Collection, 1976.*

FIGURE 8.16

Small fold of banded sulphides and tuff, Heath Steele Mines. The mining method did not permit access to the stopes to observe folds in place.

FIGURE 8.17
Lean iron formation from the hanging wall of the B Zone showing graded bedding and folding that mimics the regional pattern seen in Figure 8.9.

FIGURE 8.18
Folded syngenetic bands of pyrite and sphalerite, Lyon Lake Mine, Sturgeon Lake, Ontario. Note the sharp contacts between the beds and lack of any suggestion of remobilization. Metamorphism is greenschist and the folding is isoclinal; scale is in centimeters. *McBride Collection, 1976.*

Similarly, the gold-rich Montauban deposit of Helikian age in Quebec has the primary mineralogy preserved. It is estimated that the metamorphic temperatures were 650 °C (Bernier et al., 1987). These are two examples of deposits that retain their original character when highly deformed and metamorphosed.

Folding of quartz veins will be discussed later in this chapter and in the section on gold deposits. Two examples are included here to demonstrate that the pattern is identical to that observed in other syngenetic ores. The first sample is a banded quartz–iron carbonate vein from the Ecum Secum Map Area. It shows that the banding is tightly folded and the bands are not thickened in the fold hinge (Figure 8.20). Wes Hanson, mine geologist, informed the authors that these narrow bands can be traced for many tens of meters in the Coxheath Mine. Layering is common in many Meguma quartz bands and duplicates the pattern described in Chapter 2.

One additional folded sample is from the famous Homestake Gold Mine, South Dakota, United States; it was collected by McBride on a field trip in 1992 and is from a stope on the 6350 Level of the mine. The sample consists of a folded bed of mineralized sandy quartzite (Figure 8.21). The trend of the gold mineralization is diagonally offset to the right of the main anticlinal fold axis by about 30° and undulates across the parasitic folds, which in mine terminology are called "ledges."

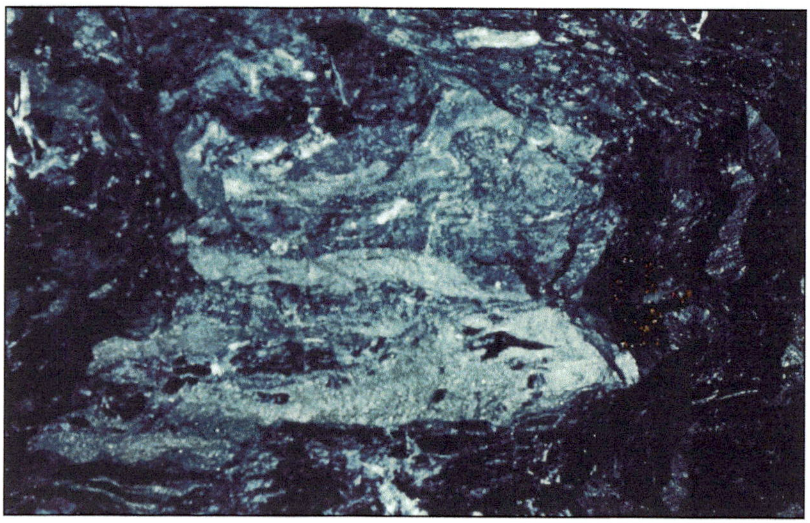

FIGURE 8.19

Folded pyrite bed with fragments from the G-Zone, 10th level, Henderson II Mine, Chibougamau, Quebec; fold is 0.5 m long. *McBride Collection, 1985.*

FIGURE 8.20

Folded band of layered quartz and iron carbonate, Miller Lake Gold District, Ecum Secum Map Area. *McBride Collection, 1975.*

FIGURE 8.21

Sample of folded auriferous quartzite from ledge 13 6350 Level, Homestake Gold Mine South Dakota; scale is in centimeters. *McBride Collection, 1992.*

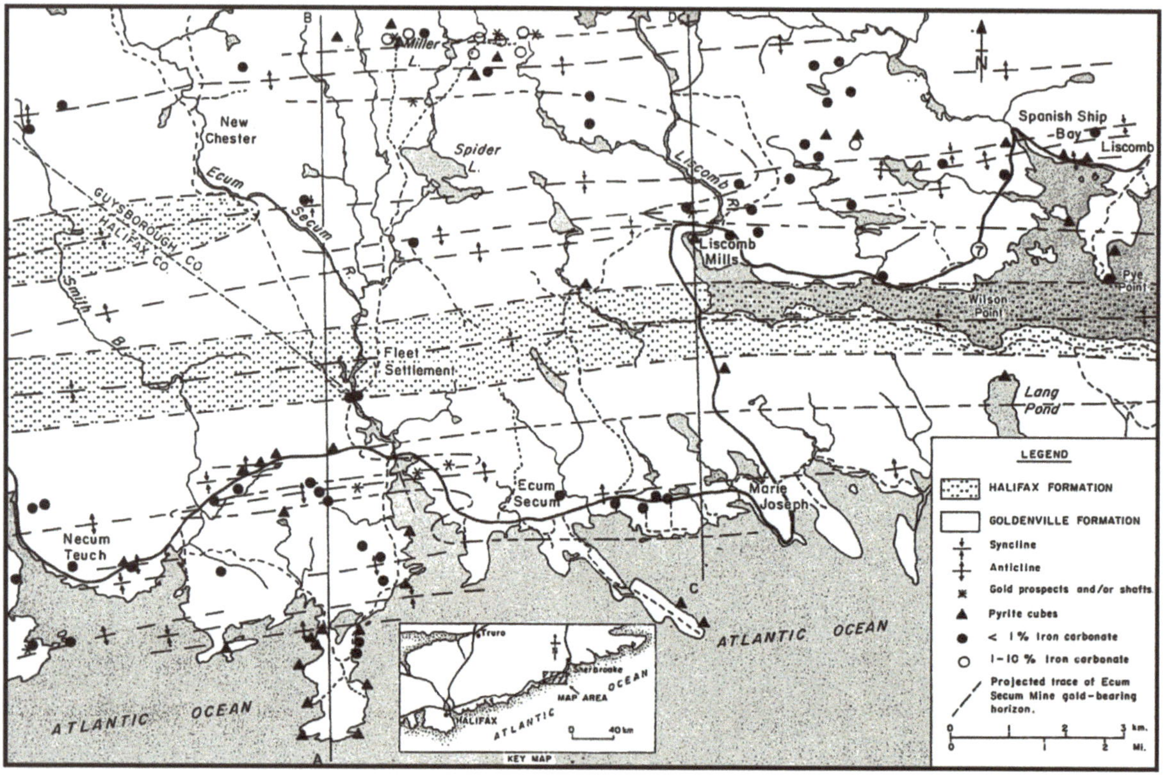

FIGURE 8.22

Map of Ecum Secum Map Area showing the distribution of pyrite cubes, disseminated iron carbonate, and gold prospects. *McBride (1978), Figure 1.*

A detailed examination of the Meguma gold deposits demonstrates a similar pattern on regional and mine scales. In the Ecum Secum Map Area, McBride (1978) demonstrated that gold was restricted to a section of the stratigraphy that was anomalous in iron carbonate and pyrite. This horizon had been subsequently folded around a series of regional folds (see Chapter 9 and Figures 8.22 and 8.23).

This interpretation is based on 400-m-spaced mapping traverses that provide a statistical database over the map area of 200 km^2. A compilation of these data is shown in Figure 8.24; it provides an example of the data required to understand the structural control of the rocks hosting the mineral deposits contained therein.

Minor folds of gold-bearing quartz bands or veins that are bedding-parallel and outline classic minor folds around a major fold are documented across the anticline at the Coxheath Mine. There is not any evidence of quartz moving out of

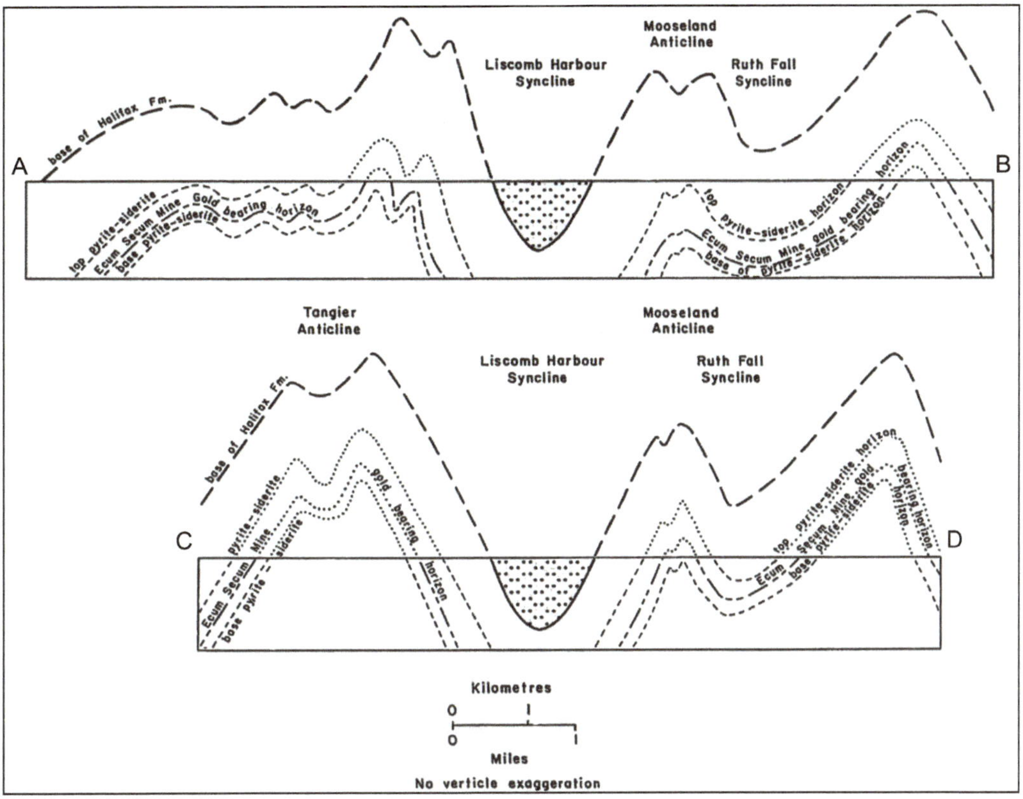

FIGURE 8.23

Sections across the Ecum Secum Map Area showing the mineralized horizons outlined by pyrite cubes and iron carbonate–rich stratigraphic section. *McBride (1978), Figure 2.*

the stratabound position; they are just parasitic folds on a major one (Figure 8.25). Figure 8.26 shows the same pattern in the Guysborough Mines Ltd. Goldenville Mine. This 1940s mine drift follows this barren vein around the hinge of the anticline, but no mining took place (Chataway, 2015, personal communication). According to the remobilization model, this part of the vein should be mineralized with gold! However, the sharpness of contacts between adjacent beds and lack of gold mineralization supports the observations of Whitehead (1973) and McBride (1976b, 1992) that any movement of metal and minerals during diagenesis and regional deformation is of geochemical magnitude only, and hence this section of the vein was never mineralized.

Malcolm (1912) shows numerous photographs of examples of these quartz bands within the host slate bands (Figures 8.27–8.30). These photos

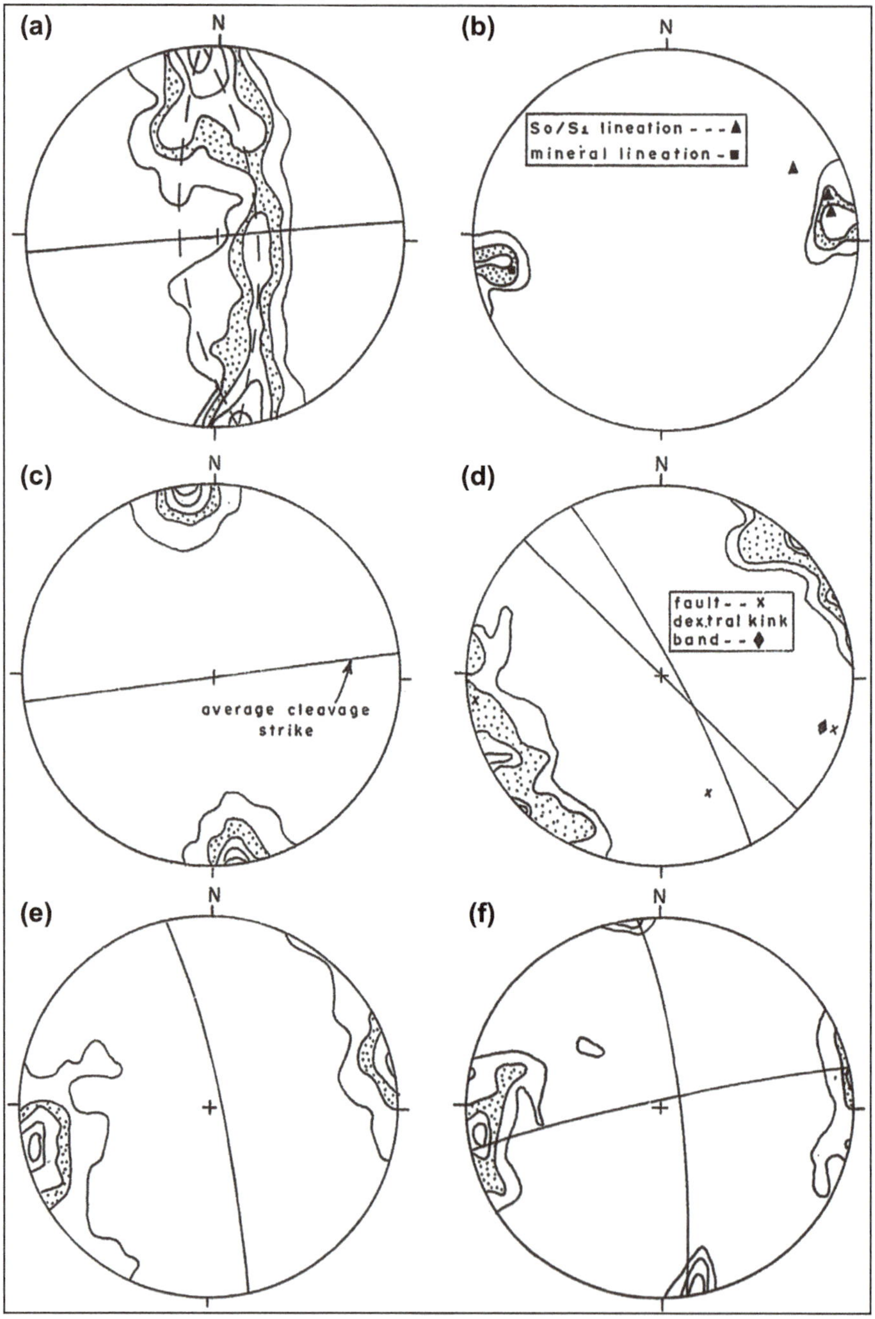

FIGURE 8.24

Lower hemisphere equal area projections of structural data from the Ecum Secum area. (a) Poles to bedding (300 points; contours: 1, 3, 5, 8, and 15%). (b) Fold Axes (31 points; contours: 7, 14, and 20%), S0 and S1 lineations and mineral lineations associated with major deformation. (c) Poles to cleavage (500 points; contours: 1, 10, 20, 30, and 35%). (d) Poles to sinistral kink band axial planes (41 points; contours: 2.5, 5, 10, and 15%), dextral kink band axial planes and faults. (e) Poles to joints (226 points; contours: 1, 5, 10, 20, and 25%). (f) Poles to quartz veins (67 points; contours: 3, 6, 9, 12%). *McBride (1978), Figure 3.*

FIGURE 8.25
Composite photo looking along hinge of the Coxheath Mine anticline, showing sharp parasitic folds of gold-bearing quartz over the major anticline. *McBride Photo, 1989.*

FIGURE 8.26
Barren quartz vein in a cross-cut around the fold hinge, Goldenville Mine, showing "M" folds. *Chataway Photo, 1989.*

FIGURE 8.27
Parasitic folds in major anticline, Moose River, Nova Scotia, "showing the cleavage approaching the corrugations." *Malcolm (1912), after p. 67.*

FIGURE 8.28
Moose River, Serpent Lead, showing laminated quartz bands and folding. *Malcolm (1912) after p. 96.*

FIGURE 8.29
Bordon lead at Mount Uniake. *Malcolm (1912) after p. 100.*

FIGURE 8.30
Thick quartz vein over the crest of the anticline Dufferin Mine. *Malcolm (1912) after p. 222.*

demonstrate that the same major-parasitic fold relationships persist throughout the gold deposits and their host rocks of Nova Scotia. They were probably taken of mineralized veins, but as Figure 8.26 illustrates, the pattern is true for all veins in the system whether or not they are mineralized.

8.6 STRUCTURAL GEOLOGY AND MINING ECONOMICS VERSUS GEOLOGICAL CONTINUITY

Many discussions that support remobilization use the history of mining to demonstrate that "saddle reef" deposits concentrate the remobilized mineralization in the fold hinges and that the mineralization vanishes down the fold limbs; but is this a true representation of the continuity of the gold-bearing veins or a simple case of mine economics? In historical mine records, the minimal payable grade for narrow vein mining was 10–15 g per ton.

In similar folding, the thickness of a bed is constantly parallel to the axial plane. It is the true thickness at the hinge of the fold. On the limbs, the axial plane thickness remains constant, but the true thickness of beds decreases as the limbs approach tight to isoclinal. Concentric folding, as seen in the Meguma deposits, produces an identical situation. In the hinge of the fold, the narrow gold-bearing vein is repeated many times (Figures 8.25–8.30). Down the limbs, the fold becomes "S" or "Z" shaped, with two long limbs and one short limb (Figure 8.1). Consider an example in which the gold vein is 10 cm thick and grades 30 g per ton. In the fold hinge, this vein is repeated, say, five times over a meter mining width in a 2-m high working place. It therefore represents 50 cm of mineralization in a 1-m width or 15 g per ton, diluted, which is ore. On the limb, the vein width averages 10 cm of the same 30 g per ton. Over the mining width of 1 m, the grade is diluted to 3 g per ton. This grade is not economic! However the vein still carries the same 30 g of gold per ton. Geologically the mineralized vein is continuous, but the economics are not. From this simple calculation, it can be demonstrated that historical mine records cannot be used to define the limits of the gold-bearing vein structure (Figure 4.5).

This relationship holds true throughout the Meguma gold deposits (Figures 8.25–8.30). The relationship in the Goldenville Anticline is demonstrated from multiple bedding-parallel veins; here the folds of the veins are sharp, stay within the host dark-gray slate band, and show clear, sharply folded contacts. Mr. Bob Chataway managed the late 1980s exploration of the Guysborough Mines Ltd., Goldenville Mine, and informed the authors that all of the mining was from bedding parallel quartz bands only and that they were not continuously mined, having gold-bearing and barren sections. These shoots crossed the main fold on the northwest–southeast diagonal Figure 8.31 Chataway also discovered records, now in the Nova Scotia Department of Mines, that show the grade control used in the mine. Apparently the miners checked the face for visible gold; as long

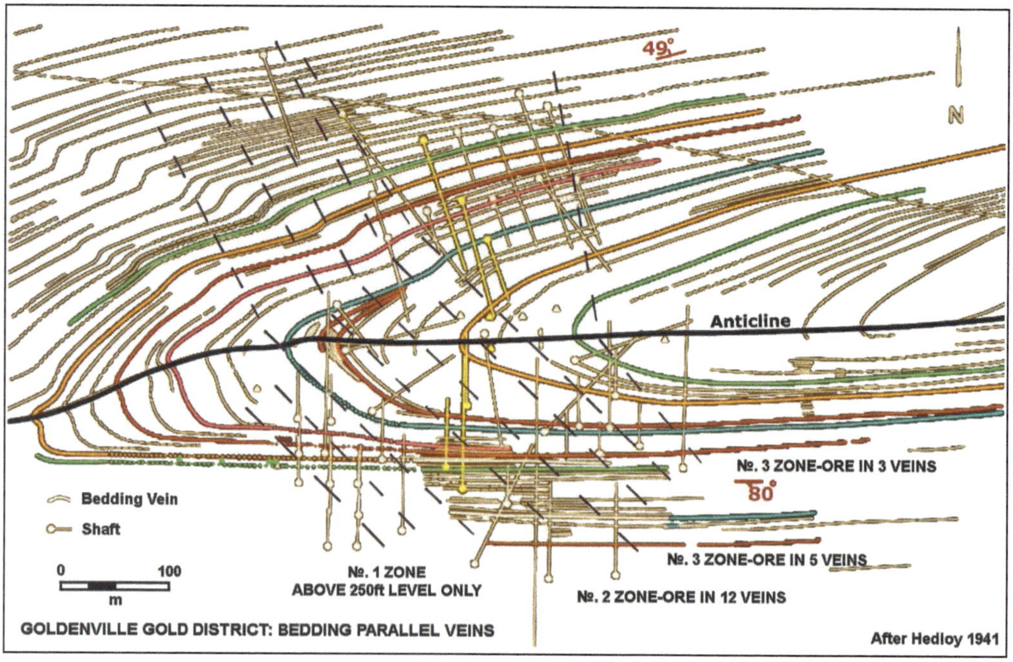

FIGURE 8.31

Distribution of bedding parallel veins, Goldenville Mine, showing limits of economic zones within them as defined by Hedley, 1941 (Figure 4).

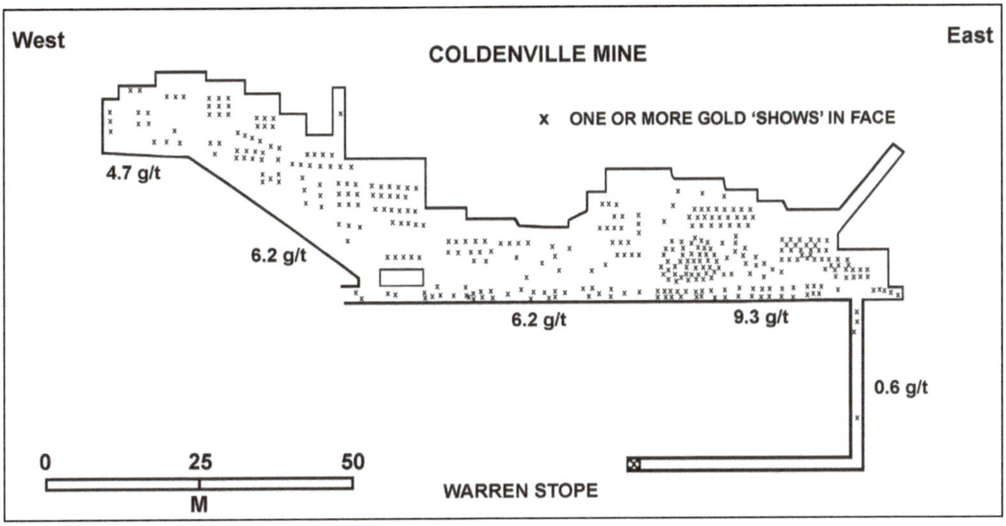

FIGURE 8.32
Grade control by "gold sight" method, 514 Warren Stope, Guysborough Mines Ltd., Goldenville Mine. *After Hedley (1941), Figure 5.*

as they got some shows of gold, the vein was considered ore. If they did not, the vein was considered waste, and the workers went to another location. The vein continued, but the economic gold content did not (Figure 8.32). All ore was hand cobbed (sorted) on surface to eliminate the cost of processing waste rock in gravity stamp mills. Often the economics of the vein was governed by the returns from the "crushing." Modern grade control was unknown.

8.7 SUMMARY

This chapter outlines the scientific parameters that control events during burial, diagenesis, and deformation of mineral deposits in mobile belts. Physical actions involving compaction, heat, and pressure will produce the same results wherever they are applied to rock sequences of similar composition such as mobile belts. In these mobile belts, deformational events produce folds and faults of all rock units present prior to deformation. Folded metamorphosed rock sequences with cleavages or schistosities represent the first structural event. Any exception to this rule must be demonstrated with sufficient evidence, especially if another structural event is interpreted to predate it and to affect the bedding. Evidence for such events has not been found despite many structural studies including those mentioned in this chapter. It is therefore concluded that the regional folding event is the first deformational event and has equally affected all of the rocks including those hosting gold deposits and the gold mineralization as well.

Metallic minerals remain within the stratigraphic sequence fixed in the natural system once they are deposited. The examples above clearly demonstrate that fact. Remobilization does not occur in economically significant amounts. The only remobilization that can be documented during diagenesis and deformation is of geochemical interest. The distribution of economically interesting metals was established prior to these events and was not affected by them in any fashion that was not expected from the regional deformation (McBride, 1991).

It has also been demonstrated that the historical mine plans define only the economic limits of the gold-bearing veins, not the geological continuity of the mineralized system. Thus they cannot be used to determine the vein economics without geological confirmation.

Shear zones have been advocated as the mineralizing conduits through which gold-bearing solutions travel from an unknown source. The scientific evidence presented herein demonstrates that shear zones are late structural events that postdate the regional deformation and the mineralization. They usually mark the boundaries between rigid continental blocks; hence they are too late for mineralization except to physically displace it. Late shear zones can also be found in mobile belts, and two have been documented in this chapter. Both fit the shear zone criteria quoted above and postdate the mineralization and its later deformation history.

A good example of a modern shear zone is the San Andreas Fault in California, which does not have any identified mineralizing centers along it of which the authors are aware.

CHAPTER 9

Lode Gold Deposits: Their Geometry and Evidence for Seafloor Vent Systems

CONTENTS

9.1 BACKGROUND FOR SYNGENETIC MODELS OF LODE GOLD DEPOSITS

This chapter describes gold deposits throughout the greenstone and sedimentary environments and concentrates on Canadian deposits. These brief descriptions provide summaries of the important relationships of mineralization and geology to establish their geological setting, history and possible modes of formation. Many comprehensive models have been proposed by Kerrich (1983) for Archean lode veins and by Haynes (1986) for Meguma veins. Combined with the work of Hutchinson and Burlington (1984), a practical, comprehensible and almost complete model for lode veins has been in existence since that time. An update was provided by Kerrich et al. (2000).

In 1987, for the Beardmore Camp east of Lake Nipigon in Northern Ontario, McBride (1987) suggested that the criteria for concentrating gold into economic deposits depended on:

1. the presence of a fairly substantial sequence of permeable mafic/ultramafic pyroclastic rocks;
2. a hiatus at the end of mafic volcanic activity to permit the development of a geothermal system;
3. the presence of a long-lived volcanic or intrusive center to act as a heat pump to drive the geothermal system; and
4. a favorable depositional environment for the hydrothermal fluids so that the gold would not be dispersed.

He also demonstrated that subsequent structural history had not had any appreciable effect on the syngenetic distribution of the mineralization.

The successful application of this model will be discussed at the end of this book (Figure 9.1).

The Metallogeny of Lode Gold Deposits. http://dx.doi.org/10.1016/B978-0-12-803222-0.00009-5

In the recent literature, there are again studies, particularly those originating in China, Russia, and with Ross Large in Hobart, that postulate a syngenetic origin for some gold (see References to syngenesis examples). In the US, it appears that a lead is being taken by Ayuso, Foley, and co-workers with their studies of deposits in the Carolina Slate Belt and Embro, Hutchinson et al. for Carlin-type deposits.

The observations and interpretations presented in Chapters 2 to 6 are based on careful field work in the Meguma and the Chester Complex and lead to the current reinterpretation of lode gold deposit genesis. By following subtle clues in outcrops and in core, and by starting with the premise that quartz is primary unless there is clear evidence to the contrary, it has been possible to gain significant new insights into lode vein and seafloor gold deposition processes. A brief recapitulation of evidence follows.

9.2 SUMMARY OF EVIDENCE FOR A PREDOMINANTLY SYNGENETIC FORMATION MODE FOR LODE VEIN DEPOSITS

Primary processes are invoked by Kretschmar (2011) based on the following:

1. the delicate nature of quartz–chlorite (pelagic–pelitic interflow) interbedding and laminations;
2. lateral and vertical mineral zoning in the veins;
3. asymmetric alteration and asymmetric geometry of interflow sediments (GCE) and quartz veins;
4. stacking of monomineralic layers of sulfides of different composition in veins, which reflect chemical gradients and secular change in fluid composition;
5. a common absence of feeder veins reflecting hydrothermal fluids that seep onto the seafloor in a diffuse manner, or are sourced laterally;
6. coxcomb and open vug textures in veins and variable NaCl content in fluid inclusions indicative of seafloor or subseafloor boiling;
7. the presence in veins, of carbon that may be biogenic or formed under Fisher–Tropsch temperature and pressure conditions in Archean Strelly Pool veins;
8. isotopic disequilibrium between veins and wall-rock, which results when fluids whose isotopic signature reflects generation at depth are deposited on the seafloor or subseafloor, with a different isotopic signature;
9. "lamprophyres" common in many worldwide gold camps, which are pelagic–pelitic sediments (Gold Cycle "E" or GCE units) and formed at the same time as the gold-bearing quartz;

10. mineralized horizons and the mineralization within them, which can be shown to predate diagenesis and regional deformation;
11. continuity of the mineralization that can be disrupted by folding to give the impression of a series of en echelon lenses when only a single lens was originally present.

The actual geometry and spacing may be derived from geological maps, cross-sections, grade-thickness contours, and other information in the literature. An excellent source of information are geological plans, mine longitudinal and drill sections that constitute geological descriptions in NI43-101 reports commonly filed by exploration companies on the SEDAR Website (www.sedar.com). Another excellent source of older data are the publications of the Canadian Institute of Mining and Metallurgy, (CIMM) specially Structural Geology of Canadian Ore Deposits Volume 1 (1948) and Volume 2 (1957). These volumes contain descriptions of Canadian deposits mined at the time or earlier, but are almost never referred to in the academic literature. Their importance is that the gold deposits were reported on by geologists who actually worked in the mines. Most modern geologists do not have the access, as these mines are now closed and inaccessible. Thanks to the present authors, these volumes will soon be available electronically from the Canadian Institute of Mining, Metallurgy and Petroleum.

Many descriptions concentrate on the mined (economic) portions of the veins, which may not represent the full extent of the mineralized system. A complication that the authors have encountered is the many deposits are folded and the economic mineralization is restricted to the long limbs of the asymmetric "S" and "Z" folds. The short limbs produced a series of intersections that could not be identified as mineable blocks of gold mineralization

It must be remembered that these volumes were prepared when all deposits were considered replacements. When they are described as being subparallel to the strata, it was considered a coincidence not a possible geological control!

9.3 EXAMPLES OF CANADIAN STRATABOUND GOLD DEPOSITS, GEOMETRY AND SPACING FROM GEOLOGICAL PLANS AND SECTIONS

9.3.1 Northern Empire Mine

In 1981–1982, McBride investigated the Northern Empire Mine (Figure 9.1) on two levels and tied this geology into his detailed field work, which mapped the favorable geological stratigraphy east of the mine for 10 km on a 100-m line spacing (McBride, 1981). In the winter of 1981–1982, the mine was pumped down 100 m and the first two levels were mapped

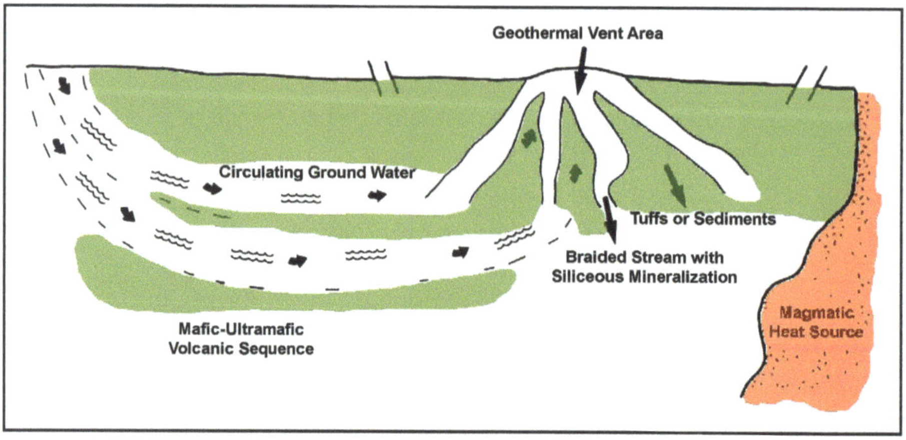

FIGURE 9.1

Diagrammatic view of a geothermal system for the Northern Empire Gold Deposit showing circulating ground water depositing gold and quartz on a paleosurface. *McBride (1987) reprinted with permission of the Canadian Institute of Mining, Metallurgy and Petroleum.*

and sampled. Mapping showed that the gold-bearing sections of the vein were spaced along the volcanic tuff horizon and were later folded and boudinaged by a single period of deformation. Figure 8.7 clearly demonstrates that in the map area only one schistosity is present. Additional data were obtained from exploration projects searching for volcanic massive sulphide deposits northeast along the greenstone belt and investigations of many local gold prospects (McBride 1980a,b). Together they provided the database for the paper on the position of economic gold deposition in the geological evolution of the Geraldton–Beardmore Greenstone Belt (McBride, 1987).

Underground mapping demonstrated that the ore shoots were in a single stratabound horizon width of 60- to 150-m-long economic lenses, separated by zones of weak and narrow mineralization. This weak gold-bearing horizon was more or less continuous but less than 0.20 m wide and much lower grade. Benedict and Titcomb (1948) produced a longitudinal section of the ore shoots (Figure 9.2). It is included here without the later diabase sill to show the radiating finger-like shoots originating from a central source at the bottom of the mine. Each of these shoots occurs in a more silty host than the vein between them and has been interpreted as a shallow channel (McBride, 1987).

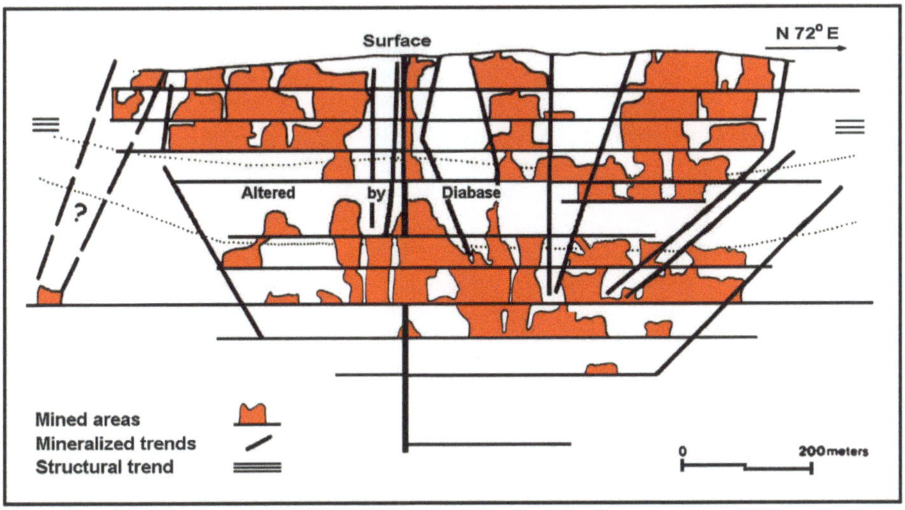

FIGURE 9.2

Long section of the Northen Empire Mine showing the radiating finger-like ore shoots; 200-m-thick diabase sill remove for clarity. *After Benedict and Titcomb (1948) Figure 3, p. 394, from CIM Special Volume 1, reprinted with permission of the Canadian Institute of Mining, Metallurgy and Petroleum.*

These veins have been deformed by a single regional deformation as illustrated in the stereographic projection of over 600 schistosity readings from the property (Figure 8.7). This deformation has boudinaged the main vein parallel to the stratigraphy and folded the cross-cutting feeder veinlets (Figures 9.3 and 9.4). These photographs and the stereographic projection of the schistosity demonstrate that the mineralization was present parallel to bedding, and folded with it by the first deformation. In addition, the pattern

FIGURE 9.3

Photograph of a typical boudin of the main vein at the Northern Empire Mine. Bedding bends into the neck of the boudin. *McBride (1987) reprinted with permission of the Canadian Institute of Mining, Metallurgy and Petroleum.*

FIGURE 9.4

Photograph of the face at the west end of the first level, Northern Empire Mine, showing a boudin similar to the one in Figure 9.3 and parasitic folding of a vein cross-cutting to the main vein horizon. *McBride (1987) Photo 7. Reprinted with permission of the Canadian Institute of Mining, Metallurgy and Petroleum.*

of ore shoots suggests a vent source below the bottom of the old mine, which supplied gold-bearing solutions to the ocean floor that flowed down slope in a "braided stream pattern."

9.3.2 Leitch Mine, Beardmore, Ontario

The Leitch Mine is located in the Beardmore Camp and was the richest mine in Canada, producing with an average grade of 0.92 oz/ton (34.6 g/T). It lies in sedimentary rocks that form a slightly overturned, west-plunging syncline. A schistosity and parasitic folds are associated. Figure 9.5 show plans of the mineralized veins. These veins are folded, but more importantly, they are not continuously mineralized. Mineralized sections of the veins can occur in the folds or on the limbs. Where the mineralization is on the limbs, it shows "lensing" that is identical to the boudinaging documented above for the Northern Empire mine in the same mining camp. Ore shoots for the No. 4 vein are shown on the longitudinal section (Figure 9.6). The pattern suggests a series of closely-spaced parallel shoots; however, a single vent system is more likely, given the folded nature of the vein. There is insufficient evidence to position the vent

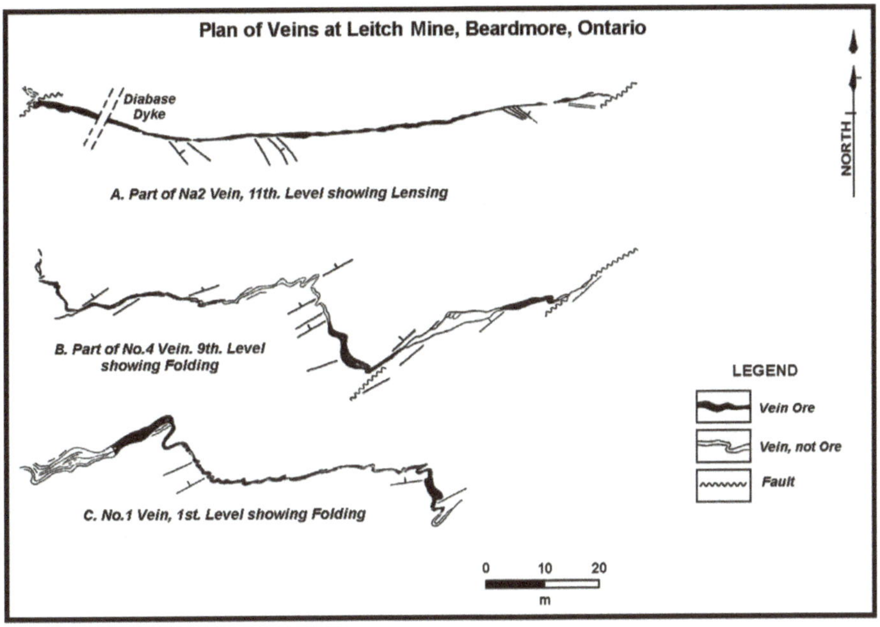

FIGURE 9.5
Plan showing veins, ore shoots, folding, and lensing, Leitch Mine, Beardmore. *McKay et al. (1948) Figure 2 on p. 386 from CIM Special Volume 1, reprinted with permission of the Canadian Institute of Mining, Metallurgy and Petroleum.*

system, but deposition before deformation in a channel way of some type, is apparent. The authors go on to state, "the quartz in all the veins is more or less banded parallel with the strike with dark streaks of mineralized schist. Ordinarily, the darker and more numerous these streaks the higher the grade of the ore. The vein walls are usually slickensided and well defined" (McKay et al., 1948). Thus the veins are well defined, internally banded, folded, associated with mineralized schist, and have far greater vertical continuity than horizontal.

The same diabase sheet that cuts the Northern Empire Mine also cuts the Leitch at from 570 to 750 m. The authors comment that "it intruded the sedimentary rocks with very little disturbance. The chief effect was baking which darkened and hardened the greywacke near the contact." They mention a 2-m dike that cuts the mineralization and provided a pillar during mining. It is clear from these observations that later intrusions had no effect on the gold mineralization.

The western extension of the Leitch vein system was mined as the Sand River Deposit.

9.3.3 Sturgeon River Gold Mine, Beardmore, Ontario

The third mine in the Beardmore camp is the Sturgeon River or Namwaminikan Mine. It exploited a narrow, high-grade vein that was consistent over a strike

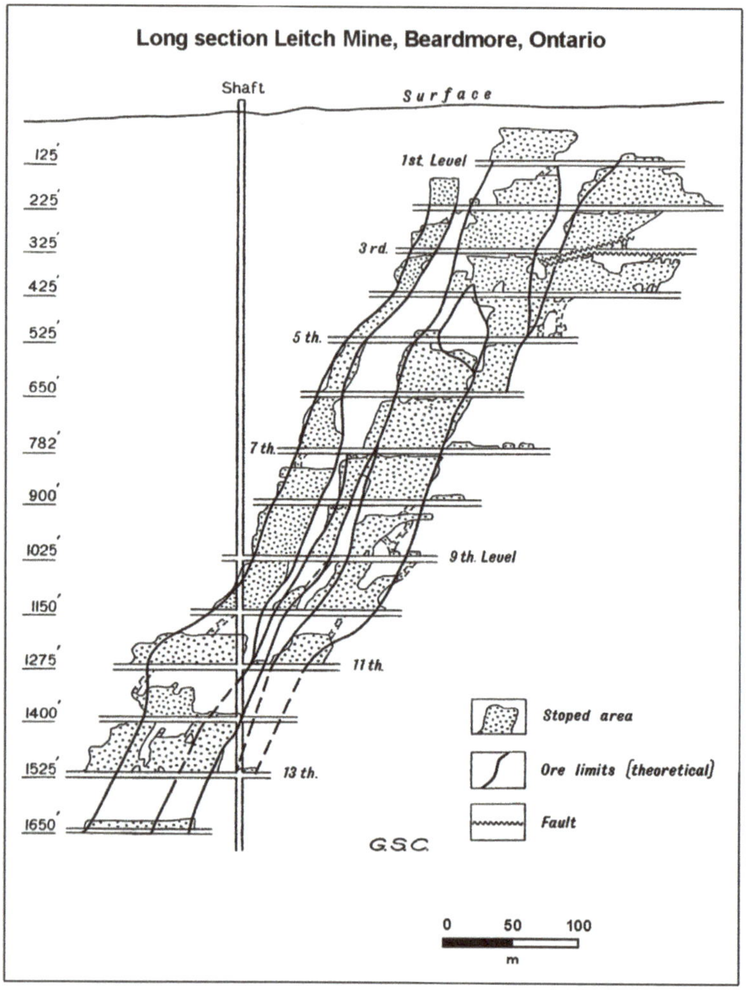

FIGURE 9.6

Longitudinal section of the ore shoots on the N. 4 vein, Leitch Mine. *McKay et al. (1948) Figure 3 on p. 388 from CIM Special Volume 1, reprinted with permission of the Canadian Institute of Mining, Metallurgy and Petroleum.*

length of 400 m and mined to a depth of 525 m Figure 9.7 is a longitudinal representation of the stoped vein. The mined limits define a wedge-shaped economic vein with the apex above the present surface and eroded away. It is likely that the vent system was in that area. Parallel zones are present in the lower part of the mine workings that could represent folded extensions. The main vein is conformable in volcanic rocks described as mafic and quartz eye tuffs; it was reported to extend into a granite to the east.

These three deposits in the Beardmore Camp share many things in common. All are folded, boudinaged or both; they were mined in the 1930s; only the

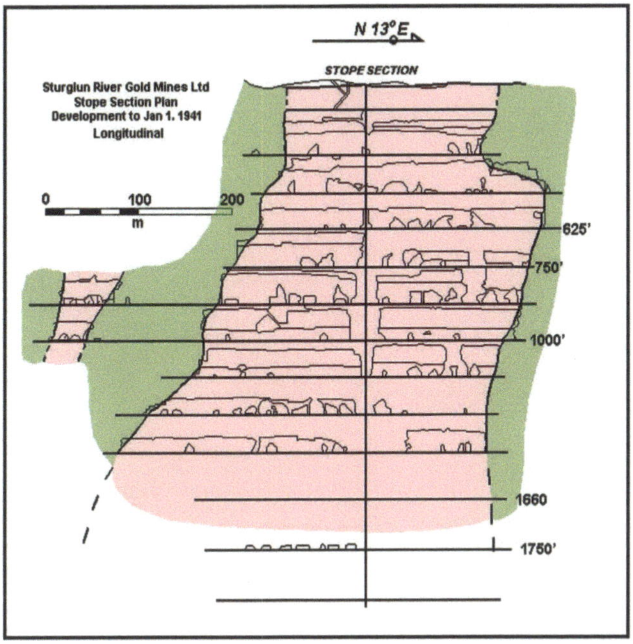

FIGURE 9.7

Longitudinal representation of the Sturgeon River Mine, Beardmore, showing the limits of the economic vein. *Compliments of Laurion Mineral Exploration Inc.*

Northern Empire has seen a recent underground evaluation; and all occur near a major geological contact. There is one interesting difference, and that is the major accessory sulphides; in the Northern Empire it is arsenopyrite, the Sturgeon River, pyrite, and the Leitch it is gold alone. This difference suggests that the source of mineralizing fluids is unique for each and that they are not derived from a single deep-seated source.

9.3.4 Cline Mine

The past-producing Cline Mine and the Island Gold Mine of Richmont Mines lie in the Goudreau–Lochalsh Greenstone Belt northeast of Wawa, Ontario. McBride has studied the Cline Mine database and walked the property (McBride, 2009). Two mineralized horizons are present that saw mining in the late 1930s. The main mine was described by Bruce (1940). His work clearly shows the vein following the folded contact between the granodiorite (felsic volcanic rocks) and andesite (Figure 9.8). Some mining took place, associated with a second band of "granodiorite." McBride's review examined the entire property using the existing database. It found that the mineralization followed felsic rocks across the property for over 7 km and around two major folds (Figure 9.9). All of the documented mineralization is associated with the two

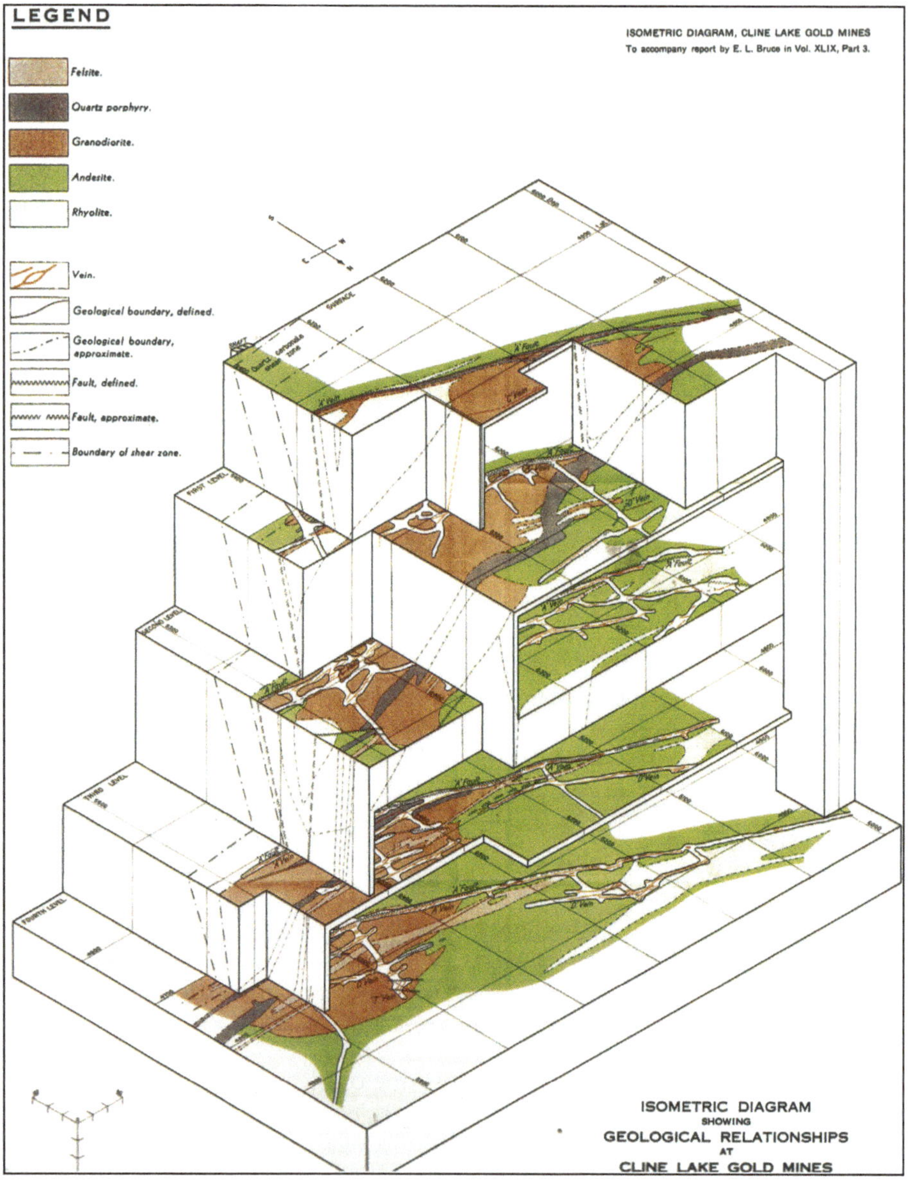

FIGURE 9.8

Block diagram from Bruce (1940), showing the main Cline Mine zone following the folded contact between felsic and mafic volcanic rocks. Most of the mining was from the A vein.

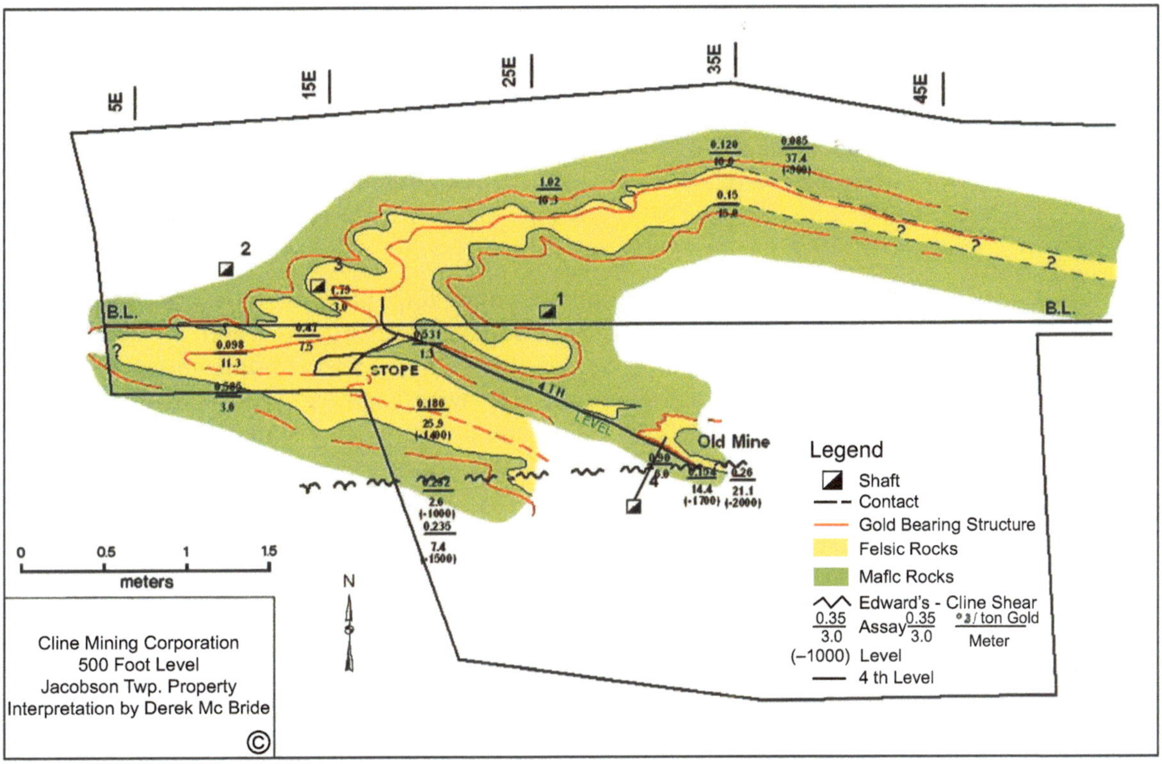

FIGURE 9.9

Level plan of the 500 ft (150-m) level, incorporating the old mine workings and recent drilling, illustrates the location of the old mine workings and the veins associated with the north felsic band. In the old mine, the A vein bends around to become the C vein. (McBride, 2009).

felsic volcanic bands, and the entire system is folded by the regional deformation. The historical data do not suggest a vent system, but the fact that two parallel felsic bands have substantial areas of gold mineralization, indicates at a vent system was present and active for a considerable time period.

Once this pattern was recognized, the recent drill intersections demonstrated that the northern felsic band had multiple parallel mineralized bands over considerable strike and dip lengths; in addition, the extensions of the favorable horizon beyond the old mine have not been explored.

Reinterpretation using modern geological knowledge has been able to redirect exploration at this past-producing mine.

9.3.5 Jerome Mine

The Jerome Mine is situated about 20 km west, northwest of the Chester Area, which has been discussed in Chapter 6 and is within the same Swayze

Greenstone Belt. It was another of those small Canadian gold mines that was exploited in the 1930s. Brown (1948) describes the mineralization as lying along the contact of sedimentary rock and granodiorite porphyry. His long section shows steep westerly raking ore shoots whose keel plunges off to the east at 30° (Figure 9.10). The mineralization is described as a composite vein. The footwall consists of quartz–carbonate with sulphides and gold; the hanging wall is a bluish, cherty silica. The main zone dips southerly at 80°.

The property remained dormant until the 1980s, when surface and underground programs were carried out on the main zone and the south zone. This latter zone had not been evaluated in the past. Drilling and underground drifting showed that these two zones were limbs of a syncline with a fairly flat easterly plunge. McBride visited the property and examined the underground workings. The hanging wall rocks are not intrusive, but are felsic volcanic tuffs with polymict clasts (Figure 9.11). Thus the mineralization lies along a volcanic–sedimentary contact.

Stope shape and drift patterns indicate that the ore has been folded as parasitic folds on the major syncline (Figure 9.12–9.14). Figure 9.12 is a plan view of the 350 West level; mining has come up to the level, and the stull-supported track layout defines the stope pattern, making the interpretation very simple.

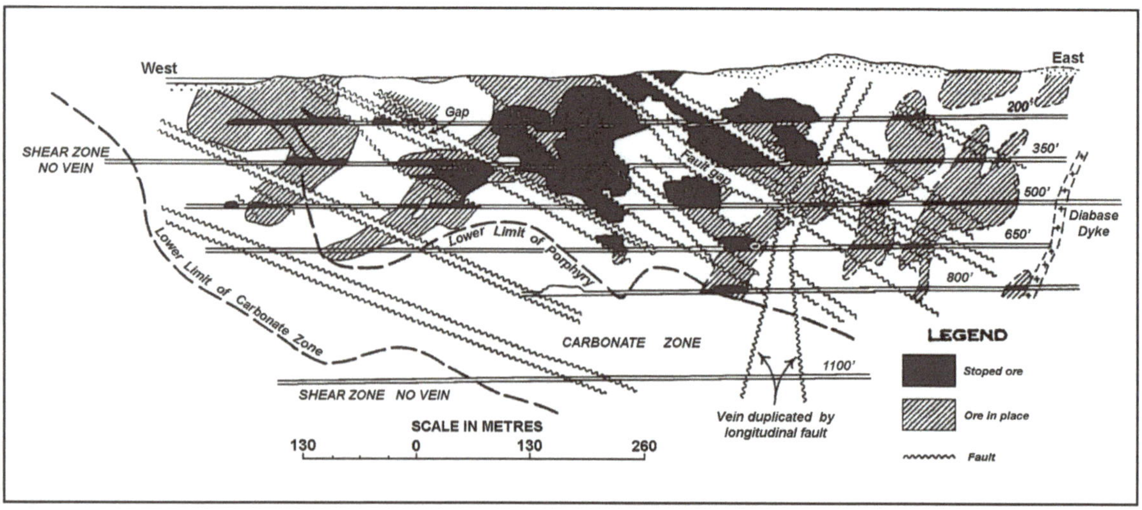

FIGURE 9.10

Brown's long section of the Jerome Mine north zone, with steep westerly raking ore shoots and an easterly dipping keel. *Brown (1948) Figure 3 on p. 440 from CIM Special Volume 1, reprinted with permission of the Canadian Institute of Mining, Metallurgy and Petroleum.*

FIGURE 9.11

Sample of the Jerome "porphyry," showing that it is really a felsic volcanic agglomerate (McBride Collection, 1989).

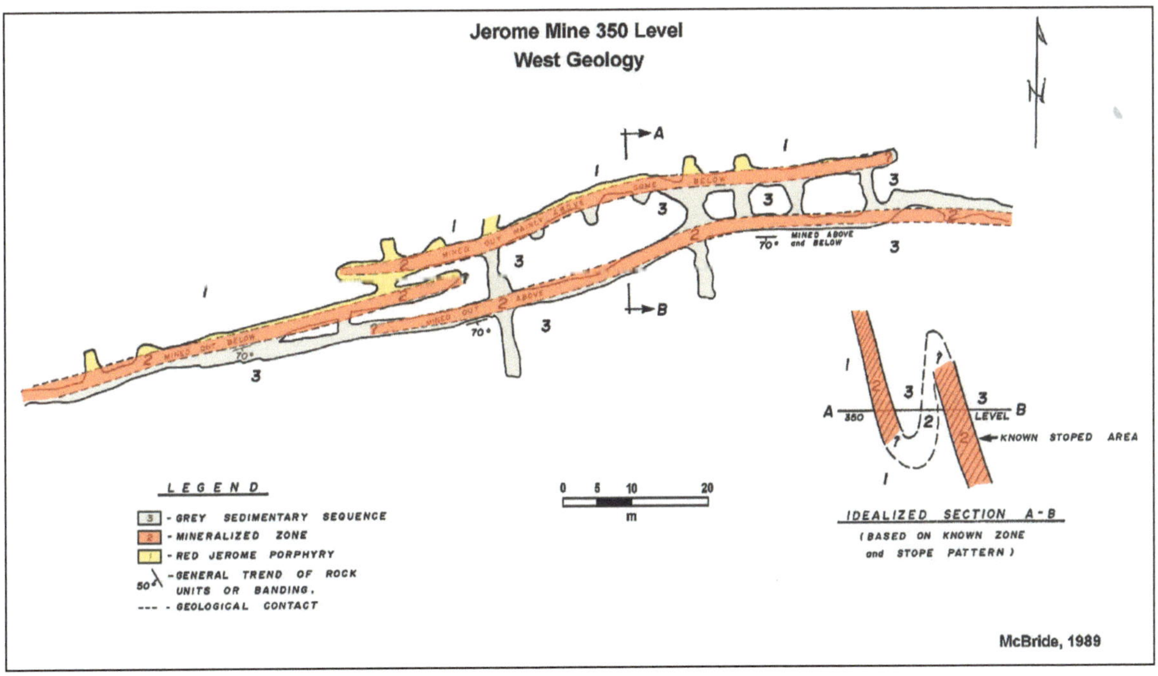

FIGURE 9.12

Jerome Mine 350 level west, showing the parallel stope outlining a fold of the mineralization as demonstrated in the included section (McBride's mapping, 1989).

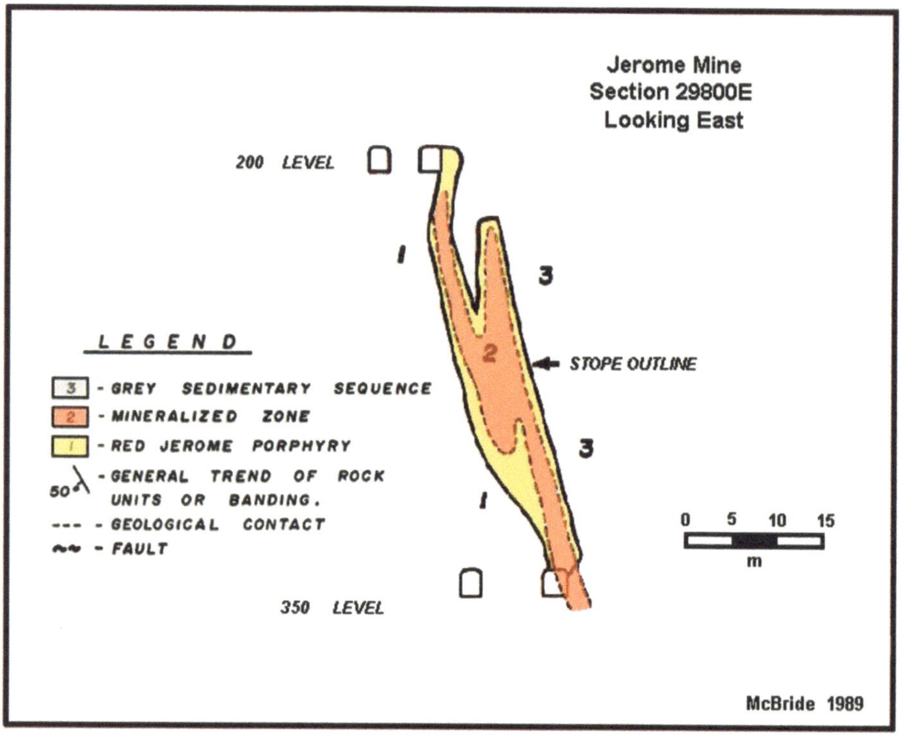

FIGURE 9.13
Jerome Mine stope outline showing an interpreted folded vein (McBride's mapping, 1989).

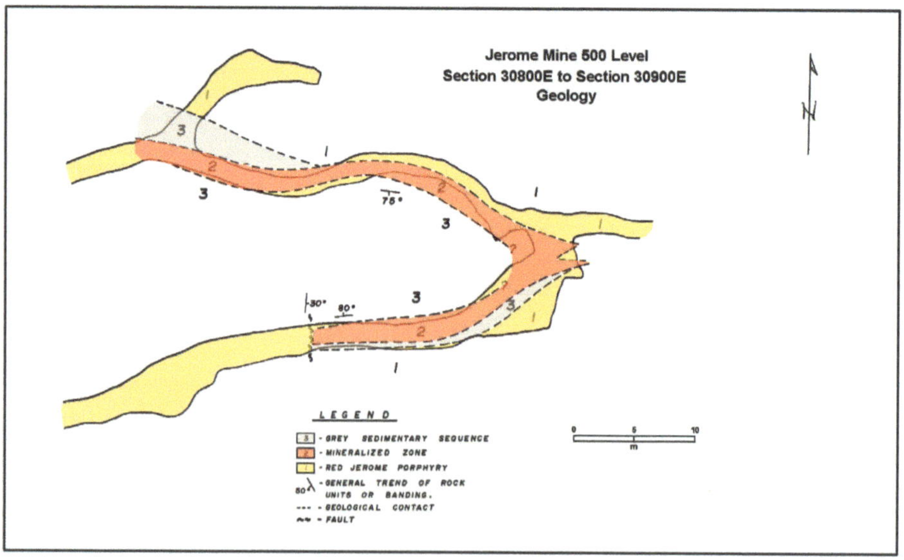

FIGURE 9.14
500 Level plan showing the mine drifts following the mineralization around the fold hinge (McBride's mapping, 1989).

Included is a section that interprets the fold form. Figure 9.13 is the interpretation of the mineralized band in a stope with the vein folding over before on the hanging wall and continuing above on the footwall side. Figure 9.14 removes any doubt that this is a folded vein system; the mine workings follow the fold around the fold nose. It is interpreted that the keel is the major synclinal fold hinge as the felsic volcanic–sedimentary contact plunges off to the east at 30°.

Together, these features show that the mineralization formed as a band on the volcanic–sedimentary contact and folded during the regional deformation. The South Zone is offset to the east so that the limits of the two zones together mark a northwest diagonal trend band across the major synclinal fold (Figure 9.15). This pattern suggests that the mineralized band is a stratum. A syngenetic gold system is the most obvious answer. It is interesting to note that the feeder system(s) could now be horizontal, probably striking perpendicular to the vein. It is not possible to say with confidence which way the mineralization flowed except to say that one would expect the better-grade material to be nearer the source, which would be on the northwest side beyond the mineralization and now eroded away. Superimposing the structural pattern of an east-plunging

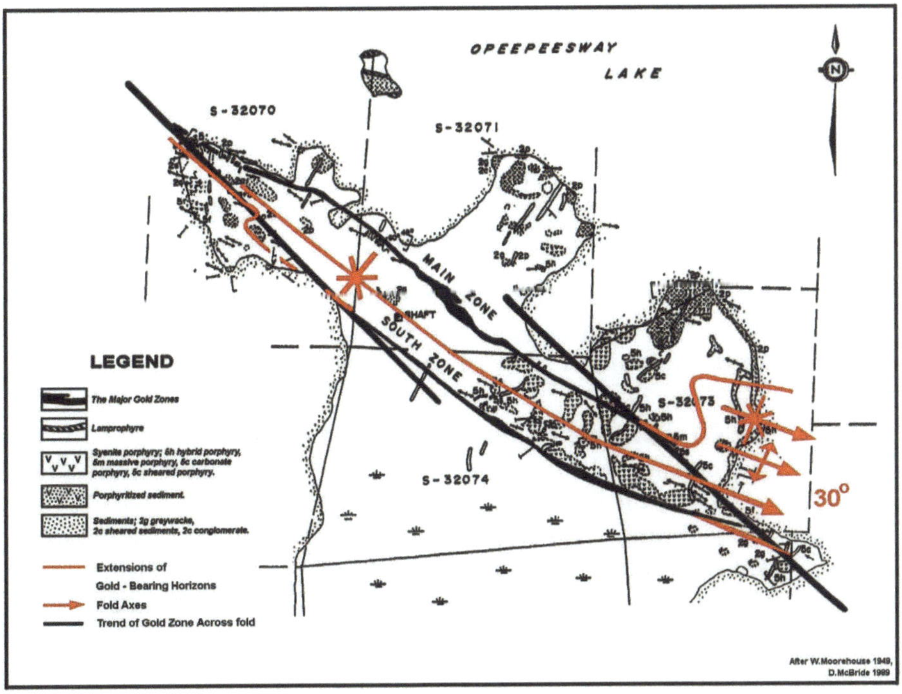

FIGURE 9.15

Plan of the north and south zones, Jerome Mine, showing the diagonal trend of the mineralized zones across the syncline within the folded contact (after Moorehouse, 1949, p. 23).

syncline on the existing database will assist in tracing the gold-bearing contact into unexplored areas, especially under Opeepeesway Lake.

9.3.6 Distribution of Gold Deposits in the Timmins Camp

The Timmins Camp hosts many past-producing gold mines and is known as one of the great gold mining camps of the world. Major gold deposits occur over a strike length of more than 40 km in the Tisdale Group volcanic rocks north of the Destor–Porcupine Fault (Figure 8.8). Regionally, these mines occur in the top section of the Tisdale volcanic rocks near the overlying clastic sedimentary rocks of the Temiskaming and Hoyle Series. Gold mineralization is found in iron carbonate–rich mafic volcanic rocks, commonly associated with the "95 flow", felsic intrusions and the adjacent sedimentary rocks. What they all share is proximity to the major break from volcanism to sedimentation. Figure 8.8 is a reinterpretation of the map of Dunbar (1948) and Pyke (1982). This version extends the fold pattern throughout the camp using the historical data; the mines are numbered. Only a few mines will be discussed here, but maps and photographs show that the mineralized horizon and mineralization is folded (Dunbar, 1948, pp. 454–455). McBride examined the "intrusive porphyries" and found that the Pearl Lake Porphyry to contains a very strong schistosity and fragments whereas the Preston East Dome Porphyry contains fragments and bombs. Many of the rocks termed porphyry historically are actually felsic crystal tuffs, agglomerates, and related rocks. It seems that much of the lower-grade gold ore found in the super pit of Dome Mines was formed by solutions percolating through porous felsic rocks. These differences are mentioned to demonstrate how a few field observations can modify the historical rock classifications, but that much work remains to be done to properly understand these gold deposits even in mature mining camps.

Most of the historical higher-grade production came from narrow quartz veins. We return to Structural Geology of Canadian Ore Deposits Volume 1 for descriptions of some of these deposits.

9.3.6.1 Buffalo Ankerite Mine, Timmins, Ontario

The Buffalo Ankerite Mine lies on the southern limb of the Edwards syncline, which has the Paymaster on the northern limb and Edward's prospect around the fold hinge. It is important because two zones were mined at the Buffalo Ankerite, each on the opposite limbs of an anticline (Figure 9.16). The geological plan shows that the ore bodies were stratabound, conformable, and occurred on the limbs of the fold. There is no remobilization into the major fold hinge. Individual ore shoots to be continuous down dip for hundreds of meters. The mine recovered over 30 tons of gold to a depth of 1200 m (Kinkel, 1948). The host rocks consist of pillow lava, chlorite–talc schist, sediments, and dacite.

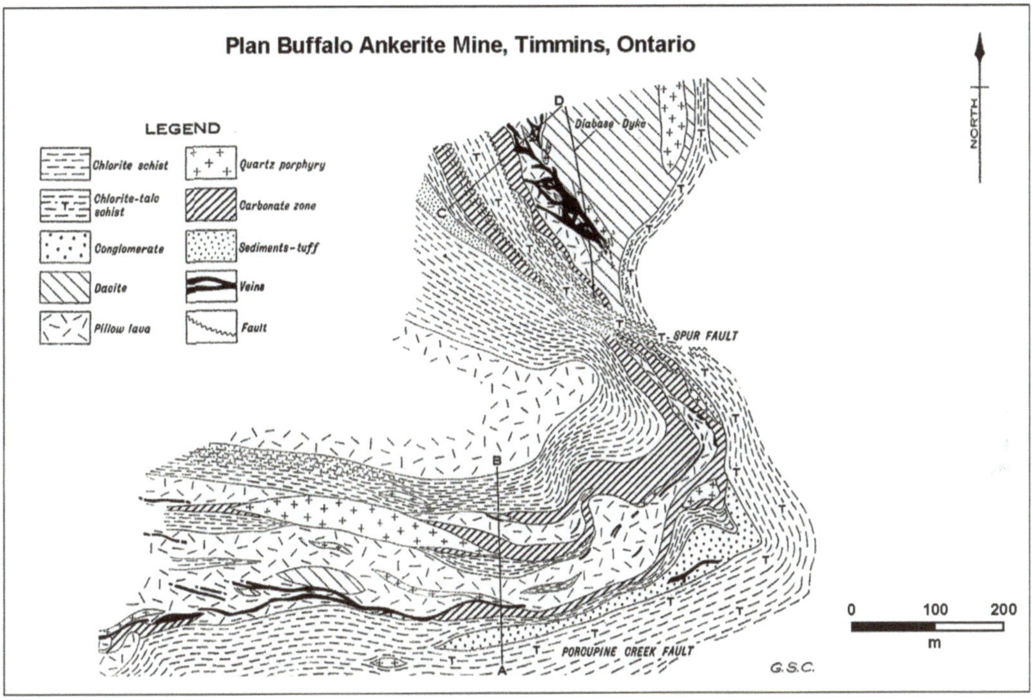

FIGURE 9.16

Geological plan of the 1050 (320-m) level, Buffalo Ankerite Mine, Timmins. *Kinkel (1948) Figure 1, p. 516 from CIM Special Volume 1, reprinted with permission of the Canadian Institute of Mining, Metallurgy and Petroleum.*

9.3.6.2 McIntyre, Hollinger, and Moneta Mines

The McIntyre and Hollinger Mines demonstrate the continuity of the gold mineralization over 500 m in the 95 flow and how the contact starts to bend around the Porcupine Syncline. As this stratigraphic package continues, it hosts the Moneta deposit before turning east through Vipond, Paymaster North Zone, and the Dome Mine then turns back west to the Preston East Dome, Paymaster, and Buffalo Ankerite described above. In the McIntyre Mine, the veins follow fragmental or tuff bands as conformable veins (Furse, 1948). Some of these vein systems can be traced down for almost 1700 m. The fold hinge of the Porcupine Syncline hosts the Moneta Mine, which strikes almost north–south and dips east at about 70°. Its footwall is a graphitic unit Figure 9.17.

Structural Geology of Canadian Ore Deposits, Special Volume 1 (1948) describes all the major mines in the Timmins Camp. In all descriptions, the conformability of the gold mineralization with the stratigraphy is apparent. It is also clear that many have been deformed by the only regional deformation. The readers are referred to the Canadian Institute of Mining, Metallurgy and Petroleum Special Volumes 1 and 11 for many more Timmins area gold mines

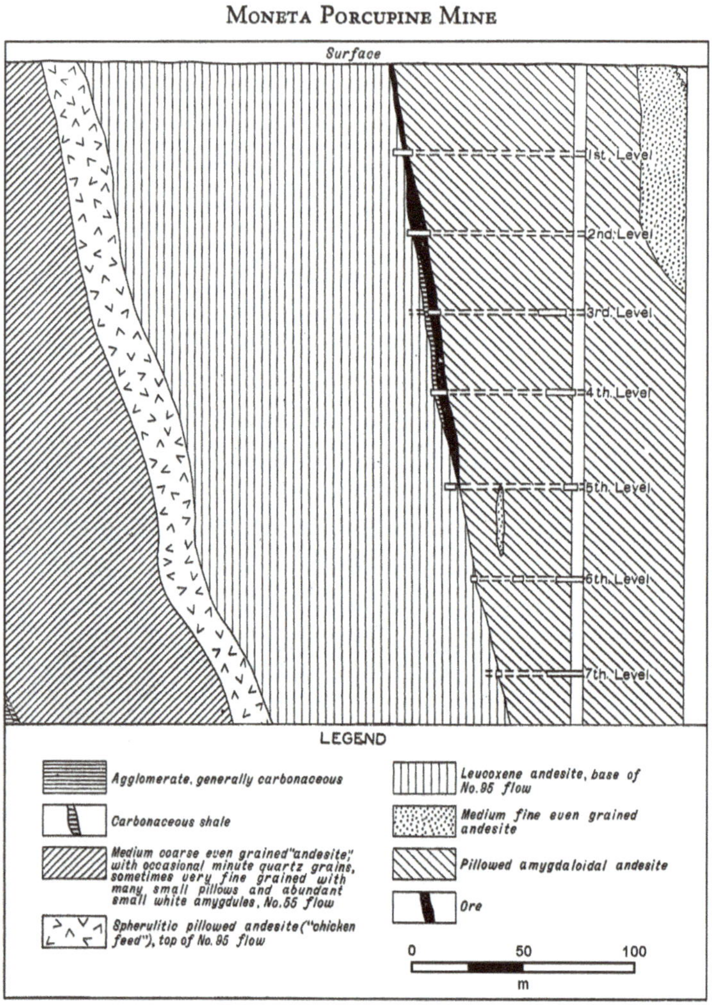

FIGURE 9.17

Vertical cross-section through the Moneta Mine, Timmins. *Buffam (1948), Figure 3, p. 463 (from CIM Special Volume 1, reprinted with permission of the Canadian Institute of Mining, Metallurgy and Petroleum.*

that further confirm the relationships described above. These special volumes are available electronically.

9.3.7 Belleterre Mine, Belleterre, Quebec

The Belleterre Mine lies in a greenstone belt, well south of the main mineral belt of Noranda-Val d'Or. It presents an interesting example because the favorable horizon can be traced around an oval, defining a doubly plunging anticline. The author McBride carried out a major mapping and re-evaluation project in

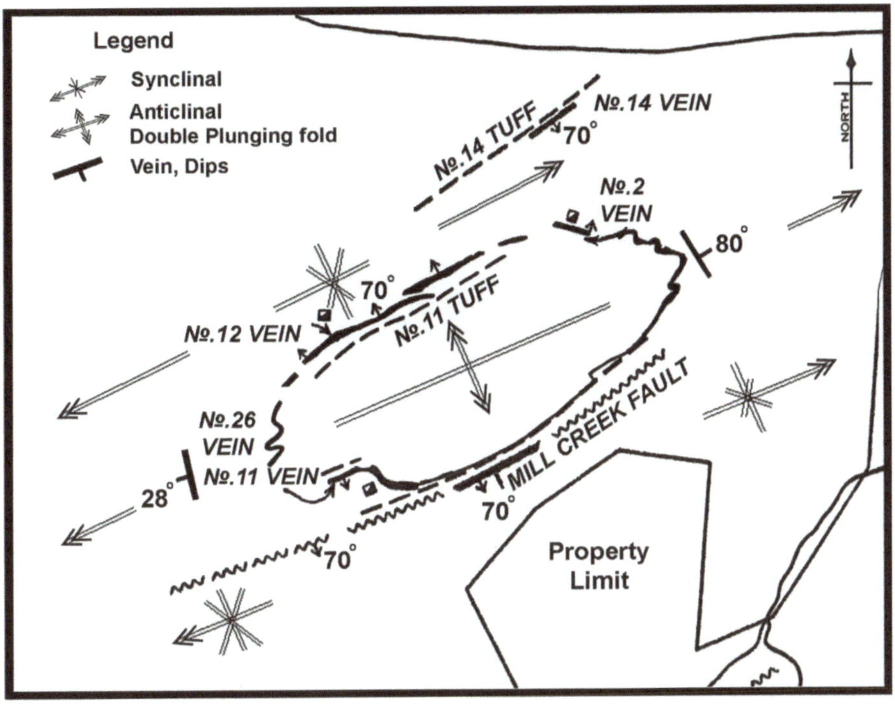

FIGURE 9.18

Surface map of the Belleterre Mine property showing the oval trace of the mineralized horizon from McBride mapping.

1992; his interpretation is shown in Figure 9.18 (revised from Trenholme, 1948). He was able to follow the gold-bearing horizon around the oval by the work of the original explorers. Most of the favorable horizon is composed of laminated quartz (Figure 9.19) in the No. 12 flow, which is more of a bedded tuff. Where the vein is not economic, it may be interbedded with the bedded tuffs (Figure 9.20). Mining was conducted in four economic shoots around the

FIGURE 9.19

Laminated and folded quartz band near old mined workings, Belleterre Mine.

FIGURE 9.20
Laminated and folded quartz band with interbedded tuffaceous sedimentary rocks beyond mine area.

main structure. The No. 12 Vein was a north dipping vein that was mined down over 450 m, and shows an easterly rake and a bending of the strike to become the No. 2 Vein east of the No. 1 Shaft. Beyond the economic shoots, the vein continues as a laminated contorted vein in a graphitic schist. It was traced continuously around to a small shoot, Vein 11, mined from the No. 3 Shaft, where it is a south-dipping vein. West of this shaft, the vein bends north through a series of parasitic folds that dip west at about 30° as the No. 26 Vein, which was mined from the No. 2 Shaft. Finally it bends east back into the No. 12 Vein.

The above detail illustrates a complex structural pattern with a very simple explanation. The regional deformation produced a series of east northeasterly trending folds. Probably because of the heterogeneity of the rock sequence, a "doming" took place, giving the appearance of a cross-fold or a doubly plunging anticline flanked by synclines (Figure 9.18). The result is that the Belleterre gold-bearing system can be traced as an oval with ore shoots distributed on its north, south, and west sides. A single source for the hydrothermal fluids could have been on the part of the dome that was weathered away.

9.3.8 O'Brien Mine, Cadillac, Quebec

The O'Brien Mine in Cadillac, Quebec, is a past producer with a new zone. Historical mining of two veins took in the main zone place from 1937 to 1956 and to a depth of 900 m. The property was originally described by Brown (1948), where he shows two essentially bedding-parallel veins: one the No. 1 or conglomerate vein and the other the No. 4 or No. 9, which is the same vein folded (Figure 9.21). Recently a new zone, the 36 Zone, was discovered and is being evaluated. Drilling has outlined this zone and intersections indicate that it continues to the depth of the old mine.

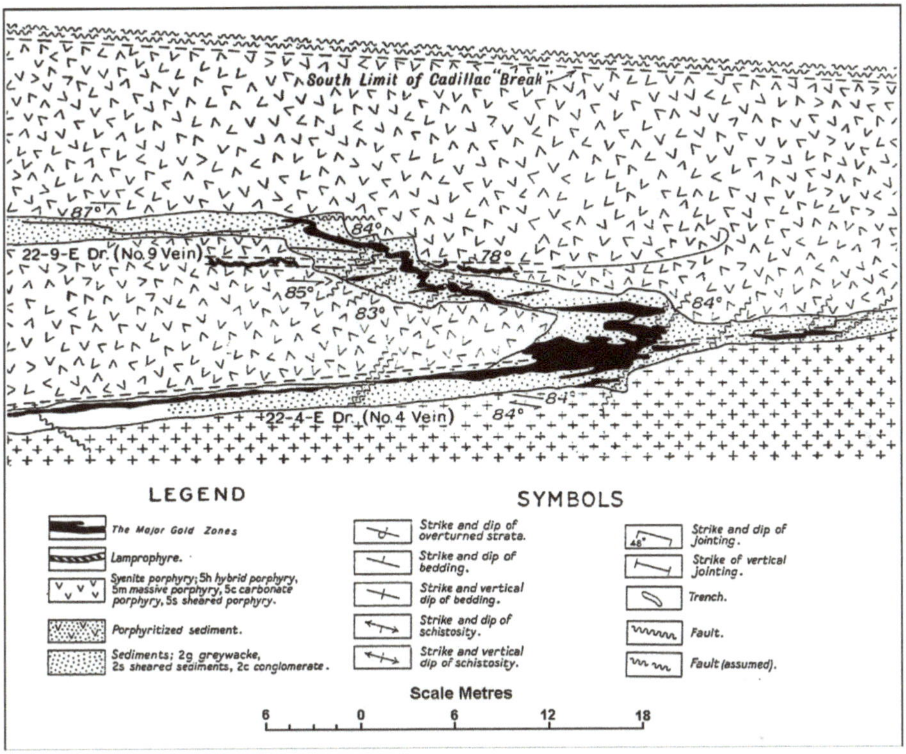

FIGURE 9.21

Plan of the 2250-ft (685-m) Level showing the No. 4 Vein and its folded No.9 Vein extension. *Brown (1948) Figure 3, p. 813 from CIM Special Volume 1, reprinted with permission of the Canadian Institute of Mining, Metallurgy and Petroleum.*

These veins are confined to the Pichee Group of ultramafic volcanic and porphyritic rocks that extend across the property to the east and host the historic Kewagama mine workings. The character and thickness of the individual beds varies along strike. In the porphyry beds, the porphyritic character decreases to the east to the point where they become fine-grained bedded tuffs. Mafic volcanic fine-grained tuffs make up most of the southerly rocks, and the northern unit is a talc chlorite rock with thin sandy layers. This rock has been referred to as a fault zone, but the clasts and relic beds indicate that it is a highly deformed thinly bedded ultramafic tuff–arenite sequence. Mine plans show the bedding conformability of the main ore shoots within the Pichee volcanic rocks (Figure 9.22).

The long section is very revealing (Figure 9.23); it shows all of the mineralized zones from the mining experience and the more recent drilling. It is apparent that the old mine and new 36 Zone are distinct zones separated by a nonmineralized section. The outline of the main zone is known from mine plans and plunges east from 70° on the west to 80° in the main mine area. Drilling on the 36 Zone suggests a very steep easterly plunge. Together, the data point to a

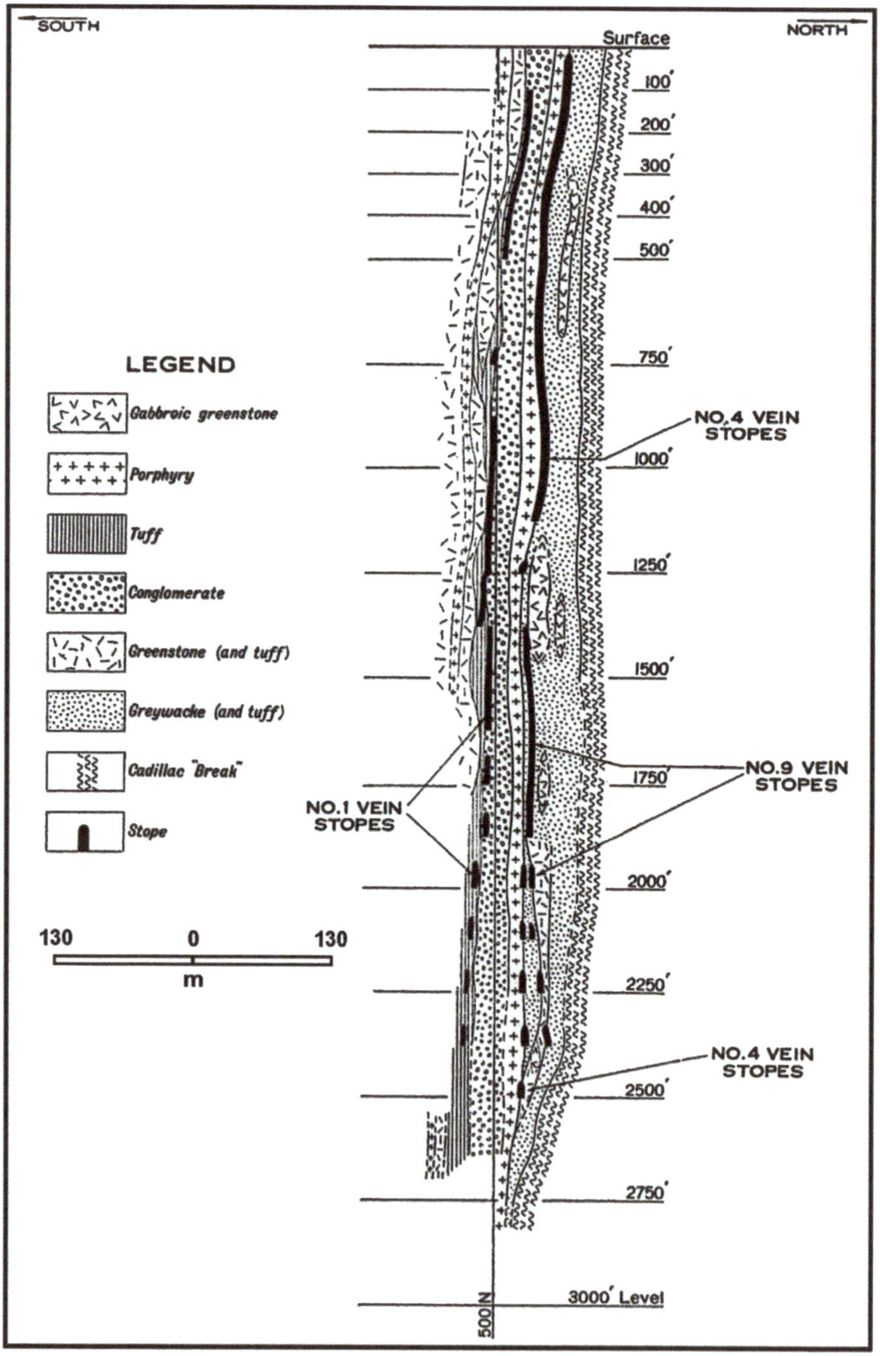

FIGURE 9.22

Figure shows the conformability of the veins with bedding over the depth of the original mining. *Brown (1948) Figure 2, p. 812 from CIM Special Volume 1, reprinted with permission of the Canadian Institute of Mining, Metallurgy and Petroleum.*

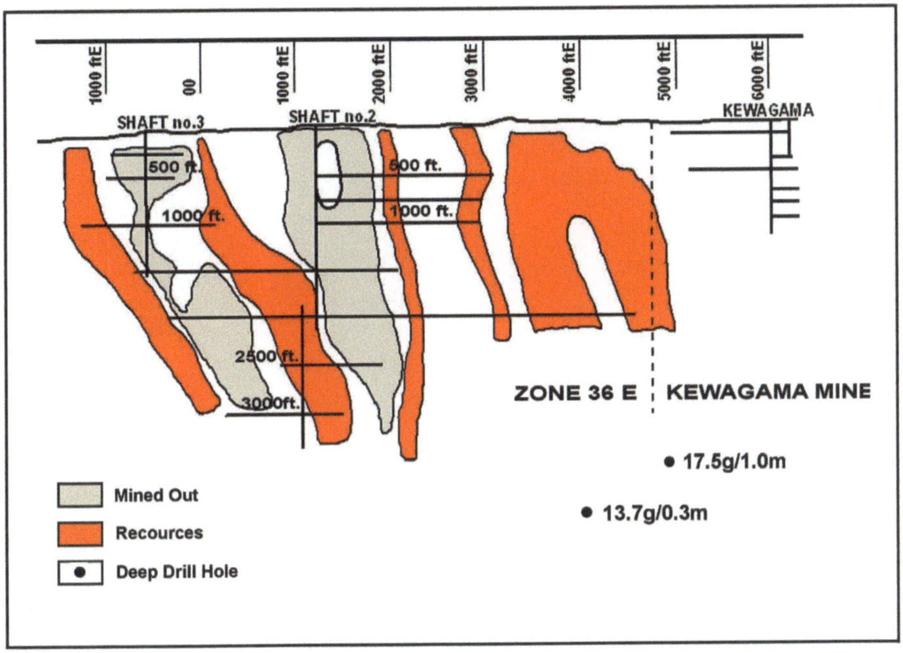

FIGURE 9.23

Long section of the old mine and new 36 Zone showing the original mined zones and the similar pattern indicated in the 36 Zone by the recent drilling. Two deep drill holes support the idea of vertical continuity below detailed drilling parallel to the mined shoots (Compliments of Radisson Mining).

system that radiates from a point some distance below or above the existing mine infrastructure or indicates that two or more vent systems are present. The authors expect the 36 Zone to be a duplicate of the mined zones. The knowledge from the known mine can be used to predict the potential of a new zone and where to concentrate the exploration efforts.

Whether or not more than a single vent system is responsible for the O'Brien mineralization remains to be determined, but a single vent system well below the mine workings could be responsible for the observed pattern.

9.3.9 Consolidated Discovery Mine, Yellowknife

The Discovery Mine is another of the small gold mines that were prominent in the 1930s to the 1960s. The vein occurs in thinly bedded greywacke (siltstone) and is conformable. As Wiwchar (1957) shows, the vein is folded, and has individual ore shoots within a much longer vein that is conformable with the stratigraphy where unfolded; the vein is 225 m long, and has a high-grade center and barren extensions (Figure 9.24). He goes on to say that the vein crosses the fold so that on the 400-m level so that all the economic shoots lie on the left limb of the anticline, which suggests that the vein rakes across the fold hinge

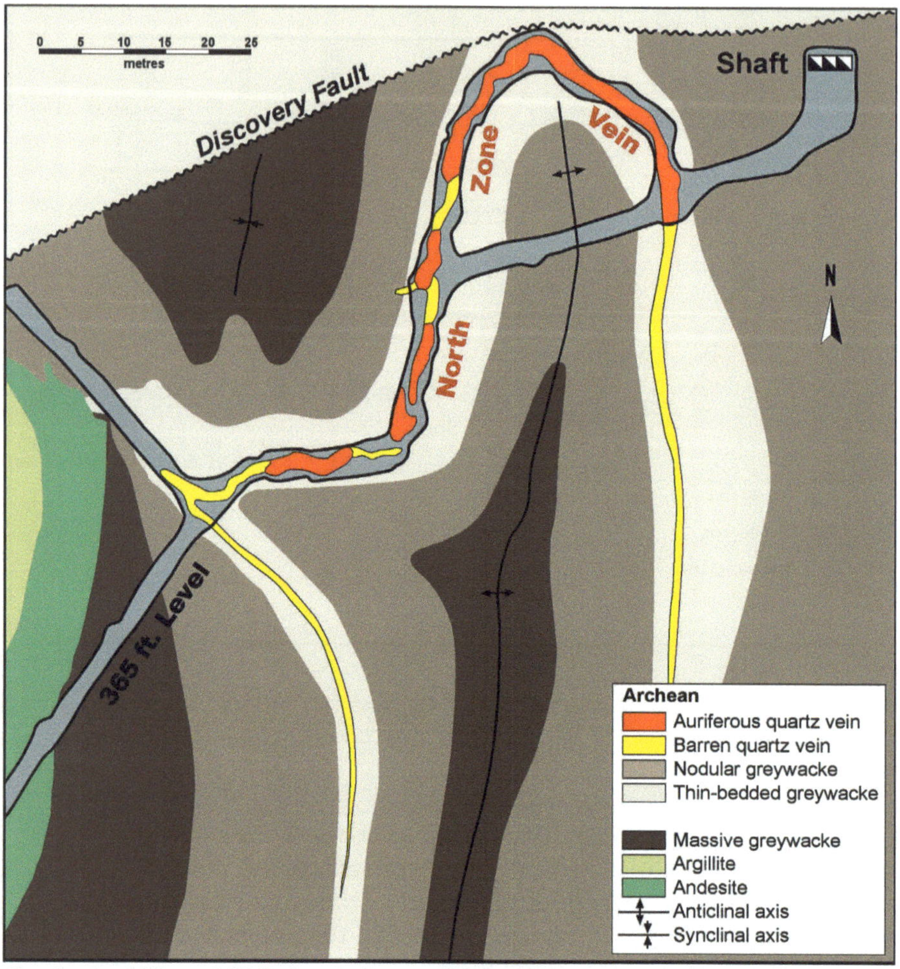

Geological Plan, 365 ft. level, Consolidated Discovery Yellowknife Mine
(after Wiwchar, 1957)

FIGURE 9.24

Consolidated Discovery Mine, Yellowknife. Geological Plan. 365 Level shows an isoclinally folded, turbidite-hosted vein. Notable is a high-grade central portion of the vein and peripheral barren quartz. *Redrawn after Wiwchar (1957), Figure 2 on p. 203 from CIM Special Volume 11, reprinted with permission of the Canadian Institute of Mining, Metallurgy and Petroleum.*

diagonally. In addition, the vein shows many parasitic folds. This combination of features can be best explained by a sheet of mineralized quartz parallel to bedding and folded during the regional deformation of the mine sequence.

All of the above examples are from Archean greenstone belts, but that does not mean that lode gold deposits are restricted to these ancient rocks. Any greenstone belt of similar evolution has the potential to host these deposits. In

1988, McBride was part of the exploration team to evaluate the Addington Gold Mine in southern Ontario, and in 2015 carried out detailed mapping and supervised a small drill program.

9.3.10 Addington Gold Mine

The Addington Gold Mine is one of the numerous small gold mines explored and exploited in eastern Ontario from the 1850s to the 1940s. All of these deposits lie in geological settings similar to the Archean greenstone belts of northern Ontario, but this greenstone belt is Grenville in age, dating from 1.2 to 1.3 Gy (Figure 9.2). The property was discovered in 1881 and exploited on a small scale; in the 1930s, it had a major underground program. Recent exploration has outlined a gold-bearing horizon over 2 km long, which is located at the top of the mafic volcanic rocks where they rapidly grade into a quartz arenite–conglomerate sequence. McBride became interested in this deposit as the geological setting is the same as the Archean Northern Empire described in Section 9.3.1. The Addington area rocks have been metamorphosed to hornblende–chlorite and micaceous schists; the conglomerate pebbles and cobbles have been flattened and stretched in the schistosity. Folding of the main contact is evident and offsets the most northerly shoot D-2 to the west (Figure 9.25).

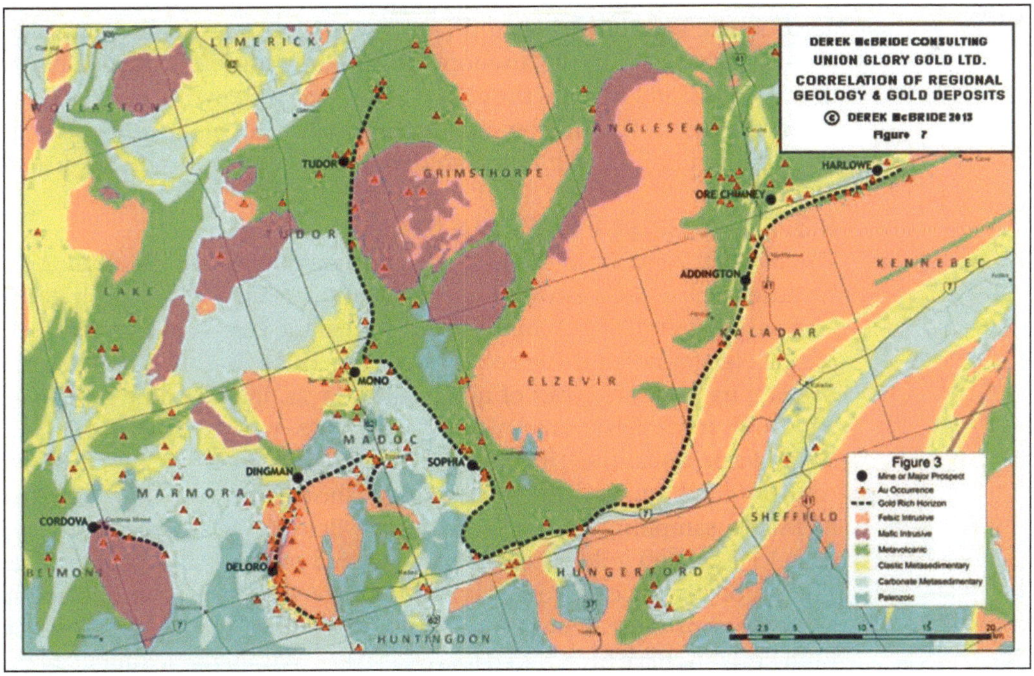

FIGURE 9.25

Map of southeastern Ontario showing the relationship of gold deposits and prospects to the major volcanic–sedimentary contacts.

FIGURE 9.26
Mode of mineralization in the Addington D Zone. The gold is in the arsenopyrite, which is confined to the host rocks. Arsenopyrite is rare in the quartz except where it contains host rock material.

Bedding-parallel and cross-cutting quartz veins are common in all of the rocks and show intense deformation; they are rarely mineralized except with tourmaline. Mineralization consists of gold in arsenopyrite in the transition zone between the two major lithologies. This arsenopyrite is present as flattened grains up to 3 mm and/or bands less than 1 cm thick that follow the bedding (Figure 9.26). All occurs in the rocks and not the quartz. Gold content is proportional to the amount of arsenopyrite A small pit was mined in 1939 that shows an "S" fold; it and the D Zones demonstrate that the mineralized horizon was deformed by the regional deformation.

Five individual mineralized shoots have been defined by drilling and underground exploration. All shoots dip parallel to the main contact and are elongated in the stretching lineation (Figure 9.27). The D Zones are as described above, having gold in the arsenopyrite and only traces of other sulphides; but in the A, B, C, and E Zones the gold is associated with pyrite, pyrrhotite, and minor chalcopyrite. This difference indicates two sources of the mineralizing fluids; are both rich in iron, sulfur and gold, but one also has significant arsenic.

9.3.11 Nugget Pond Gold Mine

The Nugget Pond Gold Mine is of particular interest to the author McBride because he led the discovery and evaluation team. This deposit lies in the Ordovician Betts Cove Ophiolite Complex on the Baie Verte Peninsula of north central Newfoundland. The complex consists of basal ultramafic volcanic rocks overlain by multiple cycles of mafic pillow lavas, intervolcanic clastic sedimentary, and some felsic volcanic rocks. It hosts two volcanic hosted massive sulphide deposits, the Tilt Cove (1854–1965) and Betts Cove (1874–1886) copper (zinc) mines, which have been shown to have high gold content. McBride's sampling of a pyrite dump at Betts Cove returned an average of over 15 g per ton gold. These volcanic hosted massive sulphide deposits lie at the top of the lower pillow lava-sheeted dike sequence. A second pillow lava sequence caps these deposits and is the footwall of the Nugget Pond deposit. Both horizons have been traced for over 10 km. Within the sedimentary horizon, the economic mineralization formed an oval raking to the southwest at 50 degrees (Figure 9.28).

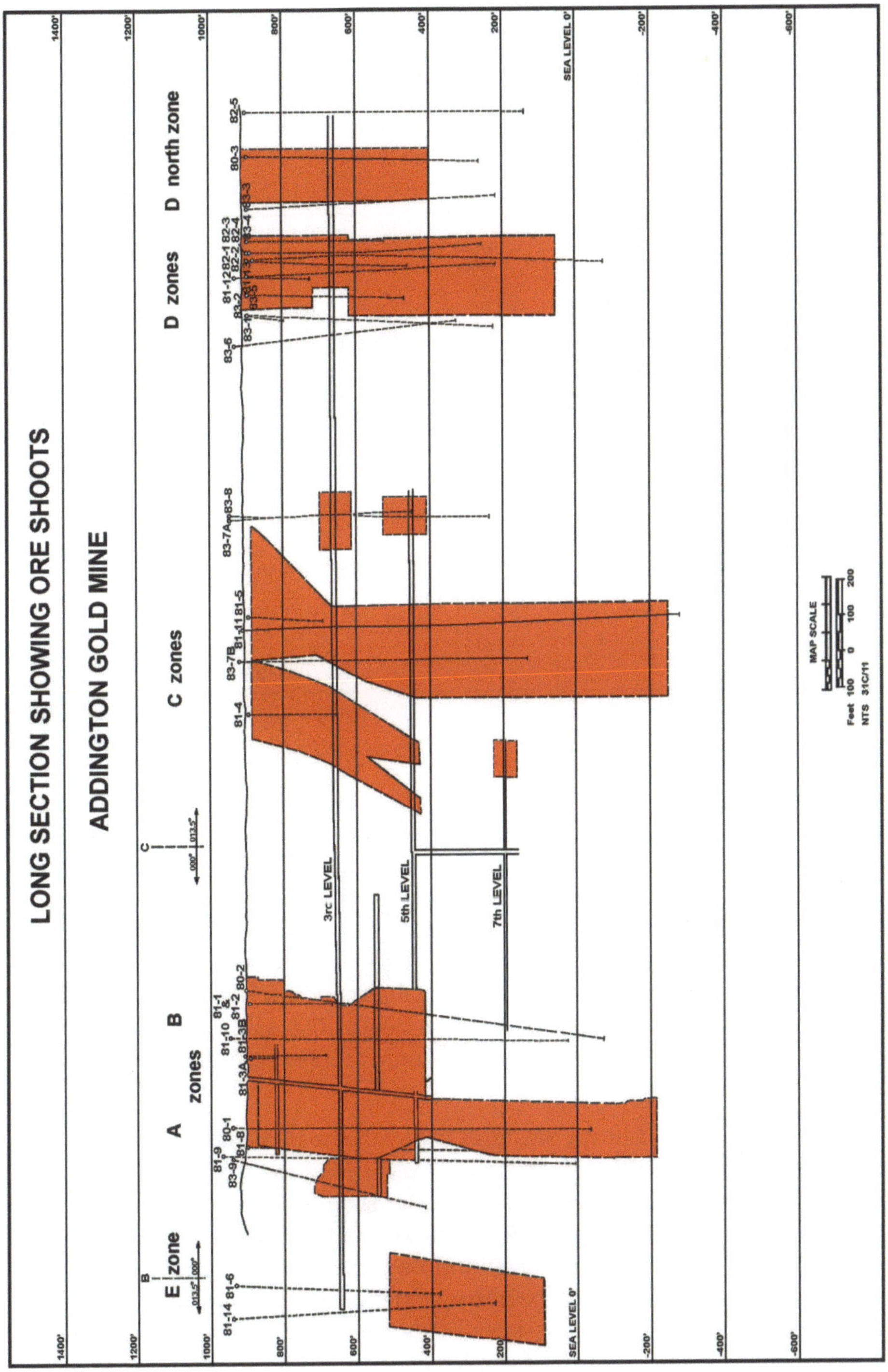

FIGURE 9.27
Long section of the Addington Deposit illustrating the individual mineralized shoots.

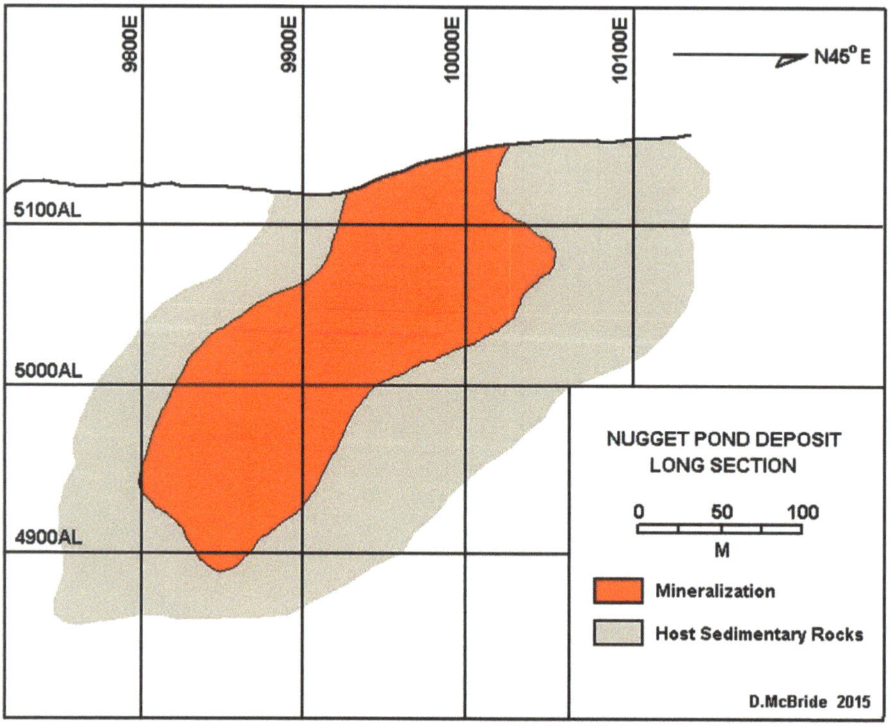

FIGURE 9.28

Longitudinal section of the Nugget Pond Deposit; the deposit dips southeast at 45 degrees and rakes southwest at 45 degrees.

Gold mineralization is associated with fine-grained pyrite and coarse pyrite crystals up to 2 cm in a thin sequence of gray-green to maroon siltstones. Gold mineralization is concentrated in the gray siltstones where they are dark gray and locally graphitic. The grade varied directly as the grain size of the pyrite with grades above 30 g per ton in pyrite cubes over 1 cm; the grade dropped to 3 g per ton in 5-mm pyrite cubes and down to geochemically anomalous in the fine pyrite peripheral to the economic deposit. A 1990 feasibility study estimated the gold resource to be 514,000 tons, grading 14.1 g per ton for 7.24 tons (294,263 oz) of gold diluted at 25% at zero grade. The deposit has been mined out.

Gold grains are found in the siltstone and with the pyrite crystals. Figure 9.29 (center photograph on the cover of this book) shows the typical mineralization and permits one to interpret the system. Across the upper section is a graded pyrite bed of fine-grained pyrite in black siltstone. Clearly this band demonstrates deposition of the mineralization in a sedimentary sequence. Diagonally in the lower-left part is a small vein that contains pyrite crystals. It is interpreted that the pyrite grew in the vent system below the surface and was transported to surface, where it was deposited. Swinden et al. (1990) noted that "there is little

FIGURE 9.29
Sample of ore from the Nugget Pond gold deposit showing a graded bed of pyrite in the upper portion, pyrite cubes in a cross-cutting feeder fracture, and the black siltstone host rock.

evidence of shearing, widespread hydrothermal alteration or metasomatism." They also state that other sulphide minerals occurred with the pyrite cubes and gold, which suggests that they all were deposited together. During the exploration program, a lithochemical study was carried out that showed a sodium depletion halo in the footwall pillow lavas, but none in the hanging wall lapilli tuffs further supporting a hydrothermal vent system below and to the mineralization.

9.3.12 Deposits of the Nova Scotia Goldfields

These Cambro-Ordovician gold deposits have been discussed in Chapter 8, and have been shown to be bedding-parallel and deformed by the regional low-grade deformational event. Most of the deposits have been described by Malcolm (1912). Their regional setting has been described by McBride (1978) for the Ecum Secum Aerea, and by Haynes (1987) for the entire Meguma goldfields. Figure 8.31 shows bedding-parallel veins in the Goldenville Mine with the diagonally cross-cutting ore shoots of Hedley (1941). He adds considerable data to the understanding of the Goldenville Mine. Superimposed on the figure are the ore shoots that rake across the anticline, with the descriptions of the number of veins in each the economic package. Where the ore shoots are not economic, the barren veins still fold around the hinge (Figure 8.26); the structural pattern of harmonic and disharmonic parasitic fold is the same for all veins, demonstrating that the economics of a vein system predates the deformation and is bedding-parallel. Hedley notes that cross-cutting veins, angulars, are mineralized only where they intersect the bedding-parallel mineralized veins. They probably represent feeder veins depositing gold when the vent system intersected the ocean floor.

9.3.13 Faille B Gold Deposit, Haiti

From 1995 to 1997, McBride oversaw an exploration program in northeastern Haiti of a major volcanic system that hosts copper and gold. A geological mapping program confirmed the presence of VMS copper deposits of the Besshi type, low-grade pyritic gold mineralization of the Kuroko type that produced low-grade gold deposits through weathering of the sulphides, and the Faille B gold deposit. The geology of Haiti is very complex; the volcanic rocks have been folded and subsequently chopped up by narrow shear zones, some of which are still geologically active.

The Faille B Gold Deposit is a bedding-parallel, laminated quartz vein that is an extension of a bedded barite deposit. It has been folded so that the vein resembles the Belleterre deposit described in Section 9.3.8. Mineralization occurs in finely bedded barite and a laminated quartz band (Figures 9.30 and 9.31), which have been folded. Drilling had established that the gold system is associated with a felsic volcanic band within a mafic volcanic sequence similar to the Cline Mine (Section 9.3.4). Mapping and trenching has defined this deposit setting in detail (Figure 9.32). The significance of this deposit is that it has the characteristics of Precambrian deposits, but is Cretaceous (70–140my) in age.

FIGURE 9.30

Faille B deposit auriferous bedded barite reported to average 3g/t gold.

FIGURE 9.31

Faille B deposit bedding parallel, laminated and folded auriferous quartz band main zone.

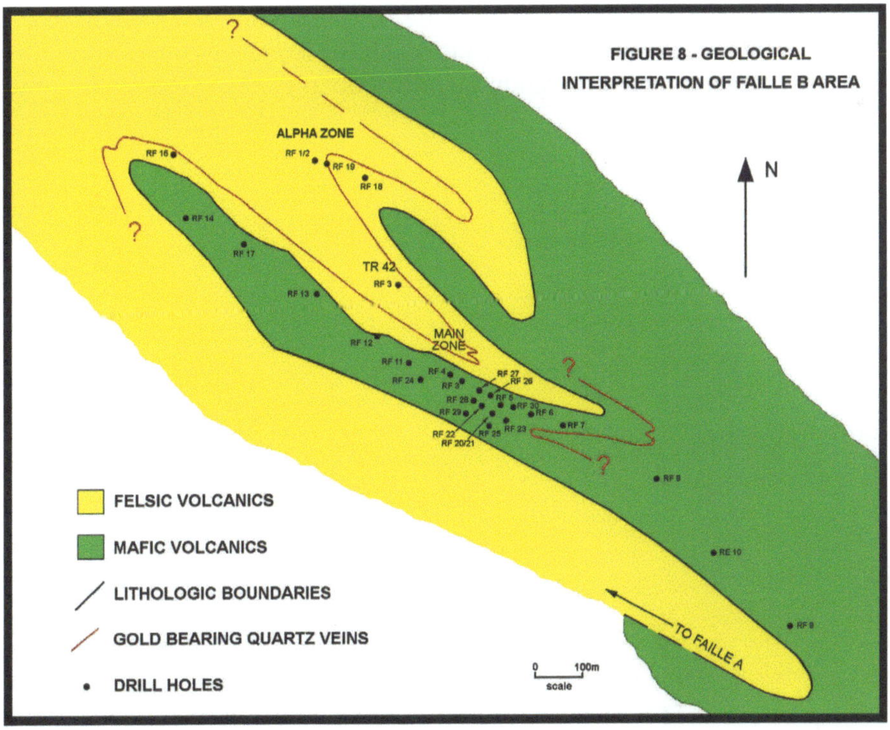

FIGURE 9.32

Plan map of the Faille B Zones.

9.3.14 Other Areas

The previous descriptions have been taken from areas in which the authors have personal experience. They have investigated other deposits, but the studies were never put into the public record. McBride spent considerable time in the Hill End Gold Camp west of Sydney, Australia, which closely resembles the Meguma. The lithologies, veins, and structural deformation are comparable; with the main difference being the north–south trend at Hill End. The stacking of bedding-parallel veins observed during the author's work is the same as Haynes (1987) described for the Meguma.

Exploration in the goldfields of Ghana revealed volcanic–sedimentary conformable mineralization similar to that described in this chapter. The same can be said for the gold deposits in the western Ukraine and throughout the Commonwealth of Independent States. McBride visited some of these deposits and has used the summaries provided by Levitan (2008). Readers can refer to his publication for more detailed information; here it is sufficient to say that the CIS gold deposits fit the summary in the next paragraph.

It is pertinent to any genetic model that all of the above-described gold deposits occur in similar volcanic–sedimentary sequences, regardless of their age, metamorphic grade, and intensity of deformation. The volcanic–sedimentary sequence is the common factor; all subsequent events do not affect the mineralization any differently than they do any other rocks in the system.

CHAPTER 10

Toward a Syngenetic Model and Vent Geometry

CONTENTS

10.1 REGIONAL TRENDS AND THEIR REINTERPRETATION

Grade-thickness contours and the spacing of ore "shoots" or lenses reflect the shape of lode veins and seafloor fracture patterns. Examples were given in Chapter 9 to illustrate the patterns and distribution observed in mined deposits. In smaller mines, such as the Leitch, Sturgeon River, Northern Empire, and Nugget Pond, single vent systems explain the observed mineralization. The Addington deposit with its two mineralogies suggests two vent systems. Mining camps such as Timmins with many deposits suggest many vent systems are present along a common stratigraphic succession. Numerous other examples in mining and exploration company literature including their Web sites provide details and publications of the Canadian Institute of Mining, Metallurgy and Petroleum especially Special Volume 1 and 11.

Spacing between the centers of ore lenses from mines and prospects yields a wide variation. Coumou et al. (2008) provide independent evidence from 3D numerical simulations. Hydrothermal cells in a mid-ocean seafloor setting self-organize into pipelike upflow zones, surrounded by narrow zones of focused and relatively warm down flow. This configuration ensures optimal heat transfer and efficient metal leaching. Concentric flow geometry results from nonlinearities in fluid properties and at high heat-flux conditions, convection naturally forms individual black smoker fields at a regular spacing of 500 m because of the nonlinear temperature and pressure dependence of hydrothermal fluid properties.

The authors conclude that the shape, thickness, and density of lode veins are related to seafloor fracture patterns, bottom topography, substrate porosity, the duration of the vent(s), and sedimentation parameters during accumulation.

Several approaches have been used to characterize the distribution of gold deposits from the perspective of structure (e.g., Carlson, 1991; Blenkinsop and Sanderson, 1999; Bierlein et al., 2006; and references therein). For a simplified

The Metallogeny of Lode Gold Deposits. http://dx.doi.org/10.1016/B978-0-12-803222-0.00010-1

model of the Ashanti Belt, numerical simulations of heat and fluid flow suggest the existence of convection cells within major fault planes, which is reflected by the location of gold mines and alteration indexes (Harcouët-Menou et al., 2009).

Rabeau et al. (2013) characterize regional distribution of gold deposits in the Abitibi Belt by using a stochastic approach. In probability theory, it is explained that random processes are collections of unknown variables and therefore could result in an infinite number of possibilities.

Similarly, Rabeau et al.'s analysis starts with assumptions that:

1. "specific characteristics of known deposits cannot serve as guides for targeting prospective sites since it is common to encounter barren structures that are identical and adjacent to gold-bearing sites;
2. a common genetic process is suggested, however, because there is a strong spatial association with major fault zones, similar metal associations, fluid chemistry, and timing; and
3. major shear zones represent a regional guide for gold deposits in Archean Greenstone Belts because they control the distribution of gold deposits by acting as conduits that channel deep-seated hydrothermal fluids."

Deposits are said to lie along the Cadillac–Larder Lake Fault Zone (CLLFZ) or "Break" (Figure 10.1). Needless to say, there is a large database on where to locate gold deposits, which makes it suitable for using their spatial distribution to help understand the underlying geological formation mechanism. To the north, a second important fault zone, the Porcupine–Destor Fault Zone (PDFZ), marks the contact between the northern and the southern volcanic zones in the Abitibi Greenstone Belt. In the Timmins area, as discussed in Chapter 8, it separates the gold deposits in the Tisdale rocks to the north from the nonauriferous Deloro rocks to the south.

Their study focuses on the eastern segment of the CLLFZ (Break), marking the boundary between the Abitibi (volcanic) and Pontiac (sedimentary) Subprovinces. This segment is about 160 km long and represents almost 80% of the zone's total length. Therefore in the study area, it marks the "break" between volcanism and sedimentation. It hosts numerous gold deposits including several world class deposits. The study focuses on the structural evolution of both the PDFZ and CLLFZs or Breaks, which have been shown in the case of the PDFZ, to postdate the regional structural deformation (Chapter 8 of the present study, Figure 8.8). They state that such crustal-scale fractures acted as conduits channeling deep-seated hydrothermal fluids and represent the main spatial control on deposit distribution within the Archean Greenstone Belt. The problem is that the mineralization has been demonstrated in Chapter 8 to predate the folding which in turn predates the fault zones or tectonic Breaks."

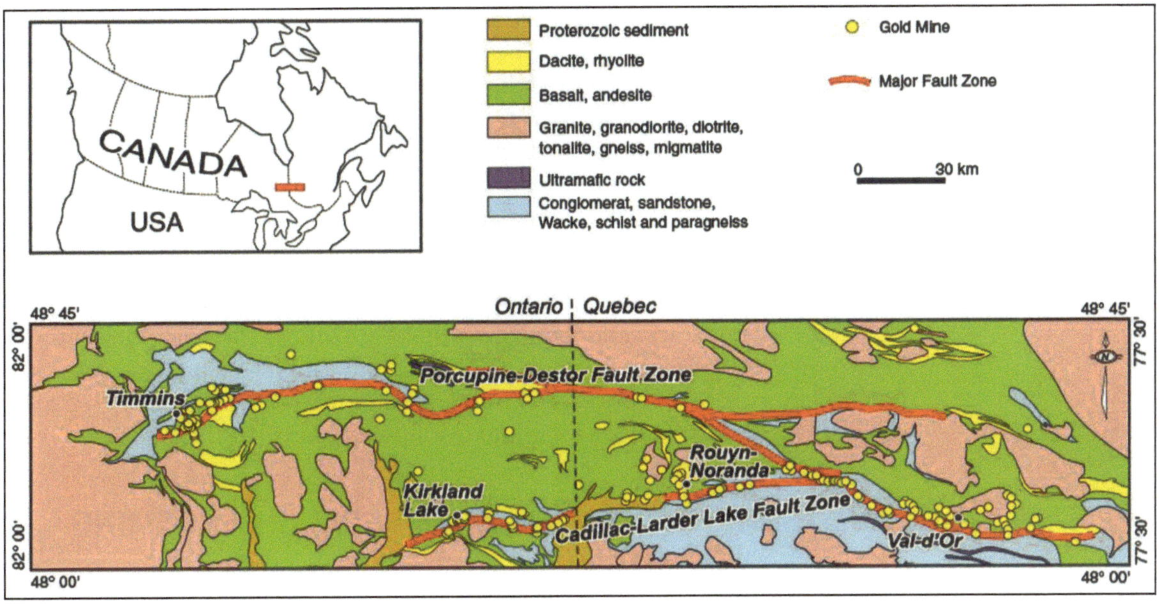

FIGURE 10.1

Location of gold deposits in a segment of the Archean Abitibi Belt of Ontario and Quebec, along the Cadillac—Larder Lake and Porcupine—Destor Breaks. *Redrawn from Rabeau et al. (2013, Figure 1) after Robert et al. (2005).*

This difference does not alter the fact that the deposits occur at intervals along the trace and in the close vicinity of the interpreted "Break" (CLLFZ and PDFZ). Deposits were chosen for which the relative timing was interpreted to be late in the overall tectonic evolution of the greenstone belt and for which the base metal content is low (indicating mineralizing fluids with low salinity).

Rabeau et al. also suggest "that the emplacement of gold deposits within a 1-km range of major fault zones could correspond to the location of structural failures along the main structures representing the mark of a pressure valve release mechanism induced by hydraulic overpressure (Sibson and Scott 1998). A 5.6-km distance obtained as the maximum range of the log-uniform distribution law in the close vicinity of the CLLFZ and DPFZ (PDFZ) might be the maximum length of segmentation along the shear zone."

The study is interesting because it shows how completely different perspectives may be brought to the same problem. It indicates that the mineralization is related to the fault, but it has been demonstrated in the present treatise that the deposits are conformable and deformed by regional tectonic events before the development of the CLLFZ and DPFZ. In the language of syngenesis, the location of deposits is most likely related to the major volcanic—sedimentary cycle, and the weakening of the volcanic activity allowing convection cells and seafloor fracture patterns to remain active for long periods of time

(McBride, 1987). These hydrothermal vents are buried by later volcanics flows, volcaniclastics, and sediments. The second- and third-order "shear zones" are stratigraphic horizons. They overlie or underlie the main ore-hosting stratigraphy and have had the schistosity (shearing) superimposed on them as part of the regional deforming event. Changes in orientation along major shear zones and geological complexity reflect primary facies changes or deformation that postdates to the mineralizing episode.

It has been shown that the CLLFZ zone is generally parallel to the steeply dipping stratigraphy. If the fault predates the regional deformation of the volcanic–sedimentary stratigraphy, to make it compatible with the field evidence presented herein, it must be a flat fault parallel to bedding in undeformed, unconsolidated volcanic, and sedimentary rocks which is a highly unlikely scenario because of the unconsolidated nature of these rocks!

10.2 SCHEMATIC MODELS OF LODE VEIN FORMING ENVIRONMENT

R.W. Boyle presented a model for the formation of lode gold deposits (Figure 10.2). This model includes sedimentation from older basement complexes as part of the basement subsidence by normal faulting. Subsidence produces folding in the volcanic–sedimentary sequence and granite intrusions pushing up along the margins of the deformed volcanic–sedimentary belt. This model is a variation of the sagduction concept, which indicates that all the features one sees in these belts are due to subsidence in a rift system.

Such a scenario is unlikely for a number of reasons. Subaqueous successions are rarely stable on slopes above 20 degrees or less, and extensive slumping of the beds would take place when active faulting occurs. Slump folds would be common in the latter sequences and would be shown to predate the regional schistosity if any developed (Stringer, 1975). Added to this situation, the marginal rocks would be crosscutting intrusive bodies with contact metamorphic aureoles. On the contrary, the Beardmore Greenstone Belt does not have a contact metamorphic aureole; rather, the bedding in the greenstone belt is parallel to the contact with the Onaman Lake Batholith (McBride, 1987), and the top of the batholith is a coarse quartz arenite with crude bedding. The field evidence clearly shows the greenstone belt was deposited on an older weathering Onaman Lake Batholith. In contrast, the late diabase sill that cuts the Northern Empire and Leitch Mines at depth has a pronounced contract metamorphic aureole east of the Northern Empire Mine in which the greenstones are baked into hornfels (McBride, 1981). This model also does not explain the regional schistosity present in all greenstone belts (see Chapter 8).

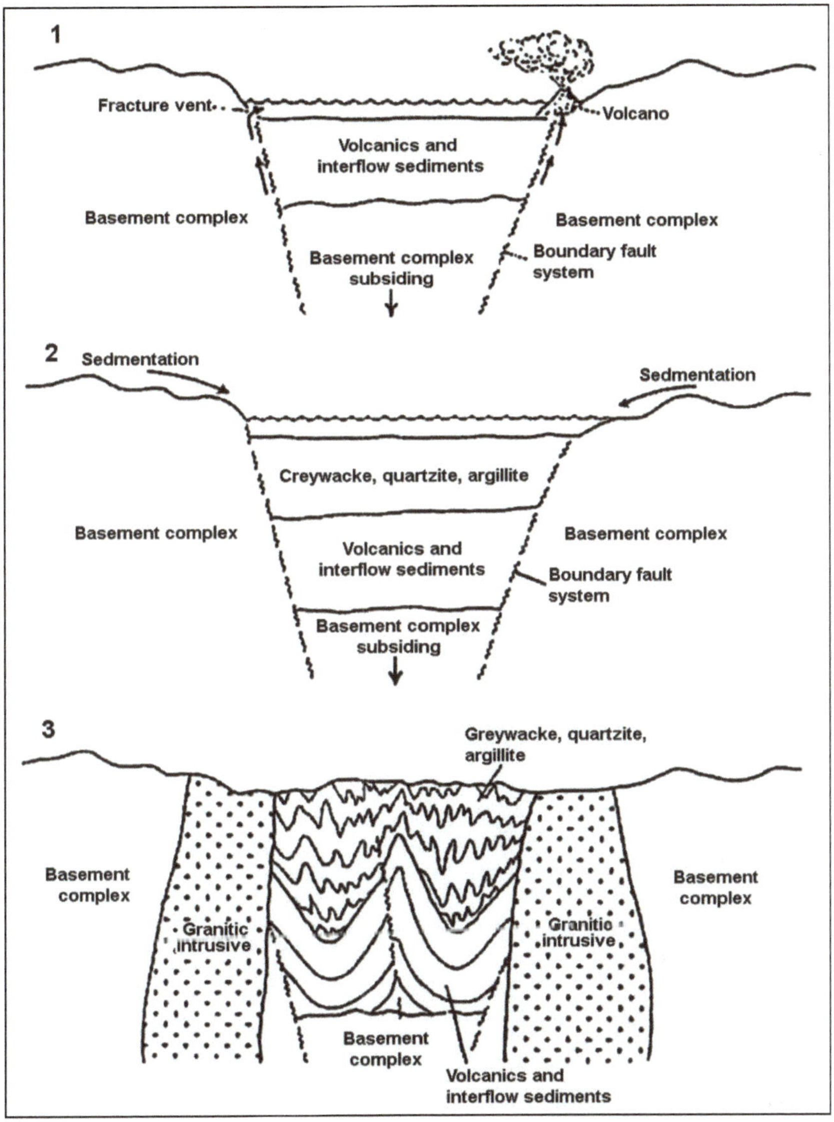

FIGURE 10.2

Development of Archean volcano—sedimentary terrains according to Boyle. The first two-stages reflect syngenetic elements of the current model. Processes illustrated by the third stage are not required and are not explained by the deformation features seen in greenstone belts. *Reproduced with the permission of Natural Resources Canada 2015.*

Alternatively, McBride (1987, see Chapter 1 and Figure 9.1) developed a model based on his experience in the Appalachian mobile belt (McBride, 1976a). It incorporates all the features observed in field mapping the rocks in the Beardmore area and northeast where the greenstone belt is very narrow, but also includes all the features of the larger parts of the belt. To differentiate between prospects and mines, a 50,000 ounce (1600 kg) production cutoff was used. As a result, the major gold deposits were found in a limited section of the greenstone belt centered on the main volcanic–sedimentary contact. The authors have been able to extend this model to include most major gold deposits in Archean Greenstone Belts.

Other authors have added details on how vent systems may be developed on the ocean floor. Turner and Gustavson's model appears in Figure 10.3.

Figure 10.4 illustrates the principle settings for major lode gold deposits: a base of predominantly mafic volcanic rocks with minor sedimentary and intervolcanic, exhalite beds overlain by more sedimentary units, exhalitic iron carbonate beds, and gold-bearing siliceous units (Fyon and Crocket 1982). These exhalitic rocks may be deposited in the upper margins of the mafic volcanic cycle or in the overlying sedimentary rocks. Felsic volcanic crystal tuffs and related rocks are commonly present and may have been porous enough to have gold deposited from percolating hydrothermal solutions from seafloor vents.

This figure covers the environments discussed for the deposits in the Beardmore area discussed in Chapter 9 and it can be applied equally well to any gold mining camp in any Canadian Greenstone Belt regardless of the age.

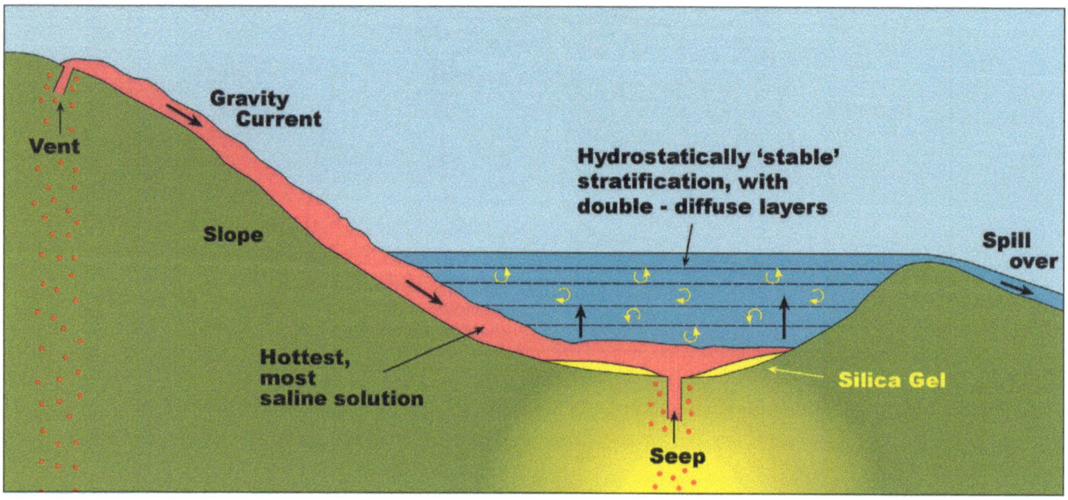

FIGURE 10.3

Seafloor deposition geometry. Dense hot saline hydrostatically stable stratified layers undergoing double-diffuse cooling pond in hollows. Silica gel (and gold) precipitate through cooling by diffusion and mixing with fresh seawater. *Redrawn from Turner and Gustavson (1978).*

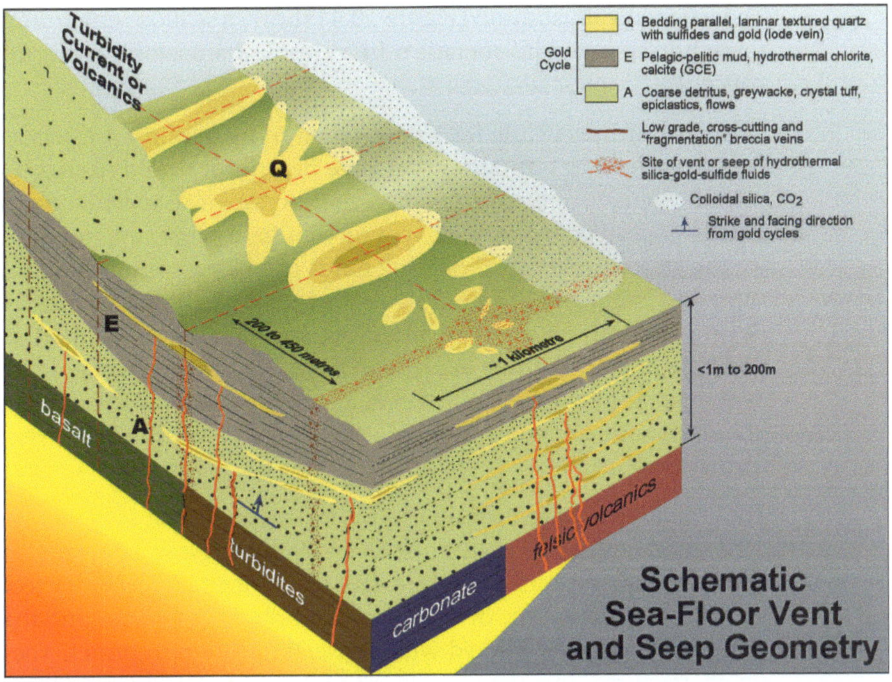

FIGURE 10.4

Schematic diagram (not to scale) illustrating the main processes responsible for formation of the majority of lode gold vein deposits (Q) in a sea-floor hydrothermal vent and seep environment. Hydrothermal fluids generated by circulating seawater deposit silica and gold onto the sea-floor substrate (A) which may be 1) sediment (in turbidite environments) 2) bi-modal felsic, where periodic basaltic and felsic volcanic/intrusive rocks can bury the hydrothermal source or 3) rarely carbonate. *Fracture spacing, lode shapes and other details are discussed by Kretschmar (2011).*

10.3 A SYNGENETIC MODEL FOR FORMATION OF LODE VEIN DEPOSITS

A syngenetic model for lode vein gold formation is based on direct observation: readily verifiable field relations and plausible P-T-X formation conditions. The model is entirely process-based and is comprised of the following basic elements:

1. lode gold veins formed in an open seafloor environment;
2. quartz and gold are everywhere generated and deposited by means of similar thermochemical processes;
3. these processes were in operation throughout much of geologic time, and despite secular changes over time are in operation today;
4. all deposits can be shown to predate the regional deformation that affected their host rocks and rotated most of them into moderately to steeply dipping deposits.

The schematic illustration (Figure 10.4) shows a seafloor environment in which these processes are in operation in a simple framework which unifies genetic concepts.

Localized prospect-scale heat-flow, when in operation for extended periods of time, may lead to formation of a large deposit or a gold district. Deformation, metamorphism, and superposition of processes lead to structural modification and geological complexity of vent or vent field geometry, but are not essential for the generation of gold-bearing fluids or to upgrade low grades. Mineral textures and chemistry of interflow sediments are frequently virtually unchanged during metamorphism, and interflow sediments may contain elevated gold and base metals (Bierlein et al., 1998b) as a ready source of gold for lode veins. Sequence stratigraphy (Thurston et al., 2008), basin analysis, paleogeography, and deciphering of volcanic architecture assumes a critical importance in a synformational origin.

The ubiquitous nature of quartz in, for instance, the Meguma terrain (up to 15% of a stratigraphic column at Sheet Harbour, Nova Scotia and 21% on Softwood Island) combined with the virtual absence of crosscutting feeders strongly suggests that the source of the silica-rich brines was diffuse and that large areas of the seafloor were covered with dense, hot silica-rich brines. The brines came up in fractures, incipient faults, porous substrate and they were ponded in depressions with dimensions ranging from millimeters to meters to kilometers. Silica precipitated episodically as the concentration and temperature of the brines fluctuated in response to diffusion, adiabatic cooling, seawater dilution, bottom currents, and fluctuations in original brine composition.

10.4 LODE GOLD DEPOSIT CLASSIFICATION

The issues underlying gold deposit classification have been well addressed by Poulsen et al. (2000). The syngenetic perspective presents a unifying process which can help our understanding by deciphering the end members present. If one considers Figure 10.4 as representing the types of economic gold deposits found in the volcanic–sedimentary setting in greenstone belts, the criteria of McBride (1987) will act to guide future investigations. It is the goal of the exploration geologist to find mineral resources for the future. One must use all the local information available to select targets, once an area of investigation is selected. It has been noted that historical data from mines have been ignored by many recent researchers. All are important. For instance, an empirical model can eliminate large areas of rock in a greenstone belt. The Red Lake Camp Greenstone Belt covers a large area, but if one looks for the economic gold deposits as a guide and draws a line from the McKenzie Island Mine to Campbell

Red Lake-Goldcorp to Starratt-Olsen-Madsen and back to Mackenzie Island, the gold production in that triangle is over 99.9% of the gold in the camp from only 15% of the greenstone belt. One would be well advised to stay within or near this triangle if he wishes to optimize his chances of exploration success.

Chapter 8 has shown that deposits in greenstone belts have common features which have been shown to predate the regional first period of deformation. It does not matter if they are veins, iron formations, low-grade porphyritic bedding crystal tuff (porphyry), massive sulfide, or gold deposits. The scope of this work does not permit wholesale examination of all deposits and can only hope that the reader will see from this treatise that future studies are approached from the perspective that, unless proven otherwise, the exploration program should contain a large syngenetic component.

10.5 OBSERVATIONS TO AID EXPLORATION APPLICATIONS

As Robert et al. (2007) point out, exploration consists of defining the footprints of known gold deposits and integrating various techniques with geology for their efficient identification and detection. It is apparent from reviews of gold deposit models (Chapter 1) that formation theories are subject to fashion and syngenetic concepts are by no means new. However, the concepts of gold cycles (GCs) and syngenetic seafloor formation may be considered new cornerstones of lode gold deposit genesis and bring into focus volcanism, sedimentation, and seafloor P-T vent discharge parameters. Each vein can be viewed as a stratum in effect, a paleo-P-T-X fingerprint that reflects fluid flow and records changes in the earth's heat budget.

The conceptual reorientation that took place when syngenetic concepts were integrating into gold formation models was discussed by Hodgson (1988) and summarized in Chapter 1. Additional new and powerful tools arising from the concept of GCs, vein asymmetry, and the recognition that many lamprophyres are Gold Cycle "E" (GCE) units (first presented by Kretschmar (2011)) carries with it exploration implications that are now presented in greater detail.

10.5.1 Top Determinations and Structure from GCs

The Maskwa Batholith and Little Bear Lake lode veins were discussed in Section 4.3. The suggestion that "lamprophyres" in the Little Bear Lake area are GCE units was discussed in Section 5.3.1. Observations provide important exploration guidelines for intrusion-hosted lode vein deposits. Equally important is the structural information provided from a sketch geological map of a small open pit from the high-grade Black Beaver vein (Figure 4.18).

The sketch shows the upper part of a "quartz—GC" equivalent to "E_C" (Figure 3.1). By being aware of the cyclical and geopetal nature of the deposition process, and keeping the lode vein formation model in mind, it is possible to tell stratigraphic tops from this map, at a single glance.

This observation is interesting considering that the veins are considered to be within an "intrusive" batholith (Duguet et al., 2007; Yang, 2013; Yang and Gilbert, 2014). The vein-hosting horizon is folded and a major fold structure is implied. As a guide to exploration, additional veins should be sought in an E-W corridor north or south of the existing veins. The 2830 Ga Maskwa Batholith may primarily consist of a series of felsic flows, exposed as older basement in an erosional window and the favorable horizon may be repeated by folding. Only detailed mapping will decipher this potential.

The presence of GCE units in the Bourlamaque Batholith was discussed by Kretschmar (2011) and presents an excellent example of the powerful tool provided by GCE asymmetry in unraveling regional-scale structures. The GCE top determination method survives high-grade metamorphism; thus, preserving the primary characteristics that can be used to direct exploration.

10.5.2 Stratigraphy, Age Dating, and Regional Exploration Guidelines

Implications of a largely syngenetic origin of lode gold veins include the use of volcanic hiatuses and correlative conformities for exploration as suggested by Thurston et al. (2008) for VMS and komatiite-associated Cu-Ni-PGE (Platinum Group Exploration) mineralization. The same criteria may be used in the search for gold deposits (McBride, 1987). Sedimentary interface zones between volcanic units represent time gaps or changes in deposition conditions. Therefore, one can correlate gold-bearing stratigraphy at the group, formation, and member-level using the sedimentary interface zones or graphitic argillite marker units at geochemical discontinuities in stratigraphy. In the Abitibi Belt, for example, felsic volcanic rock dated at 2740 Ma, the age of Côté Gold mineralization, should be sought because this age was a period of known hydrothermal activity in which proximal to intermediate felsic volcanism accompanied accumulation of economic concentrations of gold at the Jerome Mine 20 km to the west in the same greenstone belt. Just as integration of banded iron formation and komatiites as marker units into stratigraphic models permitted establishment of belt-wide stratigraphy (Thurston et al., 2008), GCEs have the potential to provide similar information at the local deposit-scale level.

On a larger scale, the maturing of the volcanic belt can be seen in the change from magnesium-rich mafic volcanic rocks to iron-rich ones in the terminal phases of the mafic volcanic activity. Beds rich in iron carbonate exhalite are common near the top of the mafic rocks in Timmins as discussed by Karvinen (1978, 1980, 1982). Karvinen (1980) noted that the ratio of gold to silver in the area varies from 5:1 near major porphyries up to 14:1 in deposits away from porphyries. He also noted that numerous thin, stratabound, ankeritic carbonate horizons throughout the stratigraphy from the Lower Member to well above the Upper Member indicate a proximal vent area where hydrothermal activity was more continuous and intense than in distal areas. He suggested using the Fe/Mg ratio of carbonate. The amount of carbonate was so obvious that it was incorporated into the name of the Buffalo Ankerite Mine and its bedding-parallel gold deposits (see Chapter 9.3.6.1). Iron carbonate was an important accessory mineral in the folded gold-bearing strata in the Ecum Secum Area of the Meguma (McBride, 1978).

10.5.3 Vent Tracing

A vent tracing technique using a proportion of free quartz (veins or layers) in a given stratigraphic column was suggested by Kretschmar in 1983. The technique involves measuring the amount of quartz (Q) and the thickness as well as the proportion of (E) (pelagic—pelitic lithologies) and (A) (basal coarse sandstone or greywacke) in stratigraphic sections or drill holes and plotting these on a triangular Q-A-E diagram. A vector may then be identified to the area of increasing hydrothermal silica layers in the stratigraphic section, which may indicate proximity and direction to a heat source or vents or seeps.

Hydrothermal quartz in turbidites can provide further insights into the turbidite deposition environment (see Shanmugam, 2006, Figure 1) as it has been investigated in detail by Fyon and Crocket (1982). They recognized two alteration environments. One was a subsurface system where the hydrothermal fluids did not reach the seafloor and did not produce economic gold production, but may define a vent system. The second was seafloor zones where the hydrothermal fluids are discharged into the hydrosphere and are spatially associated with felsic volcanic complexes and auriferous cherty dolomite. These rocks contain most of the economic concentrations of gold. They also suggest that the characteristics indicate the hydrothermal fluids were of seawater origin.

10.5.4 Stratigraphic and Sedimentary Control

For Meguma veins, stratigraphic control has been discussed by McBride (1987) and Kontak et al. (2013). For Victorian gold deposits, stratigraphic control has been discussed by Phillips and Hughes (1998). In Nova Scotia, there is a 500 m stratigraphic interval with a greater concentration of deposits than in the outside sections of the succession (Sangster and Smith, 2007, Figure 3, see summary of McBride in Sangster and Smith). A periodicity to the occurrence of gold districts might become apparent when stratigraphy is unraveled. If the Meguma Supergroup formed in a rifted basin (Waldron et al., 2009), perhaps the rifts are the loci of hydrothermal vents and seeps. Lambeck et al. (2010) successfully used sedimentary geochemistry and Nd isotopes in Northern Australia to identify favorable gold-rich stratigraphy.

10.5.5 Volcanic Facies Mapping

Volcanic facies mapping can be used to delineate distal or proximal environments. Paleovolcanic facies generally comprise large contiguous areas, and the boundaries between them are intercalated (Roobol and Hackett, 1987). Williams and McBirney (1979) define MJ as the maximum or average size of juvenile pyroclasts in each outcrop or area. Basin analysis using volcaniclastic fragment size and VMS vent tracing techniques (Fox, 1977) can be used to characterize the environment and place lode veins in a stratigraphic context. Most deposit investigations do not include systematic mapping of the deposit or its surrounds and assume that the earlier regional mapping is accurate, up-to-date and all-inclusive. McBride's (1981) mapping of the Northern Empire added considerable detail to the regional map and showed that the gold-bearing horizon had been traced by the earlier explorers for kilometers east of the mine. It was combined with geophysical surveys to demonstrate that the volcanic belt was a major anticline. Underground mapping differentiated the change in host rocks the mineralized shoots and the barren areas between them (McBride, 1987).

Exploration projects for lode gold deposits do produce some geological maps from new surveys. These are combined with geophysical and geochemical surveys to define targets. Numerous types of geophysical surveys are usually included, but the geological map is rarely upgraded with these data or fill-in mapping. The result is the loss of critical information for future investigations, especially if it remains in company files and is destroyed. Many of the authors' field investigations are lost because companies or projects cease to exist. Nevertheless, detailed field studies are required to understand the setting of the mineralization and its position in the regional geological setting (McBride, 1976a,b, 1987).

10.5.6 Using GCE and Quartz Compositions

Since the geochemistry of GCEs in a given stratigraphic column closely reflects deposition processes, as was noted in Section 3.2 and Figure 3.1 and as is also described for Carlin-type mineralization (Large, 2010), GCE geochemistry should yield information about proximity of hydrothermal vents. The major and trace element chemistry and volatile content loss on ignition (LOI) of GCE reflect marine sedimentation as well as ongoing simultaneous precipitation of elements from hydrothermal solutions. Variations in mineralogy and chemistry of GCE should provide a vector to hydrothermal vents.

10.6 AREA SELECTION

Existing gold exploration techniques, e.g., Milkereit (2007), can be supplemented and oriented toward syngenetic concepts. In a general case, the authors suggest following several steps to evaluate the exploration results before making a decision whether to continue or abandon the exploration project, which eventually should lead to the profitable mine. The steps are the following:

1. Recognition of geological structure, lithology, stratigraphy, and geological evolution of the region. It is usually summarized on the geological (lithostratigraphical) maps' scale, geological cross sections with a real depth, and explanations.
2. Development of metallogenic, mineralogenic, antracogenic etc., maps registering all mineral deposits and indices (e.g., characteristic host rock alteration, facie/zonation, gossans, geochemical and geophysical anomalies, etc.) known in the region and beyond. Definition of geological features correlated with the deposits/indices (metallotects) and statistical analysis of the results.
3. Analysis of geological potential of new discoveries based on (1) and (2) and the preparation of a map showing target areas (item 3).
4. Developing quantitative (tonnage, grade, geometric size) models of ore deposits potentially found and realize probability of finding the exploitation parameters (known from current mining practice).
5. Comparison of map (3) with a map of protected areas and ecological projects to recognize potential environmental conflicts. Categorize the areas for free/conflicting. In some countries, this applies also to social attitude.
6. Comparison of map (3) with infrastructure, water, energy, workforce availability.
7. Application exploration methods suitable for outcropping deposits/ ore bearing formations/metallotects and the blind ones.
8. Prepare criteria for distinguishing ore bodies/mineral deposits (cut-off grade thickness, depth, mineralogy, physical properties) from "barren" rock.
9. Calculation of geological resources, mean content of useful component, and other key parameters specific to minerals/industrial rocks/fossil fuels.
10. Recognition of technological ability of producing salable mineral raw material, the parameters of concentration/refining process, and finally the percentage of valuable components recovered from 1 ton/m^3 of ore. The steps 9 and 10 are parallel.
11. Recognition of hydrogeological and engineering conditions at the mine site and the approximate costs of mining, dressing,

infrastructure costs. Economic, financial availability, and risk analysis of alternative mining projects.

12. Determination of economic geological and technological criteria for chosen mining project. Calculate the ore reserves and develop optimal parameters for mine development, exploitation, processing run of mine etc., and problems with waste disposal.

Specific situations may include more steps.

1. The work of Thurston et al. (2008) provides a stratigraphic framework for gold deposits and the search for them in the Abitibi Belt and all Archean greenstone terrains. Those segments of greenstone belts that already have gold mines, significant prospects, or are of similar age are prime targets.
2. In turbidite environments, quartz veins are a general first-order criterion—evidence that the sedimentary prism reached the critical silica-dissolution temperature and those hydrothermal silica-rich fluids were channeled along fractures or emanated from diffused seeps. Within turbidite sequence, the presence anywhere, but in particular lower in gold district sediments, of sulfidic, calcareous, graphitic intervals increases the potential that the quartz will be gold-bearing (Kretschmar, 1983; Large et al., 2007).

10.7 THE NONBORING BILLIONS

The concept of boring billions originally addressed the apparent absence of life in the mid-Proterozoic (1.8 to 0.8 billion years ago). The appearance of terrestrial red beds and the disappearance of detrital pyrite beds indicate oxidative processes in the terrestrial environment after about 2.4 Ga and a permanent rise in atmospheric O_2 concentrations above the very low Archean levels ($<0.001\%$ of the present atmospheric level). Planavsky et al. (2014) and Goldfarb et al. (2001) use it to point out the apparent scarcity of economic gold deposit formation in this same period. Reasons given include the lack of preservation and changes in plate tectonic activity (Bédard and Harris, 2014).

Considering the vast increases in isotopic age dates and dating of deposits, an obvious tool for narrowing down target selection is to look for lithologies that have the same age as known deposits. We can call these the nonboring billions.

Lode gold deposits do occur in rocks of the boring time period and even much younger as demonstrated in Chapter 9. One of the reasons major deposits are lacking may have more to do with the size of the greenstone belt and its ability to generate and maintain larger hydrothermal systems.

10.8 TONALITE, TRONDHJEMITE, GRANODIORITE SUITE LITHOLOGIES AS TARGETS: EXTRUSIVE OR INTRUSIVE?

As was evident from the Little Bear Lake and Cote Lake examples, (Chapter 4.3) and further discussion of tonalites, trondhjemites and granodiorites (TTGs), the presence of GCEs in TTG suite lithologies represents a new type of large low-grade gold deposit. Evidence presented in Chapters 2 and 4 suggests an intrusive volcano–sedimentary origin, which has wide implications for TTG rocks, Archean crustal processes, the source of gold-bearing hydrothermal fluids, and in particular exploration for felsic-hosted gold deposits. Also see the discussion for the Missanabie Lake Batholith (MLB) later in this section.

GCEs may provide facing directions, as they are often asymmetric, which constitutes a previously unrecognized facing direction indicator that survives high-grade metamorphism. The local evidence for regional-scale folding in the Chester Complex, the Bourlamaque Batholith (Belkabir et al., 1993), and the Maskwa Batholith (Kretschmar, 2011) attests to its utility to unravel stratigraphy and structure.

Cross sections, longitudinal sections, and grade-thickness contours provide information on the 3D shape of lode veins, vent field geometry, and the spacing of local and district-scale fracture patterns. Much readily accessible information is contained in NI43-101 reports filed by public gold companies (see www.sedar.com).

The Ontario Geological Survey exploration recommendations for 2013 (http://www.mndm.gov.on.ca/sites/default/files/recommendations_for_exploration_2013.pdf) address the gold potential of TTG phases of the MLB (Pace, 2013).

This document states that:

> Recent gold discoveries involving brittle–ductile shear-hosted gold quartz vein deposits in trondhjemitic–tonalitic phases in (1) the Marmion Lake Batholith at the Hammond Reef project of Osisko Mining Corporation, and (2) the Côté Lake deposit within the Chester Granitoid Complex, owned by IAMGOLD Corporation, have given prospectors a new model when searching for gold deposits in Ontario. The Missanabie Lake Batholith (MLB) should be considered a target, based on the new model.
>
> The Missanabie Lake Batholith is part of the Wawa Domal Gneiss Terrain (WDGT) which represents an easterly transition from predominantly greenschist facies metavolcanics rocks of the Michipicoten Greenstone Belt to tonalitic, trondhjemitic and granodioritic (TTG) gneisses of the WDGT. Gold mineralization lies outside the Michipicoten Greenstone Belt, and is concentrated within the WDGT.

The rocks of the TTG are coarse-grained, foliated, and weathered pale-grey, and exhibit well-developed layering consisting of varying proportions of mafic minerals, generally biotite. The trondhjemitic–tonalitic phases host a series of west to northwest-trending quartz veins that cross-cut the regional metamorphic foliation as a result of brittle–ductile shearing. The vein structures provide a means of structural control for the gold mineralization in the area. Alteration within the TTG consists of hematite, giving the rock a distinctive reddish colour, as well as sericite, quartz and pyrite. Disseminated to massive pyrite, with accessory galena, molybdenite and chalcopyrite in the shear-hosted quartz veins are characteristically associated with the gold mineralization.

Sage and Heather (1991)

The TTG phases of the MLB in Leeson and Brackin townships have been known to host economically significant brittle–ductile shear-hosted gold quartz vein deposits. The past-producing Renabie gold mine operated in Leeson Township for over 40 years, producing over 1 million ounces of gold.

There are 37 known gold occurrences hosted in the TTG phases of the MLB; all are associated with west to northwest-trending quartz veins hosted in brittle–ductile shear zones. The Renabie Main Zone, C Zone, Nudulama and the Braminco zones are some examples of the larger gold mineralized zones identified in the TTG" (see Table 10.1).

An alternative model can be proposed based on the work by the authors in the Chester Complex and the work of McBride in the Renabie Camp from 1984 to 1986. His work as an exploration geologist at the Renabie Mine confirmed that the Renabie mineralization formed a distinct horizon that was folded about an axial planar schistosity (shear zone) as a north, northwest, steeply plunging anticline and that the quartz horizon had a granular (sugary) texture (Bruce,

Table 10.1 Tonnage and Grade Estimates for Some of the Larger Mineralized Zones in TTG Lithologies

Name of Mineralized Zones	Township	Tonnage Estimate (tons)	Gold (oz/ton)	Gold (g Au/T)
Renabie main zone	Leeson	1,000,000	0.2	6.2
Nudulama	Leeson	579,325	0.194	6.0
Braminco zone (comprised of #21 vein, #7 vein, and B vein)	Brackin	100,000 (#21 vein)	0.15	4.7
		23,000 (#7 vein)	0.31	9.6
		5000 (B vein)	0.26	8.1

Pace and McMillan (2012).

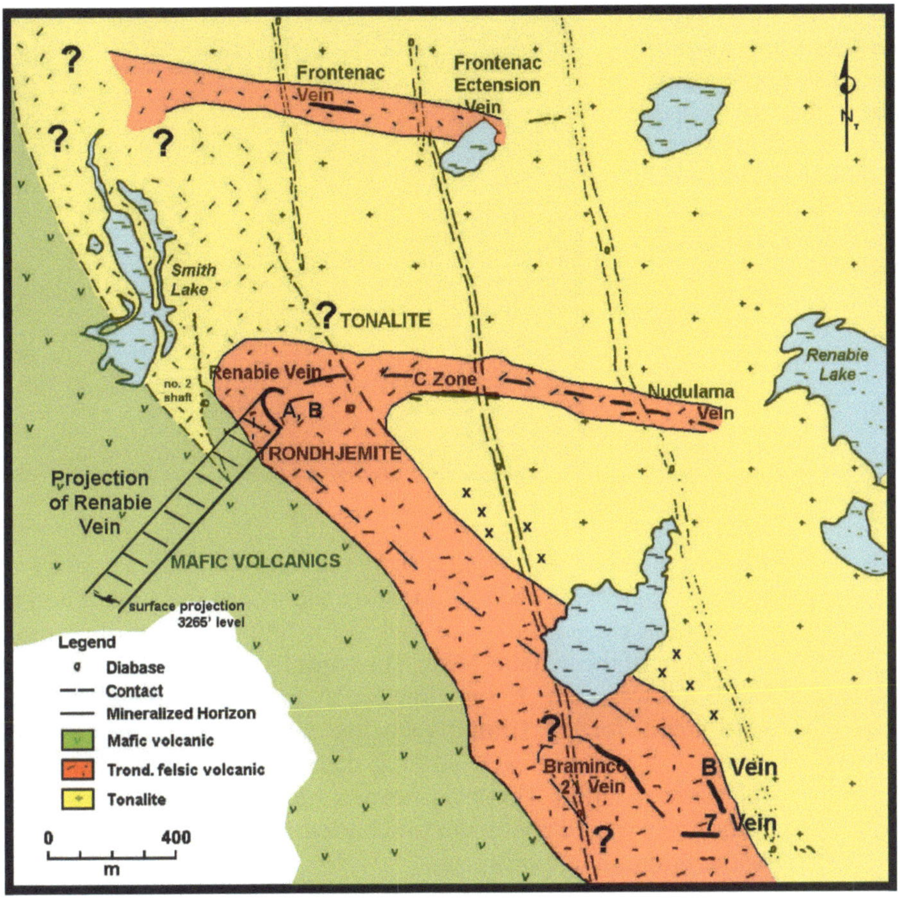

FIGURE 10.5

This revised map of the Renabie Area that shows the gold zones and incorporates the observations on the main mine and prospects by McBride during the period 1984–1986. At this time all the main prospects were accessible underground or stripped clean. A single folded horizon interpreted that hosts all the main gold zones. *After Ferguson (1968) and Callan (1991).*

1948). Mineralization in all the larger zones occurs with a sericite schist host; the limits of which are show on Figure 10.5. Gold mineralization was directly proportional to disseminated pyrite content in the quartz. None of the geologists, past or present, have mentioned any contact metamorphic aureole in the adjacent mafic volcanic rocks, which lie conformably on the MLB contact. The Campbell Vein lies conformably 90 m above (west) of this contact as a bedding-parallel quartz band approximately 1 m wide (Ferguson, 1968). Equally important is the presence of a pebble conglomerate strata of MLB rocks just above the base of the mafic sequence (Figure 10.6).

FIGURE 10.6
A photograph taken in the 3100 level exploration drift, Renabie Mine in 1984, showing a basal conglomerate, folded tuff beds, and fine-grained tuffs of fine-grained trondhjemite which host the nearby gold zones. These bands form one of many beds in the mafic volcanic tuffs a few metres above the contact with the MLB. The wire square is approximately 10 cm.

Beyond the Renabie property to the east, the Nudulama deposit is a continuation of the Renabie geology; McBride supervised the surface and underground exploration program on it as well as the Braminco and other prospects to the southwest of the Renabie Mine. The Nudulama deposit is identical to the Renabie, but lower grade. It was mined in the mid-1980s for a smelting flux, with a gold credit, for the Timmins smelter. Gold is associated with disseminated pyrite. A 30 m northwest striking diabase dike has had no visual effect on the mineralization. This gold-bearing horizon continues east to the A and B veins prospect and appears to fold back through the No. 7 vein to the northwest as the Braminco 21 prospect.

The Braminco area was drilled and explored underground under McBride's supervision. Three mineralized zones form an echelon pattern with the Braminco Zone being the most important. It was mined by a small open pit which had exposed the polymetallic gold-bearing mineralization disseminated in a conformable quartz zone. These zones were interpreted as a parasitic Z fold on the west limb of the north plunging Renabie anticline. This horizon continues southwest, parallel to the mafic contact through five more gold prospects to the Boundary showing 7.5 km beyond the Braminco Prospect (Bennett, 1978). Primary layering in the TTG on Crooked Lake (similar to that in Bennett's Photo 5 of layering in the MLC) was interpreted by McBride in 1984 to be of tuffaceous origin.

The continuity of a mineralized horizon parallel to the conformable contact with the adjacent mafic volcanic rocks for such a distance suggests stratigraphy in a continuous succession of rocks that share extrusive characteristics.

The ability to accurately interpret areas within the deformed TTG phases of the MLB can be paramount in identifying further exploration opportunities. Older maps must be updated, and more detailed mapping is required in the mineralized areas to develop effective exploration programs.

West of Thunder Bay, Ontario, the Tower Mountain Complex (TMC) has many of the Chester Complex or MLB characteristics. McBride has worked extensively on this complex including mapping the various old showings. He concluded that the TMC was an intrusive–extrusive complex with zones of disseminated gold mineralization and some high-grade veinlets. Recently, his work had been confirmed by Gelinas (2015). More information on this prospect is available on the Ontario Government Web site (Chataway, 2005, 2007).

These are low-grade gold deposits that share many common features, suggesting that they can be treated as "type of gold deposit" with its own exploration potential. They contain both intrusive and extrusive parameters.

10.9 "SHEAR ZONES" AND FAULTS

Examination of "shear zones" is important in order to see whether they have the characteristics of GCE division, i.e., are they bedded units, foliated tuff, or if they show evidence of graded bedding. Stratigraphic principles can then be used. While it is obvious, evidence for faults is that they must displace lithological units! Otherwise, they are better designated as lineaments and their significance must be further investigated.

The influence of shear zones on a mineralization must be determined by a structural analysis similar to those mentioned in Chapter 8. Differentiation from regional deformation as described in Chapter 8 and its influence on previously deformed rocks, including mineralized ones, is essential to understanding the genesis of lode gold deposits.

10.10 A SYNGENETIC CLASSIFICATION OF LODE DEPOSITS AND PROSPECTS

A classification of ore deposits that provides an essential framework for resource-assessment studies, designing exploration strategies, and evaluating prospects should be a coherent classification of gold deposits that will take into account massive amounts of new information. Some of the considerations in the classification of deposits for the generation of an exploration philosophy should contain as many as possible of the parameters in Table 10.2.

Table 10.2 A Syngenetic Classification of Lode Gold Deposits and Prospects

Type	Source of Heat	Primary Source of Metals and Fluids	Primary Deposition Environment	Prominent Examples
Turbidite	Basin	Shales, and quartz arenites host sediments	Open seafloor	Meguma, Ballarat
Tuff/ sediment hosted	Volcanic pile	Volcanic pile	Open and sub-seafloor	Hemlo
Sill or flow	Internal magmatic	Strata above or below	Open seafloor or internal	Kalgoorlie, San Antonio, Rompas komatiite hosted
Mafic volcanic	Magma	Circulating meteoric –magmatic water	Open ocean floor	Timmins, Beardmore area, O'Brien mine
Felsic or bimodal	Magma	Circulating meteoric –magmatic water	Porous crystal tuffs on or just below the seafloor	Bourlamaque Pierre Beauchemin

10.11 USING THE SYNGENETIC MODEL FOR TARGET SELECTION

The above table shows the basic observed relationships in host rocks and the gold deposits they contain. Steps to assist in area selection:

1. take an area with regular fracture spacing and regular spacing of volcanic vents (gold prospects)
2. assume 5 km depth of crust, which represents a maximum mining depth
3. assume each volcanic vent has a signature that can be recognized within a radius of 1 km and depth of 250 m
4. fold this surface by lateral compression to produce isoclinal folds
5. erode the surfaces in (4) by taking horizontal slices
6. count the theoretical deposits that occur within the practical mining depth
7. search for areas within the favorable rocks that have the desired characteristics, but lack significant exploration.

Targets can further be defined within the general area selection by researching these excellent sources of information. One should look for local geological similarities, alteration patterns, extent of overburden, and gold prospects or indications. The discovery of the Nugget Pond Deposit (Chapter 9) is an example of the success of this approach.

CHAPTER 11

Comparisons, Conclusions, Suggestions for Further Work and Application of the Syngenetic Model

11.1 INTRODUCTION

The beginnings of this book and the syngenetic concepts that form its basis were formulated when the authors first saw excellent glaciated seashore and ridge outcrops in Nova Scotia. The field relationships between quartz and turbidite sedimentation in the Cambro-Ordovician Meguma sediments suggested that they were deposited simultaneously. This conclusion seemed simple, as well as intuitively obvious and pointed to a model for lode gold vein formation in which hydrothermal silica and gold-rich fluids were deposited in seafloor hollows from vents and seeps. The authors were not aware at the time of the extreme polarization of views on gold vein genesis.

Perhaps the reader will consider that the evidence presented is plausible, even though a syngenetic model is at variance with much historical and current literature. A "syngenetic mindset" (i.e., the willingness to consider syngenetic concepts) permitted the conclusion that graded bedding in "intrusions," e.g., Chester and other TTG (tonalite–trondhjemite–granodiorite) complexes means that parts of them are water lain. The concept of "Gold Cycles" (GCs) is a natural consequence of commonly observed asymmetric alteration, horizontal and vertical mineral changes, layering, obvious zoning of veins, and regional setting. It is necessarily gravity induced. An excellent exposure, fortuitously unearthed during stripping and washing of outcrop, the "Two Lamps" outcrop constitutes essential and solid proof that so-called lamprophyres in many gold camps are interflow sediments (Gold Cycle "E" or GCE of the present work). This interpretation requires that the host rocks are extrusive, which turns attention to the nature of felsic volcanism.

In terms of existing concepts for lode vein formation, the model proposed in this work is revolutionary. It adds to the concept presented by Kuhns (1986), which determined that mineralization occurred prior to the peaks in

CONTENTS

The Metallogeny of Lode Gold Deposits. http://dx.doi.org/10.1016/B978-0-12-803222-0.00011-3

deformation and metamorphism. This treatise has demonstrated that lode gold deposits can be shown to predate all deformation and metamorphism.

Even for the single gold deposit type of lode veins, there is lack of agreement because theories are commonly based on reviews of reviews or data that lump deposit types together and reuse stacks of assumptions that need revision. There is a paucity of comprehensive deposit and camp-scale studies using syngenetic ideas to distinguish between cause and effect. One reason for confusion, however, is that existing models are faith based and the geologist's interpretation of the outcrop and drill core will therefore be based on tenets of that faith.

Perhaps the pendulum is swinging in the syngenetic direction again. At the least, syngenetic concepts are worth reexamining in light of current developments in the earth sciences. The evolution of new earth science concepts and difficulties faced by the innovative thinker, which were well addressed by Woodall (1983), are perhaps even more applicable today.

Each of these points has been addressed in detail throughout the book.

The interpretations presented herein follow from detailed outcrop mapping, careful logging of core, underground, mining property, and regional geological mapping and the resulting increase in skepticism toward a dominant role for epigenetic processes. GCs are observational, field-based, and readily verifiable. Their integration into models of lode vein formation in felsic rocks, however, requires navigating a hinterland of preconception.

This is more difficult because quartz units (QU), common in the Archean, rarely have been detected on the modern seafloor. While there is much robust, thermodynamic fluid inclusion and isotope data on quartz, the understanding of the physical–chemical conditions of trapping, and precipitation of gold-bearing hydrothermal fluids in an open or sub-seafloor system is incomplete.

Several aspects of syngenetic concepts for gold formation are discussed by Hodgson (1990) who points to an element of fashion in formation theories. The seminal work of Dick Hutchinson, summarized in Hutchinson (1993) and the geochemical bases, in essence completely described by Kerrich (1983), presented a simple, comprehensive model for lode veins in a mafic volcanic environment. In particular, the suggested passive discharge of fluids over broad areas, the formation of auriferous exhalites (GCE of the current work), and their dilution by clastic or pyroclastic detritus is integral to a complete model of lode vein deposit formation. McBride's (1987) regional study puts the formation of the GC into the evolution of a greenstone belt, thus making it part of the "the system" and not a postdepositional addition.

Equally important to the argument are the many common features gold deposits share with volcanic-hosted massive sulfide (VMS) deposits and their history, which was discussed in Chapter 8, and the similarities between gold deposits of all ages, as discussed in Chapter 9.

11.2 ANALYSIS AND COMPARISON TO EPIGENETIC CONCEPTS

GCs and a syngenetic seafloor formation mode for lode gold veins extend existing formation models and permit unification of genetic interpretations. Additional tools to aid in the search for lode gold are inherent. Focus is on sedimentation, seafloor P-T, and vent discharge parameters. Each vein can be viewed as a stratum—in effect, a paleo-P-T-X fingerprint—that reflects fluid flow and marks changes in the earth's heat budget. Syngenetic concepts permit narrowing the gaps between disparate genetic models by stressing the following parameters:

1. GCs systematize the description of lode gold veins in graded volcanic and clastic rocks and constitute evidence for subaqueous formation.
2. The presence of GCs in the TTG suite is proof of volcano—sedimentary origin, as part of the body with wide implications for the origin of TTG, Archean crustal processes, and the source of gold-bearing hydrothermal fluids.
3. GCs provide previously unrecognized facing direction indicators that can be used to unravel stratigraphy and structure, even in high-grade metamorphic terrain.
4. Most lode veins are silica crusts (QU) deposited from hydrothermal solutions emanating from faults and seeps on tuffaceous/pelagic/pelitic seafloor substrate which is subsequently buried by periodic influx of detritus.
5. Cross sections, longitudinal sections, and grade-thickness contours provide information on the 3D shape of lode veins, vent field geometry, and the spacing of local and district-scale fracture patterns.

11.2.1 Discussion

11.2.1.1 Nomenclature and Classification

In Chapter 1, arguments are presented as to why some geologists feel the syngenetic model cannot be valid, including the following:

1. The term "orogenic" accepts that the premise that orogenic deformation processes are essential for gold deposit generation.
2. There may be veins that crosscut stratigraphy and formed later than the enclosing rocks, but this must be clearly evident from mapping including a structural analysis and drilling. Age relations should not be assumed.

3. The scale of crosscutting is important. Small-scale remobilized veins commonly result from compaction and deformation. Soft-sediment deformation leads to local crosscutting features. In proximity to vents, vein orientation increases in complexity.
4. The similarities of the environment of lode gold deposition to that of VHMS deposition are not important.
5. Detailed structural analysis is not important to the understanding of source and timing of fluid generation and deposition.

11.2.2 Conclusion

The present observations provide a new approach in the quest for understanding lode vein formation. At least the authors' data and its analysis rekindle discourse on syngenesis. At best, the simplicity and predictive aspects inherent in GCs may provide the prospector another key to Pandora's box, as suggested by McBride (1987).

The debate of lode vein models would benefit from efforts to sort out primary seafloor processes from later overprinting, deformation, and metamorphism. Such investigations must start with careful field work, core logging, and study of the geology, including detailed field mapping of the deposit and its surrounds.

One must reexamine the tenets of the creed of epigenesis, which predisposes conclusions in many studies. For instance, the vertical changes in mineral chemistry and textures described for the Bogosu and Presea deposits and interpreted in terms of the crustal continuum model by Mumin et al. (1994) can be equally interpreted as horizontal and lateral zoning from seafloor vents and evolving hydrothermal fluids.

Once a primary syngenetic component for lode vein genesis is accepted, new insights are brought to other debates. The homogeneous character and distinct oxygen isotopic composition of metamorphic hydrothermal fluids (Ridley and Diamond, 2000) reflect host rock substrate from which constituents are leached under conditions of elevated geothermal gradient. A magmatic isotopic signature is entirely consistent with exsolution of fluids and hydrothermal liquids in a proximal felsic volcanic environment (e.g., Renabie Mine, Ontario as well as other examples discussed by Spooner (1991), Hagemann et al. (2006), and the current study of Côté Gold, Kontak et al., 2013). The concept of fertile upper-mantle source region (Hronsky et al., 2012) is also accommodated. Syngenesis may well be the elusive Higgs Boson of vein-dominated gold deposit formation.

11.2.3 Common Features of Present Models

The syngenetic and orogenic gold models share all the same scientific databases. These measurements will not change because of the interpretation, but the interpretation can change using the same data. Chapter 7 shows the variability of the temperatures and pressures of solutions emerging from seafloor vents. Many of these vents are precipitating mineralization in their proximity. These temperatures span the range of metamorphic temperatures given to the "shear zone" hydrothermal solutions. In the case of the Mattabi Mine in Ontario, precursor compositions give rise to andalusite and kyanite in the footwall alteration vent system when the regional metamorphic grade is lower greenschist (Mumin et al., 2007). In Chapter 7, seafloor conditions were presented that gave rise to massive sulfide deposition; it is apparent that similar conditions could give rise to other deposits that share the same general geological setting. Chapter 9 shows how many gold deposits share these common features. It becomes apparent that all aspects of the deposit and surrounding geology must be fully understood to decipher the origin of the deposit and use this information to guide future exploration.

11.2.4 Uncertainties

Various uncertainties come into the equation and understanding their derivation and how they form part of the overall picture is essential. Some of them are discussed below:

1. Vein formation pressures from fluid inclusion data. How do they fit into various models and can they be explained by syngenesis?
2. Age difference between veins and host rock: it is difficult to ascertain what is actually being dated or if there is good evidence that metamorphism or proximal intrusion has reset the ages. The best tool is the deformation textures (See Chapter 8).
3. Alteration, where the widespread replacement/sulfidation presents a mineralization style in which it may be difficult to distinguish between syngenetic features, epigenetic features, or metamorphic recrystallization. Studies in VMS deposits have shown that primary features are preserved during epigenesis and metamorphism as illustrated in Chapters 8 and 9.

11.2.5 Environment and Hydrothermal Streaming

In the candy-stripe outcrop (Figure 6.12) and along the Chester Rd, we see thin pyrite-silica veins or alteration selvages to felsic flows (with 1 g Au/T assays); other bands carry erratic gold values on the south side of the trend (Figure 4.14). These may lie in the bedding plane or appear to crosscut stratigraphy. We see them both underlying and overlying the Côté Lake stratigraphy, so they reflect a process that accompanied felsic flow extrusion.

11.2.6 Timing of Vein Formation and Emplacement

Foliated wall rock inclusions in the quartz feldspar volcanic rocks have been used as evidence that they postdate the regional penetrative deformation. Many of these inclusions have the regional schistosity superimposed on them as the same rocks outside the vein and in the same direction, which suggests that they predate the deformation. Many of the same veins show folding and boudinaging (Figures 2.8(a,b) and 9.3). The "serrated" parasitic folded contacts between the wall rock and the QV (Figure 2.19), together with folding (see Chapter 8) of the QV, indicate that the veining took place prior to the deformation that produced the schistosity and illustrated features noted above. These features are a product of the observed deformational and depositional fabrics. Any feature between the two would be preserved as well. Since none are recorded, none exist. The observed deformation is the first. It can be found throughout the deformed succession; thus, it represents a regional event and not a zonal one. Chapter 8 describes shear zones and shows that they postdate the regional deformation and everything that goes before it.

11.2.7 Aspects for Which a Syngenetic Origin is better

As discussed in the previous sections and in detail in Chapters 8 and 9, the vein timing in gold deposits has been demonstrated to predate the deformation or does not appear to have been significantly impacted by deformation. Alteration of the surrounding rocks, as seen on porphyry copper deposits is lacking; Asymmetric footwall alteration (Valliant et al., 1982) is similar to that seen in VMS deposits, which suggests a similar origin for the deposits. The lack of isotopic equilibrium between quartz veins and their host rocks, cited by Bierlein and Crowe (2000), also argues for syngenesis. A syngenetic origin to lode vein formation is internally consistent, coherent, simple, and allows insights into the following:

1. Vein-scale to kilometer-scale ankerite-siderite spotting e.g., Bierlein et al. (1998a),
2. Evolution of fluids from oxidizing to reducing in Archean gold camps and hydrothermal systems (Golding et al., 2011),
3. Nonequilibrium of stable isotope signature between wall rock and lode veins (Bierlein and Crowe, 2000), Kerrich (1993),
4. Asymmetric veining and alteration (Crocket et al. (1980), Dziggel et al. (2009), and Wulff et al. (2010)),
5. The single premise, which may be considered entirely subjective, that quartz textures in Meguma turbidites reflect interplay of hydrothermal quartz deposition during coincident turbidite sedimentation has led to new insights on lode vein formation,
6. Without what might be termed a "syngenetic mindset," recognition of the important, but very subtle, distinction between lamprophyre and

pelagic–pelitic interflow sediments (GCE units) may have continued to be misinterpreted.

The common occurrence of GCE and graded-bedding interflow sediments has led to unsuspected details about earth history, which must be integrated into gold deposition models. GCE is a new diagnostic tool that opens windows to felsic volcanism, quartz in lode gold deposits, seafloor hydrothermal systems, and TTG suite lithologies. The relative roles of structure, stratigraphy, and large-scale geologic processes in the formation of lode gold deposits as discussed by e.g., Barley and Groves (1992), Groves et al. (2003), and Kerrich et al. (2000) are bound to benefit from syngenetic concepts.

Bierlein et al. (2000) question why rich lode gold deposits are virtually absent from the Caledonides and the Alpine fold belts, and their answer is that these fold belts are the result of continent–continent collisions which do not favor the development of major lode gold metallogenic provinces. Lithospheric delamination to initiate devolatilization of hydrated ocean crust is limited, and largely transpressional tectonics are some of the possible causes that account for the deficiency in lode gold mineralization in internal orogens. The answer may be as simple as "the rocks are wrong." Most lode deposits occur in mobile belts consisting of large volumes of oceanic volcanic rock with lesser amounts of associated clastic sedimentary rocks. As demonstrated by Bilibin (1968), the host rock controls the deposit type.

11.3 DISTRIBUTION OF DEPOSITS THROUGH TIME

Meyer (1981) discussed the apparent absence of orogenic gold lodes dating between 2.4 and 0.3 Ga and suggested that the 2 Ga years gap in ore formation may reflect:

1. lack of gold-enriched source areas in host terrains or
2. the lack of the enriched source areas in host terrains or
3. the lack of the required type of hydrothermal process throughout most of the Proterozoic and into the Phanerozoic.

Chapter 9 shows that mobile belts of all ages have similar lode gold deposits. The authors argue that the age of the mobile belt is not as important as its size. Gold deposits, albeit small, occur in the same geological setting in the Grenville rocks of southern Ontario, which are dated at 1.2–13 Gy. Included are the gold deposits at Eldorado, Deloro, Bannockburn, and the Golden Fleece (Addington) which is discussed in Chapter 9. In the Ukraine, similar gold deposits are shown to be bedding parallel in rocks of 1.6 Ga, Maiskoye Deposit, and 1.7 Ga, Klintsy Deposit (Levitan, 2008). None of these deposits have been exploited, but all were drilled and have significant resources. In eastern

Canada, the Nugget Pond deposit occurs in Ordovician rocks. The observations of Swinden et al. show that the deposit is tilted but does not contain any "shear zone" characteristics. A similar lode gold deposit is described from Haiti in Chapter 9, which confirms that auriferous producing systems produced similar deposits in seafloor rocks of all ages. The Faille B deposit in Haiti is a very young deposit (see Chapter 9). One may add that the exploration and development of these younger deposits have a lot to do with the local political situation, the Ukraine and Haiti being examples.

The common feature in these deposits is the host rocks! The most likely controlling factor on the size of the lode gold deposit is the size of the volcanic belt and its ability to supply the required metals. Most of these deposits formed on the ocean floor as seen from their association with fine-grained volcanic and sedimentary rocks; they would have been rapidly buried and were protected from erosion. Only later by regional deformation were they subjected to uplift and erosion. By this time the lode gold system was part of the deforming sequence as demonstrated in Chapters 8 and 9.

According to Goldfarb et al. (2001), the development of major lode gold deposits center on tectonics, source rocks, thermal regimes, and structure for understanding the distribution of orogenic gold deposits. They point out that apparently, orogenic gold shows a spatial association with collisional orogens and wonder whether Phanerozoic style collisional plate tectonics continue back into the Archean. They state that the paucity of "orogenic gold" in the Proterozoic may be due to:

1. a lack of trans-crustal structures, the conduits for large-scale ore fluid flow,
2. a scarcity of dissolved sulfur and gold in the ore-forming fluids (Loucks and Mavrogenes, 1999) and lower overall fluid volumes (Fyfe and Kerrich, 1984),
3. the absence of higher thermal gradient needed to drive crustal fluid migration,
4. a consequence of nonpreservation of gold mineralized orogens at particular times in Earth history.

All these parameters are explained by the relative sizes of the Archean Greenstone Belts relative to the more recent ones such as in the Grenville of Ontario, Central Mobile Belt of Newfoundland, or the mobile belt in Haiti. One can go further and show that the large Abitibi Belt of Ontario and Quebec has produced many more and larger lode gold deposits than the much smaller Beardmore-Geraldton Belt north of Lake Superior.

A summary of the investigated parameters and their various interpretations is shown below in Table 11.1.

Table 11.1 Simplified Comparison of Key Model Characteristics Illustrating How the Same Observation May Be Interpreted from Epigenetic and Syngenetic Perspectives

Interpretation Characteristic	Epigenetic Interpretation	Syngenetic Interpretation
Geology		
Geological contacts	Structural transposition of different lithologies or locus of shearing and deformation	Stratigraphy: key defining variable
Features that may be hundreds of kilometers long and several hundred meters wide	first order or transcrustal structures	Mainly stratigraphic horizons
Fault zones	Transcrustal fault zones, ore bodies are situated along second-order and third-order subsidiary faults	May be time-equivalent stratigraphic horizons
Lower order faults	Lower order faults are dominated by brittle–ductile shear zones	Ore generation not related to brittle–ductile deformation or faults
Structure		
Steep dip of veins	Due to contemporaneous deformation	Due to later, regional folding
Foliation	"Shearing" or differential movement	Superimposed and due to deformation
"Deformation" zone	Closely spaced faults reflect cumulatively broad zone of deformation	Faults must offset stratigraphy "deformation zones" may be disconformities or strata or part of regional deformation
Quartz		
"Ribbon" texture veins	Parallel to foliation or XY plane of finite-strain ellipsoid	Bedded quartz (QU), parallel to bedding, reflect sedimentation and hydrothermal deposition conditions
"Ladder" texture veins	Reflect compressional deformation and brittle failure of banding	Elongation during deformation and small tension gashes in stratabound unit
Laminated vein texture	Crack seal origin due to structural, brittle–ductile regime	Stratigraphic layering due to hydrothermal, geochemical, and sedimentary variables
Foliated wall rock inclusions in veins	Implies that quartz postdates regional deformation	May be contemporaneous sediments or clasts
Laminated quartz-carbonate with breccia fragments	reflect repeated fracturing and sealing in hydrothermal systems over several million years	May be crosscutting feeder veins, or primary synsedimentary beds (QU) with rip-up clasts
Quartz	Forms as quartz	Mainly forms as gel, which recrystallizes to quartz, some quartz forms from hydrothermal solutions

Continued

Table 11.1 Simplified Comparison of Key Model Characteristics Illustrating How the Same Observation May Be Interpreted from Epigenetic and Syngenetic Perspectives continued

Interpretation Characteristic	Epigenetic Interpretation	Syngenetic Interpretation
Ore Formation		
Deposition depth	3–30 km lithostatic depth	Fluids form in deep reaction zones but deposition is on the seafloor under hydrostatic conditions
Seismicity	Correlates with majority of gold-forming events	Has no necessary role
Association with felsic or porphyritic intrusions	Intrusives supply heat which leach gold from host rock	Intrusives may supply heat engine, or intrusives may in part extrusive as flows or more likely crystal tuffs
Timing		
Timing of vein formation	Later, epigenetic	QU are beds formed at same time as host rock unless they are clearly crosscutting veins
"Veins"	Gold-bearing extensional veins	Commonly synformational quartz beds (QU); crosscutting veins often feeders
Graded bedding	Difference competency with adjacent unit	Indicates subaqueous formation and deposition
Relationship to orogeny	Key variable	If related must be deposit specific
Extensional or compressional environment	May be either	Mainly extensional but interpretation must be deposit specific
Rheological heterogeneities serve as favorable ore host rock	Structural	Incidental to ore forming processes
Similar or Same		
Mineralization down plunge continuity	1.0–2.5 km	Variable up to 6 km, commonly 1–2 km depends on original shape and deformational elongation
Fluid source	metamorphic or magmatic	Magmatic and/or meteoric
Gold source	Black shales, fertile crust, felsic intrusions	Black shales, fertile crust, exsolution from felsic volcanics, leaching from host volcanic pile
Overprinting or telescoping	Exactly the same	Vein complexity increases with proximity to vent and introduction of lithic material from exterior sources
Fault thrusting of high grade over lower grade rocks	Inherently characterized by abundant auriferous quartz veins	Much later tectonic event; must be deposit specific to be valid
Small vein systems, scattered through greenschist facies rocks	If no relationship to first-order faults, ore bodies are small and subeconomic	The economics of quartz vein system is related to gold in the hydrothermal system, deposition and erosion/preservation parameters
Relationship to metamorphic grade	Ore deposits form subsequent to development metamorphic mineral show overprinting features	Ore deposits formed at same time as enclosing rocks. Show same metamorphic features. Primary textures are generally preserved and assist of timing each event

Table 11.1 Simplified Comparison of Key Model Characteristics Illustrating How the Same Observation May Be Interpreted from Epigenetic and Syngenetic Perspectives continued

Interpretation Characteristic	Epigenetic Interpretation	Syngenetic Interpretation
Uncertain or Equivocal		
Vein age dating	If age younger than host, considered proof of epigenetic origin	Must be deposit and technique specific, many dating methods have large errors associated with them and it is important to know what is dated
Metamorphic grade	Key variable	Not important, greenschist facies reflects primary seafloor alteration and deposits shown to not vary over large ranges of metamorphic grades
Dilational jogs	Key localizing variable for major concentrations of mineralization	Not important, may reflect vents or feeders for ore solution or a later deformation pattern

11.4 FURTHERING SYNGENETIC CONCEPTS

11.4.1 Gold Camps

Gold camps offer the opportunity to investigate gold deposits. Much has been written and deposit studies undertaken. Many have been reviewed in previous chapters, but much remains to be done. Most of the systematic field mapping is old; it was good work in its time and remains an important resource today. It needs to be updated by experienced field geologists who are capable of introducing modern geological knowledge. Some of the fields that should be applied are stratigraphy, structural geology, regional tectonism, and analysis of basic field relationships. When one looks at some modern geological interpretations, a critical eye should be sure to present a complete picture of the deposit and its setting.

Using gold cycles (CGE) is a good way to see if the gold mineralization can help decipher the stratigraphy and structure. Sampling of the mineralization for dating requires that the sample is representative of the vein or the host rock if that is the sample medium. The CGE thickness may be used to indicate a measure of accumulation and is probably related to the distance from a hydrothermal center. In the sedimentary environment, the sand/shale ratio in a given section may be used to determine the paleocurrent direction's determination of the turbidites, and possibly the veins, to establish palaeoslopes and the sources of sedimentary material into the basin as well as the direction of gold-bearing solutions. Carrier et al. (2000) provide an excellent description of a large hydrothermal vent deposit with variations of mineralogy, stacking, and changes in fO_2 as a function of stratigraphy. Work on the Bendigo ore system is an excellent study using the above approach (Bull and Large, 2014).

11.4.2 Lamprophyres

Chapter 5 described the lamprophyres in the Chester complex and determined that some of their field relationships were that of intervolcanic layers. Chemical analyses demonstrated that some "lamprophyres" in gold camps are pelagic–pelitic marine or interflow sediments that could form a GC. These observations demonstrate that careful field mapping is required to determine the relationship of these layers to determine if they are conformable bands, crosscutting later dikes feeding younger hanging-wall volcanic beds or much later, postdeformation, intrusions similar to the common northwest striking diabase dikes.

11.4.3 Quartz Chemistry

Chapter 7 discusses seafloor mineralizing systems and the role of quartz in some of those systems, like the cauliflower garden in the Galapagos spreading center. The cauliflower garden is not forming a gold deposit; however, it has similar characteristics to those forming sulfide deposits and there is not any reason to think that gold deposition would not be precipitated, if gold was available to the hydrothermal fluids. Tomilenko et al. (2010) present a comprehensive study which measures over 5000 homogenization temperatures from auriferous and barren quartz from Sovetskoye, Russia, which has produced over 100 tonnes of gold (Levitan, 2008). This study showed vein formation over a range from 225 to 400 °C, with no discernible temperature difference between auriferous and barren veins. Even knowing the formation temperature of a deposit is of limited use. To find a mineral deposit, one needs to recognize differences between proximal and distal samples and mineralized and background samples and this typically requires a spatial array of data points, including remote and background samples. But almost no research includes an adequate number of these important distal and background reference samples.

The same point is discussed by Bodnar and Yardley (2012), who state that the fluid inclusion community bemoans that fluid inclusions are not used routinely in exploration for mineral deposits because most academic inclusionists cannot separate the theoretical or experimental basis for the technique from its practical application in the field. Fluid inclusions by themselves cannot prove the existence of economic mineralization. They may be able to direct you to a part of the system where mineralization is likely to occur, but drilling, assays, and the current economic environment will determine if that occurrence is economic or not.

The genetic link between boiling and gold deposition was established by careful experimental and theoretical studies during the 1970–1990s. However, it is not necessary to understand the underlying phase equilibria and thermodynamics to apply the results in exploration; one only needs to know how to identify the former presence of boiling fluids which can be identified, coexisting liquid-rich and vapor-rich fluid inclusions, and distinctive mineral textures.

11.4.4 Vent Tracing

Tracing the location of vents for various gold deposits is described in Chapter 9. It is based on the shape of the mineralized zones and where they appear to apex. Valliant et al. (1982) have shown that the footwall of the Bousquet deposit, shares the same sodium depletion/potassium enrichment as volcanogenic massive sulfide deposits described in Chapter 8 and which is well recognized throughout the exploration industry. These two parameters can help guide vent tracing.

11.4.5 Metallogeny

The distribution of mineral deposits in mobile belts has been shown to be consistent over time and between belts as first established by Bilibin (1968) and subsequently demonstrated for the northern Appalachian Mobile Belt of northeast United States and eastern Canada (McCartney and Potter, 1962; McCartney, 1965).

In a wider context, the geologic environment appears to be extensional, similar to a mid-oceanic ridge environment. A very recent article by Webber et al. (2013) is a systematic study of gold concentration in oceanic basalts and shows that basalts were enriched up to 13 times by specific plume components of the Iceland plume along the mid-Atlantic Ridge. For the Archean, implications are that crustal recycling can enrich basalts in gold, which may have exploration implications. For Chester Township, it is worth examining the gold and trace element composition of the gabbro–diorite for element depletion and possible vectors to hydrothermal vents.

11.4.6 Detailed Study of Ore Deposits

11.4.6.1 Dome Mine, Timmins

The Dome Mine "veins" have been recently interpreted as sedimentary carbonate deposits (Fryer et al, 1979; Karvinen, 1980). At other mines, where they have been examined by the present authors they have all the characteristics of chemical sediments (Figure 3.4). They are conformable with the enclosing strata and are commonly layered; often showing an asymmetrical distribution of minerals and in some places some crude "graded bedding" is apparent. At the Pamour and Aunor, these beds are found intercalated with massive green carbonates or carbonatized pillowed magnesium-rich basalts of the Lower Member, whereas at the Dome and Pearl Lake areas they are found throughout the succession from the Lower Member to the above Upper Member. These veins generally make good ore, but can be barren in places. Native gold is common, as is pyrite. Where ladder veins are well developed, they are often restricted to a single stratum.

11.4.7 Summary

While the authors' findings are unlikely to be unique, their significance does not appear to be fully reflected in current lode vein formation models. A syngenetic/synsedimentary origin has wide-reaching implications for exploration, development, and gold deposit models.

Inherent are plausible explanations for lack of stable isotopic equilibrium between vein and host rock, the timing of vein formation, and the origin of hydrothermal gold-bearing fluids.

The key observations on which a new model of lode vein genesis hinges is that the veins formed in a cyclical deposition mode. This is considered irrefutable proof that they formed on the seafloor as silica crusts.

In turn, this constrains physical and chemical formation parameters and provides the basis of a comprehensive formation model that incorporates much of what is known about these deposits, but also provides the basis for a reexamination of these deposits, which may clarify certain previously unexplained points.

Finally, the debate of lode vein models would benefit from efforts to sort out primary seafloor processes from later overprinting, deformation, and metamorphism. Such investigations must start with careful field work, core logging, and study of the geology of the deposit. One must reexamine the tenets of the creed of epigenesis, which predisposes conclusions in many studies.

For instance, the vertical changes in mineral chemistry and textures described for the Bogosu and Presea deposits, and interpreted in terms of the crustal continuum model by Mumin et al. (1994), can be equally interpreted as horizontal and lateral zoning from seafloor vents and evolving hydrothermal fluids. This becomes easier and makes eminent sense if one rotates the stratigraphy to its original horizontal position.

Once a primary syngenetic component for lode vein genesis is accepted, new insights are brought to other debates. The homogeneous character and distinct oxygen isotopic composition of metamorphic hydrothermal fluids (Ridley and Diamond, 2000) reflect host rock substrate from which constituents are leached under conditions of elevated geothermal gradient.

A magmatic isotopic signature is entirely consistent with exsolution of fluids and hydrothermal liquids in a proximal felsic volcanic environment (e.g., Renabie Mine, Ontario) as well as other examples discussed by Spooner (1991), Hagemann et al. (2006), and the current study of Côté Gold (Kontak et al., 2013).

The concept of fertile upper-mantle source region (Hronsky et al., 2012) is also accommodated.

11.5 THE STATE OF CURRENT LODE DEPOSIT RESEARCH

The findings of this treatise are timely for enhanced understanding at the deposit scale. An urgent need for innovative approaches in earth sciences and exploration concepts has been stressed by former Chief Scientist Jim Franklin (Lopez-Pacheco, 2011). Innovation is also a subject of ongoing concern, e.g., Woodall (1983), Williams (2008), McBride (2009), McCuaig and Hronsky (2000), Sillitoe (2011), Dubé et al. (2011), and others.

Modern research seems to suffer from a lack of inclusion of old geological databases, a lack of mapping experience by many early career geologists, and use of assessment work submitted to government agencies as part of exploration projects. In current projects, the author, McBride, is using geological maps produced before many modern geological understandings were recognized. Remapping by experienced geologists familiar with regional geological concepts is an essential base to any understanding of the geological setting.

11.6 CONCLUDING STATEMENT

11.6.1 "Gute Ansichten sind wertlos. Es kommt darauf an, wer sie hat"

When the authors first contemplated the seashore and ridge exposures in the Cambro-Ordovician Meguma of Nova Scotia, they wondered about how quartz and turbidite sedimentation can occur at the same time, because it seemed to be obvious that is what the outcrops showed.

The journey toward a better understanding has been circuitous and expanded significantly from what seemed like initial simple questions. To get to this point, the authors have found it necessary to follow a path outlined by Karl Kraus, the Austrian writer of aphorisms and satire who said: "I have often been asked to be fair and view a matter from all sides. I did so, hoping something might improve if I viewed all sides of it. But the result was the same. So I went back to viewing things only from one side, which saves me a lot of work and disappointment. For it is comforting to regard something as bad and be able use one's prejudice as an excuse."

This has been necessary, since virtually every observation, from textures to local and regional have been interpreted differently in the literature. From an academic perspective, divergence is entirely normal and, by using the scientific process, may even lead to progress. For the exploration geologist, at a time in history when a large number of surface ore deposits have been found, it is necessary to widen our perspective, use all the tools at our disposal and constantly seek those from any or all fields of investigation, whether generally accepted or not, which provide a competitive advantage.

In his review of Ore Deposit Geology and its influence on mineral exploration, Jowett (1986) states that one must keep in mind that much gold has been found using syngenetic and epigenetic models and that geologic associations and controls rather than genetic models may be the key.

While much research is clearly needed, understanding the role of GCE and GCs may revitalize understanding of lode gold deposit genesis and exploration, and perhaps lead to a much discussed paradigm shift, a concept articulated by Kuhn (2012).

Simple explanations for this and other apparently contrary facts emanate from silica-rich hydrothermal seafloor vent fields and seeps, which, since the beginning of time, have accompanied cooling of the crust.

Bedding parallel quartz veins and their economically interesting varieties provide clues to fundamental geological processes.

The authors' contention, which forms the thesis of this book, is simply that most quartz veins formed at the same time as the enclosing rocks and a compelling observation pertaining to lode gold genesis as the current study shows, is that most lode veins occur in complex, frequently graded chemical sediments. They are clearly stratabound and therefore must have formed subaqueously.

Detailed geological mapping over several years and careful logging of thousands of meters of core to a scale of millimeters from several gold camps underlie the current work. The conclusions, while iconoclastic, are not entirely new. They are closely geologically constrained and contravene no fundamental physical laws. Lines of geological evidence converge and further extend syngenetic concepts into the Archean, thereby forming a closer link to a seafloor hydrothermal origin for lode gold deposits.

This study of GCE units brings a different perspective to continuing controversies—expertly summarized by e.g., Kerrich (1993), McCuaig and Kerrich (1998), Robert et al. (2007), or Hronsky et al. (2012)—such as the timing of lode vein mineralization and the source and composition of gold-bearing fluids.

The authors findings corroborate the multistage and multiprocess genetic hypothesis for greenstone-hosted Au deposits developed by Hutchinson (1993). The current work extends these concepts into felsic-hosted deposits exemplified by the Côté Gold deposit.

A final note of caution, as James Lovelock has stated (Lovelock, 2009) in The Vanishing Face of Gaia, A Final Warning: this Pygmalion fate—falling in love with your model—is all too easy. Gradually, the world of science has evolved to the dangerous point where model building has precedence over observation and measurement, especially in the Earth and Life Sciences.

11.7 THE PROOF IS IN THE SUCCESS OF THE MODEL IN ITS POSITIVE APPLICATION

Theoretically, it is the work of government and university researchers, not the explorationist, to provide a guide for finding and evaluating gold deposits. Bottril (1990) said of the shear zone hosted gold model that it was an economic and academic failure. Throughout this book, data are presented and interpreted to develop such a model. McBride (1987) did produce a syngenetic model summarized in Chapter 1 based on the Beardmore section of the Archean Beardmore-Geraldton Greenstone Belt and Northern Empire past producing gold mine. This model shows that time is not important; only the geological cycle. In 1986, while working in the Ordovician of the Newfoundland Central Mobile Belt, an area with very similar geology to Beardmore was encountered. It was the Betts Cove Ophiolite Complex. These rocks represented a target over 2000 million years younger than the Archean and 2000 miles (3200 km) east of the reference area. Claims were staked in early 1987 on the available ground to cover the favorable mafic volcanic–sedimentary rocks. The adjacent ultramafic rocks had been staked by two major companies. The Beardmore model had suggested that soil sampling on a half kilometer spacing line and a 50 m sample interval would detect any significant gold deposit; the survey carried out in 1987. Two samples ran over 500 ppb gold. Follow up returned a soil anomaly with values to 30 g per tonne. Trenching of the anomaly led to the discovery of the Nugget Pond deposit. This deposit lies in clastic sedimentary rocks above ultramafic and mafic volcanic rocks and is buried under intermediate volcanic rock (see Chapter 9). It fulfills the diagnostic criteria in the 1987 paper thus supporting the syngenetic model. It is not deformed, only moderately tilted. It represents the first gold find in the Bett's Cove Ophoilite Complex which is known for its VHMS deposits.

The discovery cost was $144,000 Canadian and it was brought to feasibility study confidence for 3.5 million dollars Canadian (McBride, 1991). It has since been mined out. The model successfully predicted the gold potential of this old copper VMS camp and then evaluated it for a minimal cost. Subsequent drilling and mining did not significantly change the numbers in the original feasibility study. This success demonstrates that the syngenetic model is an important exploration tool.

Hopefully, this book has cracked open doors, which will lead to as yet unimagined vistas.

References

Adams, M., Lentz, D.R., Shaw, C., Williams, P., Archibald, D.A., Cousens, B., 2005. Eocene lamprophyre dykes intruding the monashee complex, B.C.: petrochemical to petrogenetic relationships with the Kamloops group volcanic sequence. Canadian Journal of Earth Sciences 42, 11–24.

Agricola, G., 1556. De Re Metallica (H.M. Hoover, Trans.). Mag, London, 1912.

Akande, S.O., 1982. Mineralogy and genesis of three veins systems, Ross mine, Holtyre, Ontario. In: Hodder, R.W., Petruk, W. (Eds.), Geology of Canadian Gold Deposits, vol. 24. Can. Inst. Min. Metall., pp. 94–97.

Alt, J.C., Lonsdale, P., Haymon, R., Muehlenbachs, 1987. Hydrothermal sulfide and oxide deposits on seamounts near 21°N, East Pacific Rise. Geological Society of America Bulletin 98, 157–168.

Anderson, G.M., 2008. The mixing hypothesis and the origin of Mississippi Valley-type ore deposits. Economic Geology 104, 1683–1690.

Anhaeusser, C.R., 1976. The nature and distribution of Archean gold mineralization in southern Africa. Minerals Science and Engineering 8, 46–84.

Anhaeusser, C.R., Mason, R., Viljoen, M., Vilojoen, R., 1969. A reappraisal of some aspects of Precambrian shield geology. Geological Society of America Bulletin 80, 2175–2200.

Ashley, P.M., Cook, N.D.J., Hill, R.L., Kent, A.J.R., 1994. Shoshonitic lamprophyre dykes and their relation to mesothermal Au-Sb veins at Hillgrove, New South Wales, Australia. Lithos 32, 249–272.

Ayer, J.A., 2012. Report on the Lithogeochemical Analysis of Lamprophyre and Basalts from the Watershed Property of Golden Scarab Corporation for Dr. Kretschmar (Unpublished).

Ayer, J.A., Trowell, N.F., 2002. Geological Compilation of the Swayze Area, Abitibi Greenstone Belt. Ontario Geological Survey. Preliminary Map P.3511, scale 1:100.

Ayuso, R.A., Wooden, I.I., Foley, N.K., Seal II, R.R., Shinhz, A.K., 2005. U-Pb zircon ages and Pb isotope geochemistry of gold deposits in the North Carolina Slate Belt of South Carolina. Economic Geology 100, 925–952.

Barley, M.E., Groves, D.I., 1992. Supercontinent cycles and the distribution of metal deposits through time. Geology 20, 291–294. http://dx.doi.org/10.1130/0091-7613(1992)020<0291:SCATDO>2.3.CO;2.

Barrett, T.J., Fralick, P.W., 1989. Turbidites and iron formations, Beardmore-Geraldton, Ontario: an application of a combined ramp/fan model to Archean clastic and chemical sedimentation. Sedimentology 36 (2), 221–234. http://dx.doi.org/10.111/j.1365.1989.tb00604.x.

Barrett, T.J., MacLean, W.H., 1994. Chemostratigraphy and hydrothermal alteration in exploration for VHMS deposits in greenstones and younger volcanic rocks. Geological Association of Canada Short Course Notes 11, 433–467.

Barton Jr., P.B., 1970. Sulfide Petrology. Mineralogical Society of America, pp. 187–193. Special Paper 3.

Barton Jr., P.B., 1991. Ore textures: problems and opportunities. Mineralogical Magazine 55, 303–315.

Barton Jr., P.B., Skinner, B.J., 1979. Sulfide mineral stabilities. In: Barnes, H.L. (Ed.), Geochemistry of Hydrothermal Ore Deposits, second ed. Wiley, New York, pp. 278–403.

Barton, P.B., Toulmin III, P., 1964. The electrum-tarnish method for the determination of the fugacity of sulfur in laboratory sulfide systems. Geochimica et Cosmochimica Acta 28, 619–640.

Barton Jr., P.B., Toulmin III, P., 1966. Phase relations involving sphalerite in the Fe-Zn-S system. Economic Geology 61, 815–849.

Bastin, E.S., 1959. Interpretation of ore textures. Geological Society of America Memoirs 45, 101.

Bateman, A.M., 1950. Economic Mineral Deposits, second ed. John Wiley and Sons, New York. 916 p.

Beakhouse, G.P., 2007a. Structurally Controlled, Magmatic Hydrothermal Gold Deposits for Archean Lode Gold Deposits: A Working Hypothesis. Ontario Geological Survey. Open File Report 6193, 133 p.

Beakhouse, G., 2007b. Gold, Granite and Late Archean Tectonics: A Superior Province Perspective. Geoscience Australia. Record 2007/14, pp. 191–196.

Beaudoin, G., Mercier-Langevin, P., Dubé, B., Taylor, B., 2014. Low-temperature alteration at the world-class LaRonde Penna Archean Au-rich volcanogenic massive sulfide deposit, Abitibi subprovince, Quebec, Canada: evidence from whole-rock oxygen isotopes. Economic Geology 109, 167–182.

Bédard, J.H., Harris, L.B., 2014. Neoarchean disaggregation and reassembly of the Superior craton. Geology 42, 951–954. http://dx.doi.org/10.1130/G35770.1.

Belkabir, A., Robert, F., Vu, L., Hubert, C., 1993. The influence of dikes on auriferous shear zone development within granitoid intrusions: the Bourlamaque luton, Val d'Or District, Abitibi greenstone belt. Canadian Journal of Earth Sciences 30, 1924–1933.

Benedict, P.C., Titcomb, J.A., 1948. Northern Empire Mine; Structural Geology of Canadian Ore Deposits, a Symposium. CIMM, pp. 398–399.

Bennett, G., 1978. Geology of the Crooked lake Area, District of Sudbury. In: Ont. Geol. Sur., Geol. Rept., 172, 46 p.

Berger, B., 2011. Project Unit 11-001. Geological Investigations South of Gogama. Ontario Geological Survey. Summary of Field Work and Other Activities. Open File Report 6270, pp. 3-1–3-7.

Bergman, S.C., 1987. Lamproites and other potassium-rich rocks: a review of their occurrence, mineralogy and geochemistry. In: Fitton, J., Upton, B.G.J. (Eds.), Alkaline Igneous Rocks. Geologic Society, pp. 103–190. Special Publication 30.

Bernard, S., Papineau, D., 2014. Graphitic carbons and biosignatures. Elements 10 (6), 435–440.

Bernatchez, R., 1997. A Summary Report on the Little Bear Lake Property Cat Lake - Bernic Lake Area Lac du Bonnet Mining Division, Lac du Bonnet, Manitoba. Manitoba Assessment Report #73282.

Bernier, L.R., Pouliot, G., MacLean, W.H., 1987. Geology and metamorphism of the Montauban north gold zone: a metamorphosed polymetallic exhalative deposit, Grenville Province, Quebec. Economic Geology 82, 2076–2090.

Berry, W.B.N., Wilde, P., 1978. Progressive Ventilation of the oceans-an explanation for the distribution of the Lower Paleozoic black shales. American Journal of Science 278, 257–275.

Bierlein, F.P., Anne, D.C., Broome, J.M.N., Ramsay, W.R.H., 1998a. Metatholeiites and interflow sediments from the Cambrian Heathcote greenstone Belt, Australia: sources for gold mineralization in Victoria? Economic Geology 93, 84–101.

Bierlein, F.P., Christie, A.B., Smith, P.K., 2000. A comparison of orogenic gold mineralization in central Victoria (AUS), western South Island (NZ) and Nova Scotia (CAN): implications for variations in the endowment of Palaeozoic metamorphic terrains. Ore Geology Reviews 25, 125–168.

Bierlein, P., Crowe, D.E., 2000. Phanerozoic orogenic lode gold deposits in gold 2000. Reviews in Economic Geology 13, 103–137.

Bierlein, F.P., Fuller, T., Stuewe, K., Arne, D.C., Keays, R.R., 1998b. Wallrock alteration associate with turbidite-hosted gold deposit: examples from the Palaeozoic Lachlan Fold Belt in central Victoria, Australia. Ore Geology Reviews 13 (1), 345–380.

Bierlein, F.P., Hughes, M., Dunphy, J., McKnight, S., Reynolds, S., Waldron, H., 2001. Tectonic and economic implications of trace element, 40Arr39Ar and SmBNd data from mafic dykes associated with orogenic gold mineralization in central Victoria, Australia. Lithos 58, 1–31.

Bierlein, F.P., Murphy, F.C., Weinberg, R.F., Lees, T., 2006. Distribution of orogenic gold deposits in relation to fault zones and gravity gradients: targeting tools applied to the Eastern Goldfields, Yilgarn Craton, Western Australia. Mineralium Deposita 41, 107–126.

Bilibin, Y.A., 1968. Metallognic Provinces and Metallogenic Epochs. trans. by Alexandrov, E. A. Queen's College Press, Flushing, NY, 36 p.

Binns, R.A., Scott, S.D., 1993. Actively forming polymetallic sulphide deposits associated with felsic volcanism in the eastern Manus back-rc Basin, Papua New Guinea. Economic Geology 88, 2226–2236.

Binns, R.A., Scott, S.D., Bogdanov, Y.A., Lisitzin, A.P., Gordeev, V.V., Gurvich, E.G., Findlyson, E.J., Boyd, T., Dotter, L.E., Wheller, G.E., Murevyev, K.G., 1993. Hydrothermal oxide and gold-rich sulphate deposits of Franklin seamount, Western Woodlark Basin, Papua New Guinea,. Economic Geology 88, 2122–2153.

Blenkinsop, T.G., Sanderson, D.J., 1999. Are Gold Deposits in the Crust Fractals? A Study of Gold Mines in the Zimbabwe Craton. Geological Society, London, pp. 141–151. Special Publication 155.

Bodnar, R., Yardley, B., 2012. Selling our science (but not at all costs). Triple Point Column. Elements 8 (4), 245.

Bohlke, J.K., Irwin, J.J., 1992. Laser microprobe analyses of Cl, Br, I, and K in fluid inclusions-implications for sources of salinity in some ancient hydrothermal systems. Geochimica et Cosmochimaica Acta 56, 203–225.

Bottrill, T.J., 1990. The shear-zone hosted gold model – an economic and academic failure. In: Paper Presented at the Toronto Geological Discussion Group; Nov. 26, 1990, 14 p.

Bouchot, V., Milesi, J.P., Aledru, P., Lerouge, C., Roig, J.Y., Bellot, J.P., Becq-Giraudon, J.F., Truffert, C., 2000. Orogenic gold veins and W, LI-F mineralization related to "specialized" granites: two markers of the crustal scale Au-W-Sb metalliferous peak at 310-305 (French Variscan Belt). Documents du bureau de Rescherches Geologiques et Minieres 297, 53–55.

Bouma, A.H., 1962. Sedimentology of Some Flysch Deposits: A Graphic Approach to Facies Interpretation. Elsevier, Amsterdam, 168 p.

Bowers, T.S., Von Damm, K.I., Edmond, J.M., 1985. Chemical evolution of mid-ocean ridge hot springs. Geochimica et Cosmochimica Acta 49, 2239–2252.

Boyle, R.W., 1955. The geochemistry and origin of the gold-bearing quartz veins and lenses of the Yellowknife greenstone belt. Economic Geology 50, 51–66.

Boyle, R.W., 1959. The geochemistry, origin and role of carbon dioxide, water, sulfur, and boron in the Yellowknife gold deposits, N.W.T., Canada. Economic Geology 54, 1506–1524.

Boyle, R.W., 1979. The Geochemistry of Gold and Its Deposits. Geological Survey of Canada. Bulletin 280, 584 p.

Boyle, R.W., 1982. Gold depositės: a review ot their geological and geochemical setting. In: Geology of Canadian Gold DepositsCIMM Sp., 24. pp. 1–524.

Boyle, R.W., 1986. Gold deposits in turbidite sequences: their geology, geochemistry and history of theories of their origin. In: Keppie, J.D., Boyle, R.W., Haynes, S.J. (Eds.), Turbidite-Hosted Gold Deposits. Geological Association of Canada, pp. 1–13. Special Paper 32.

Boyle, R.W., 1987. Gold: History and Genesis of Deposits. Van Nostrand Reinhold Company. ISBN: 0-442-21161-7, TN420.B63.

Boyle, R.W., Davies, J.L., 1964. Geology of the Austin Brook and No. 6 sulphide deposits, Gloucester County, New Brunswick. Geological Survey of Canada Paper 63–24.

Branney, M.J., Suthren, R.J., 1988. High-level peperitic sills in the English Lake District: distinction from block-lavas and implications for Borrowdale Volcanic Group stratigraphy. Geological Journal 23, 171–187.

van Breemen, O., Heather, K.B., Ayer, A.J., 2006. U-Pb Geochronology of the Neoarchean Swayze Sector of the Southern Abitibi Greenstone Belt. GSC, p. 32. Current Research 2006–F1.

Brown, R.A., 1948a. O'Brien Mine; Structural Geology of Canadian Ore Deposits, a Symposium. CIMM, pp. 809–816.

Brown, W.L., 1948b. Jerome Mine; Structural Geology of Canadian Ore Deposits, a Symposium. CIMM Sp. vol. 1. pp. 438–441.

Bruce, E.L., 1940. Geology of the Goudreau-Lochalsh area. Ontario Department of Mines Annual Report. vol. 49 (Pt III), 33–42.

Bruce, E.L., 1948a. Cline Lake Mine. Canadian Institute of Mining and Metallurgy 1, 433–435.

Bruce, E.L., 1948b. Renabie Mine. In: CIMM Sp. vol. 1, Structural Geology of Canadian Ore Deposits; A Symposium, pp. 436–438.

Buffam, B.S.W., Allen, R.B., 1948. Chesterville Mine. Canadian Institute of Mining and Metallurgy 1, 662–671.

Bull, S.W., Large, R.R., 2014. Setting the stage for the genesis of the giant Bendigo ore system. In: Jenkin, G.R.T., Lusty, P.A.J., McDonald, I., Smith, M.P., Boyce, A.J., Wilkinson, J.J. (Eds.), Ore Deposits in an Evolving Earth. Geological Society, London. Special Publication 393. http://dx.doi.org/10.1144/SP393.10.

Burchfiel, B.C., February 2004. New technologies: new geological challenges. GSA Today.

Burlinson, K., 1991. Decrepitation in gold exploration. A case history from the Cotan prospect; N.T. Journal of Geochemical Exploration 42, 143–156.

Burlinson, K., 2007. Acoustic decrepitation as a means of rapidly determining CO_2 (and other gas) contents in fluid inclusions and its use in exploration, with examples from gold mines in the Shandong and Hebei provinces, China. Acta Geologica Sinica 23, 65–71.

Burlinson, K., 2013. A discussion to point out errors in H_2 analyses of fluid inclusion volatiles by mass spectrometry. Economic Geology 108, 1211.

Burrows, D.R., Jemielita, R.A., 1989. Lithophile-element systematics of Archean greenstone belt Au-Ag vein deposits: implications for source processes: discussion. Canadian Journal of Earth Sciences 26, 2741–2743.

Burrows, D.R., Spooner, E.T.C., 1989. Relationship between Archean gold quartz vein-shear zone mineralization and igneous intrusions in the Val d'Or and Timmins areas, Abitibi subprovince, Canada. Economic Geology. Monograph No. 6, 424–444.

Burrows, D.R., Wood, P.C., Spooner, E.T.C., 1986. Carbon isotope evidence for a magmatic origin for Archean gold-quartz vein deposits. Nature 321, 851–854.

Buryak, V.A., 1964. In: The Process of Regional Metamorphism Influencing Development of Gold-Sulfide Mineralization in the Central Part of the Lena Goldfield: Physical-Chemical Conditions of Magmatism and Metasomatism, All-Union Petrographic Symposium, 3rd, Moscow, Collected Papers, pp. 184–189.

Buryak, V.A., 1975. A Metamorphic-Hydrothermal Type of Economic Gold Mineralization. Nauka, 45 p.

Buseck, P.R., Beyssac, O., 2014. From organic matter to graphite: graphitization. Elements 10 (6), 421–426.

Cabanis, B., Lecolle, M., 1989. The La/10-Y/15-Nb/8 diagram: a tool for distinguishing volcanic series and discovering crustal mixing and/or contamination. In: Comptes Rendus de l'Academie des sciences, serie 2. Sciences de la Terre, vol. 309, pp. 2023–2029.

Callan, J.N., 1991. Syn-deformational Shear Zone-hosted Au-quartz Vein Mineralization in TTG Host Rocks, Renabie Mine Area, North Ontario: Structural Analysis, Microstructural Characteristics and Vein Paragenesis. Ontario Geological Survey. Open File Report 5759.

Cameron, E.M., 1993. Precambrian gold: perspectives from the top and bottom of shear zones. Canadian Mineralogist 31, 917–944.

Cameron, E.M., 1988. Archean gold; relation to granulite formation and redox zoning in the crust. Geology 16, 109–112.

Cameron, E.M., Hattori, K., 1987. Archean gold mineralization and oxidizing hydrothermal fluids. Economic Geology 82, 1177–1191.

Campiglio, C., Darling, R., 1976. The geochemistry of the Archean Bourlamaque batholith, abitibi, Quebec. Canadian Journal of Earth Sciences 13, 972–986.

Card, K., Poulsen, K.H., 1998. Geology and mineral deposits of the Superior Province of the Canadian Shield. In: Lucas, S. (Ed.), Geology of the Precambrian Superior and Grenville Provinces and Precambrian Fossils in North America. Geological Survey of Canada, Geology of Canada. No. 2, pp. 12–194.

Cargill, D.G., Gow, N.N., 2010. Technical Report on the young=Shannon Property. Chester township, Ontario a NI 43-101 report prepared for Trelawney Mining and Exploration (2009). 57 p. www.sedar.com. Aug. 20, 2010.

Carlson, C.A., 1991. Spatial distribution of ore deposits. Geology 19, 111–114.

Carrier, A., Jébrak, M., Angelier, J., Holyland, P., 2000. The Silidor deposit, Rouyn-Noranda district, Abitibi belt: geology, structural evolution and paleostress modeling of an Au quartz vein-type deposit in an Archean trondhjemite. Economic Geology 95, 1049–1065.

Cas, R., 2013. Palaeovolcanology and Palaeoenvironmental Evolution of the Archean Eastern Goldfields Province. Tectonic and Economic Implications. Monash Volcanology Research Group (MONVOLC), Yilgarn Craton, Western Australia. PowerPoint presentation to Sudbury Geological Discussion Group, November 2013.

Cathelineau, M.M., Boibon, M.C., Holleger, P., Marion, P., Denis, M., 1989. Gold in arsenopyrite crystal chemistry, location and state, physical and chemical conditions of deposition. Economic Geology Monographs 6, 328–341.

Cerny, P., Trueman, D.L., Ziehlke, D.V., Goad, B.E., Paul, B.J., 1981. The Cat Lake-Winnipeg River and the Wekusko Lake Pegmatite Fields, Manitoba. Manitoba Energy and Mines, Mineral Resources Division. Econ. Geol. Rept. ER80–1, 216 p.

Chamberlain, C.M., 2003. Geology and Genesis of the Butyanhulu Gold Deposit, Sukumaland Greenstone Belt Tanzania (Ph.D. thesis). Imperial College (Unpublished) 410 p.

Chataway, R., 2005. 2004–2005 assessment report on the Tower Mountain Property. Ontario Geological Survey Assessment Report.

Chataway, R., 2007. 2007 assessment report on the Tower Mountain Property. Ontario Geological Survey Assessment Report.

Chattopadhyay, A., 2010. A review of the structural characteristics of orogenic gold deposits, with special reference to Indian gold Fields. In: Mihir, D., Goldfarb, R.J. (Eds.), Gold Metallogeny, India and beyond (Chapter 7).

Claoué-Long, J.C., King, W., Kerrich, R., 1990. Archaean hydrothermal zircon in the Abitibi greenstone belt: constraints on the timing of gold mineralization. Earth and Planetary Science Letters 98, 109–128.

Colvine, A.C., 1989. An empirical model for the formation of Archean gold deposits: products of final cratonization of the Superior Province, Canada. In: Keys, R.R., Ramsay, W.R.H., Groves, D.I. (Eds.), The Geology of Gold Deposits: The Perspective in 1988. Economic Geology. Monograph 6, 37–53.

Colvine, A.C., Andrews, A.J., Cherry, M.E., Durocher, M.E., Fyon, A.J., Lavigne Jr., M.J., Macdonald, A.J., Marmont, S., Poulsen, K.H., Springer, J.S., Troop, D.G., 1984. An Integrated Model for the Origin of Archean Lode Gold Deposits. Ontario Geological Survey. Open File Report 5524.

Colvine, A.C., Fyon, J.A., Heather, K.B., Marmont, S., Smith, P.M., Troop, D.G., 1988. Archean Lode Gold Deposits in Ontario. Ontario Geological Survey. Miscellaneous Paper 139.

Compston, D.M., Mcdougall, I., 1994. 40Ar-39Ar and K-Ar age constraints on the early Proterozoic Tennant Creek Block, northern Australia, and the age of its gold deposits. Australian Journal of Earth Sciences 41, 609–616.

Condie, K.C., 2014. How to make a continent: thirty-five years of TTG research (Chapter 7). In: Dilek, Y., Furnes, H. (Eds.), Evolution of Archean Crust and Early Life, Modern Approaches in Solid Earth Sciences, vol. 7. http://dx.doi.org/10.1007/978-94-007-7615-9_7.

Cook, R.B., Roscoe, W.E., 2011. Trelawney Mining and Exploration Inc., Technical Report on the Coté Lake Deposit, Ontario, Canada. NI-43-101 Report, Chester Property, 148 p. www.sedar.com.

Corliss, J.B., Lyle, M., Dymond, J., Crane, K., 1978. The chemistry of hydrothermal mounds near the Galapagos rift. Earth and Planetary Science Letters 40, 12–24.

Côté Gold Deposit. Iamgold Inc. http://www.iamgold.com/English/Operations/Development-Projects/Cote-%20Lake-Ontario/default.aspx (accessed 01.09.13.).

Coumou, D., Driesner, T., Heinrich, C.A., 2008. The structure and dynamics of mid-ocean ridge hydrothermal systems. Science 321, 26.

Cox, S.F., 2000. Coupling between deformation, fluid pressures, and fluid flow in ore-producing hydrothermal systems at depth in the crust. Economic Geology 100, 39–75.

Craig, J.R., Vaughn, D.J., 1994. Ore Microscopy and Ore Petrography, second ed. John Wiley & Sons, Inc., ISBN 0-471-55175-9 Downloaded from: http://www.minsocam.org/%20MSA/Open%20Access_publications.html?p=Craig_Vaughan.

Craw, D., Chamberlain, C.P., 1996. Meteoric incursion and oxygen fronts in the Dalradian metamorphic belt, southwest Scotland: a new hypothesis for regional gold mobility. Mineralium Deposita 31, 365–373.

Crocket, J.H., Cowan, P., Kusmirski, R.T.M., 1980. Gold content of volcanic-hosted interflow sedimentary rocks in the Red Lake Area: Implications on ore genesis at the Dickenson mine; in genesis of Archean. In: Volcanic-Hosted Gold Deposits, Symposium Held at the University of Waterloo, March 7, 1980. Ontario Geological Survey, MP97, pp. 128–143.

Crocket, J.H., Fueten, F., Clifford, P.M., Kabir, A., Henderson, J., 1986. Distribution of gold and arsenic in turbidites at Harrigan Cove, Nova Scotia: Implications on gold mineralization. In: Keppie, J.D., et al. (Eds.), Turbidite-Hosted Gold Deposits. Geological Association of Canada, pp. 149–159. Special Paper 32.

Daigneault, R., Gaboury, D., 2007. Systèmes filoniens aurifères: Analyse et implications pour l=exploration. Atelier, Québec Exploration.

David, J.B., Bell, B.R., 2007. How do you grade peperites? Journal of Volcanology and Geothermal Research 159, 409–420.

Davies, J.L., Smith, R.N., 1966. Geology of the Brunswick No. 6 and No. 12 sulphide deposits. In: Poole, et al. (Eds.), Geology of Parts of the Atlantic Provinces, Geol. Assoc. Can., Guidebook. Halifax, Nova Scotia, pp. 45–57.

Davison I., Jeffcoate, A., Qing H., 200. Geometry of chevron folding and shortening estimates at Hartland Quay, north Cornwall, UK, and some regional implications for Culm Basin development; Proceedings of the Ussher Society 11, Part 1, pp. 42–50. See: http://eurekamag.com/research/019/060/geometry-chevron-folding-shortening-estimates-hartland-quay-cornwall-uk-regional-implications-culm-basin-progress.php.

Deb, M., Goldfarb, R.J., 2010. Gold Metallogeny: India and Beyond. Alpha Science International Ltd. ISBN: 978-84265-646-4, 320 p.

Decartes, R., 1664. Principia Philosophiae.

Degens, E.T., Ross, D.A., 1969. Hot Brines and Recent Heavy Metal Deposits in the Red Sea. Springer-Verlag, New York.

De Gregorio, B.T., Sharp, T.G., 2006. Possible abiotic formation of kerogen-like carbon in the Strelley Pool Chert. Lunar and Planetary Science XXXVII, 2318.

De Rosen-Spence, A., Provost, G., Dimroth, E., Gochnauer, K., Owen, V., 1980. Archean subaqueous felsic flows, Rouyn-Noranda, Quebec, Canada and their Quaternary equivalents. PreCambrian Research 12, 43–77.

Derry, D.R., 1939. The geology of the Canadian Malartic gold mine, N. Quebec. Economic Geology 34, 495–523.

Dewey, J.F., Holdsworth, R.E., Strachan, R.A., 1998. Transpression and transtension zones. Geological Society of London 135, 1–14.

Dove, P.M., Rimstidt, J.G., 1994. Silica. In: Heaney, P.J., Prewitt, C.T., Gibbs, G.V. (Eds.), Min. Soc. America Reviews in Mineralogy, vol. 29, pp. 259–307.

Downes, M.J., Hodges, D.J., Derweduwen, J., 1984. A free carbon- and carbonate-bearing alteration zone associated with the Hoyle Pond occurrence, Ontario, Canada. In: Foster, R.P. (Ed.), Gold '82. The Geology, Geochemistry and Genesis of Gold. A.A. Balkema, Rotterdam, pp. 435–448.

Downey, W., Lentz, D.R., 2006. A review of deep submarine silicic pyroclastic volcanism. Geoscience Canada 33, 5–24.

Dubé, B., Gosselin, P., 2008. Lode gold: greenstone-hosted quartz carbonate vein deposits (orogenic, mesothermal, lode gold, shearzone-related quartz-carbonate or gold-only deposits). In: Mineral Resources of Canada: A Synthesis of Major Deposit-Types, District Metallogeny, the Evolution of Geological Provinces, and Exploration Methods, vol. 5. Geological Association of Canada, Mineral Deposits Division, pp. 49–74.

Dubé, B., Mercier-Langevin, P., Castonguay, S., McNicol, V.J., Pehrsson, S.J., Bleeker, W., Schetselaar, E.M., Jackson, S., 2011. Targeted Geoscience Initiative 4. Lode Gold Deposits in Ancient, Deformed and Metamorphosed TerranesCFootprints and Exploration Implications: A Preliminary Overview of Themes, Objectives and Targeted Areas. Ontario Geological Survey. Summary of Field Work and Other Activities, 2011. Open File Report 6270, pp. 38.1–38.10.

Dubé, B., Mercier-Langevin, P., Kjarsgaard, I., Hannington, M., Becu, V., Cote, J., Moorhead, J., Legault, M., Bedard, N., 2014. The Bousquet 2-Dumagami world-class Archean Au-rich volcanic massive sulfide Deposit, Abitibi, Quebec: metamorphosed submarine Advance argillic alteration footprint and genesis. Economic Geology 109, 121–166.

Dubé, B., Williamson, K., McNicoll, V., Malo, M., Skulski, T., Twomey, T., Sanborn-Barrie, M., 2004. Timing of gold mineralization at Red Lake, northwestern Ontario, Canada: new constraints from U-Pb geochronology at the Goldcorp high-grade zone, Red Lake mine, and the Madsen mine. Economic Geology 99, 1611–1641.

Dubuc, F., 1966. Geology of the Adams Mine. CIM Bulletin 59, 101–104.

Duggan, M.B., Jaques, A.L., 1996. Mineralogy and geochemistry of Proterozoic shoshonitic lamprophyres from the Tennant Creek Inlier, northern territory. Australian Journal of Earth Science 43, 269–278.

Duguet, M., Lin, S., Davis, D.W., Corkery, M.T., McDonald, J., 2009. Long-lived transpression in the Archean Bird River greenstone belt, western Superior Province, southeastern, Manitoba. Precambrian Research 174, 381–407.

Duguet, M., Lin, S., Gilbert, H.P., Corkery, M.T., 2007. GS-13. Structural Geology and Kinematic Evolution of the Bird River Greenstone Belt, English River Subprovince, Manitoba (NTS 52L5,6); in Report of Activities 2007, Manitoba Science, Technology, Energy and Mines. Manitoba Geological Survey, pp. 144–154.

Dunbar, W.R., 1948. Structural relations of the Porcupine ore deposits. Structural Geology of Canadian Ore Deposits CIMM 1, 442–456.

Dziggel, A., Wulff, K., Kolb, J., Meyer, F.M., 2009. Processes of high-T fluidBrock interaction during gold mineralization in carbonate-bearing metasediments: the Navachab gold deposit, Namibia. Mineralium Deposita 44, 665–687.

Edmunds, C., 2009. Technical Report on Underground and Open Pit Mineral Resource Estimates, Young-Davidson Property, Matachewan, Ontario. Prepared for Northgate Minerals Corporation. Jan 23, 2009. Report under NI 43-101 as filed on. www.sedar.com (under AuRico Gold Corporation).

Edwards, A.B., 1947. Textures of the Ore Minerals. Australian Inst. of Mining and Metallurgy, Melbourne, Australia, 242 p.

Einaudi, M.T., Hedenquist, J.H., Inan, E.E., 2003. Sulfidation state of fluids in active and extinct hydrothermal systems: transitions from porphyry to epithermal environments. Society of Economic Geologists Special Publication 10, 285–313.

Emmons, W.H., 1937. Gold Deposits of the World. McGraw Hill, New York, 562 p.

Emsbo, P., 2005. The Society of Economic Geologists 2004 Awards R.A.F. Penrose Gold Medal for 2004. Citation of richard W. Hutchinson. Economic Geology 100, 1277–1278.

Emsbo, P., Hutchinson, R., Hofstra, A.H., Volk, J.A., Bettles, K., Boschuk, G.J., Johnson, C., 1999. Syngenetic Au on the Carlin Trend: implications for Carlin-type deposits. Geology 27, 59–62.

Etheridge, M.A., Wall, V.J., Cox, S.F., Vernon, R.H., 1984. High fluid pressures during regional metamorphism and deformation; implications for mass transport and deformation mechanisms. Journal of Geophysical Research 89, 4344–4358.

Etheridge, M.A., Wall, V.J., Vernon, R.H., 1983. The roll of the fluid phase during regional metamorphism and deformation. Journal of Metamorphic Geology 1, 205–226.

Faribault, E.R., 1899. The gold measures of Nova Scotia and deep mining. Journal of the Mining Society of Canada 119–128.

Farquhar, R.M., Haynes, S.J., 1986. Lead isotope data for gold-bearing veins and their host metasedimentary rocks of the Goldenville Formation, eastern Nova Scotia. Maritime Sediments and Atlantic Geology 22, 89–99.

Ferguson, S.A., 1968. Renabie Mines Limited, Nudulama Mines Limited and Adjoining Properties; Surface Geology, Parts of Rennie, Leeson and Brackin Townships, District of Sudbury. Ont. Dept. Mines. Preliminary Map P492.

Ferguson, S.A., Buffam, B.S.W., Carter, O.F., Griffis, A.Y., Holmes, T.C., Hurst, M.E., Jones, W.A., Lane, H., Longley, C.S., 1968. Geology and ore deposits of Tisdale township, District of Cochrane. In: Ont. Dept. Mines, Geol Report, No. 58, 177 p.

Ferry, J.M., 1981. Petrology of graphitic sulphide schists from southern Maine: an example of desulfidation during prograde regional metamorphism. American Mineralogist 66, 908–930.

Fletcher, H., Faribault, E.R., 1887. Report on the Geological Surveys and Explorations in the Counties of Guysborough, Antigonish, Pictou, Colchester and Halifax, Nova Scotia from 1882–1886, vol. 2. Geological and Natural History Survey of Canada, Ann. Rept., 163 p.

Foley, S.F., Venturelli, G., Green, D.H., Toscani, L., 1987. The ultrapotassic rocks: characteristics, classification, and constraints for petrogenetic models. Earth-Science Reviews 24, 81–134.

Fournier, R.O., 1983. A method of calculating quartz soluabilities in agueos sodium chloride solutions. Geochimica et Cosmochimca Acta 47, 579–586.

Fournier, R.O., 1985. The behavior of silica in hydrothermal solutions. Reviews in Economic Geology 2, 45–62.

Fournier, R.O., 1999. Hydrothermal processes related to movement of fluid from plastic to brittle rock in the magmatic-epithermal environment. Economic Geology 94, 1193–1212.

Fox, J.S., 1977. Rapid Pyroclastic Mapping in Base Metal Exploration. CIM Bulletin, pp. 173–178.

Francheteau, J., Needham, H.D., Choukroune, P., Juteau, T., Seguret, M., Ballard, R.D., Fox, P.J., Normark, W., Carranza, A., Cordoba, A., Guerro, J., Ranging, C., Bougeault, H., Cambon, P., Hekinian, R., 1979. Massive deep sea sulphide ore deposits discovered on the East Pacific Rise. Nature 277, 523–528.

Franklin, R.E., 1951. Crystal groth on graphitizing and non-graphitizing carbons. Proc. Royal Soc. London, Series A 209, 196–218.

Frimmel, H.E., 2007. Earth's continental crustal gold endowment. Earth and Planetary Science Letters 267, 45–55.

Fryer, B.J., Kerrich, R., Hutchinson, R.W., Peirce, M.G., Rogers, D.S., 1979. Archean precious metal hydrothermal systems, Dome Mine, Abitibi greenstone belt: 1. patterns of alteration and metal distribution. Canadian Journal of Earth Science 16, 421–439.

Furse, G.D., 1932. Geology of the Swayze area. Ontario Department of Mines Annual Report 41 (Pt 3), 35 59.

Fyfe, W.S., Henley, R.W., 1973. Some thoughts on chemical transport processes in particular reference to gold. Minerals Science and Engineering 5, 295–303.

Fyfe, W.S., Kerrich, R., 1984. Gold natural concentration processes. In: Foster, R.P. (Ed.), Gold '82: The Geology and Genesis of Gold Deposits, 753 p.

Fyfe, W.S., Price, N.J., Thompson, A.B., 1978. Fluids in the Earths' Crust. Elsevier, Amsterdam, 383 p.

Fyon, A., Crockett, J.H., 1984. Gold exploration in the Timmins District using field and lithogeochemical characteristics of carbonate alteration zones. In: Hodder, R.W., Petruk, W. (Eds.), Geology of Canadian Gold Deposits, vol. 24. CIMM, pp. 113–129.

Gaboury, D., 2013. Does gold in orogenic deposits come from pyrite in deeply buried carbon-rich sediments? Insight from volatiles in fluid inclusions. Geology 41, 1207–1210 doi:10.1130.

Garrels, R.M., MacKenzie, F.T., 1971. Evolution of Sedimentary Rocks. Norton, New York.

Gelinas, B.R., 2015. Characterization of the Mineralization and Alteration at Tower Mountain, Commee Township, Ont (B. Sc. thesis). Dept. of Geology, Laurentian Univ.

Gilbert, P., 2013. Bird River Greenstone Belt in Manitoba-Neoarchean Arc Magmatism in the Wester Superior Province; Powerpoint Presentation. Manitoba Geological Survey.

Glasson, M.J., Keays, R.R., 1978. Gold mobilization during cleavage development in sedimentary rocks from auriferous slate belt in central Victoria Australia: some important boundary conditions. Economic Geology 73, 496–511.

Goldfarb, R.J., Ayuso, R., Miller, M.L., Ebert, S.W., Marsh, E.E., Petsel, S.A., Miller, L.D., Bradley, D., Johnson, C., McClelland, W., 2004. The Late Cretaceous Donlin Creek Deposit, Southwestern Alaska—controls on epizonal formation. Economic Geology 99, 643–671.

Goldfarb, R.J., Baker, T., Dubé, B., Groves, D.I., Hart, C.J.R., Robert, F., Gosselin, P., 2005. World distribution, productivity, character, and genesis of gold deposits in metamorphic terranes. In: Hedenquist, J.W., Thompson, J.F.H., Goldfarb, R.J., Richards, J.P. (Eds.), 1905–2005: Society of Economic Geologists. Economic Geology 100, 407–450.

Goldfarb, R.J., Groves, D.I., Gardoll, S., 2001. Orogenic gold and geologic time: a global synthesis. Ore Geology Reviews 18, 1–75.

Goldfarb, R.J., Hart, C., Miller, M., Miller, L., Farmer, G.L., Groves, D.I., 2000. The Tintina gold belt—a global perspective. In: Tucker, T.L., Smith, M.T. (Eds.), The Tintina Gold Belt—Concepts, Exploration, and Discoveries, British Columbia and Yukon Chamber of Mines, vol. 2, pp. 5–34.

Goldfarb, R.J., Leach, D.L., Miller, M.L., Pickthorn, W.J., 1986. Geology, metamorphic setting, and genetic constraints of epigenetic lode-gold mineralization within the Cretaceous Valdez Group, south-central Alaska. In: Keppie, J.D., Boyle, R.W., Haynes, S.J. (Eds.), Turbidite-Hosted Gold Deposits. Geological Association of Canada, pp. 87–105. Special Paper 32.

Goldfarb, R.J., Leach, D.L., Pickthorn, W.J., Paterson, C.J., 1988. Origin of lode-gold deposits of the Juneau gold belt, southeastern Alaska. Geology 16, 440–443.

Goldfarb, R.J., Miller, L.D., Leach, D.L., Snee, L.W., 1997. Gold deposits in metamorphic rocks of Alaska. Economic Geology. Monograph 9, 151–190.

Goldfarb, R.J., Snee, L.W., Miller, L.D., Newberry, R.J., 1991. Rapid de-watering of the crust deduced from ages of mesothermal gold deposits. Nature 354, 296–299.

Golding, S.D., Duck, L.J., Young, E., Baublys, K.A., Glikson, M., Kamber, B.S., 2011. Earliest seafloor hydrothermal systems on earth: comparison with modern analogues. In: Golding, S.D., Glikson, M. (Eds.), Earliest Life on Earth: Habitats, Environments and Methods of Detection. http://dx.doi.org/10.1007/978-90-481-8794-2_2.

Golding, M.J., Neufield, J.A., Hesse, M.A.Q., Huppert, H.E., 2011. Two phase gravity currents in porous media. Journal of Fluid Mechanics 678, 248–270.

Goldstein, S.T., 1988. Decoupled Rb-Sr and Sm-Nd isotopic evolution of the continental crus. Lunar and Planetary Institute 65–67. Rep. 88-02.

Graves, M.C., Zentilli, M., 1982. A review of the geology of gold in Nova Scotia. In: Hodder, R.W., Petruk, W. (Eds.), Geology of Canadian Gold Deposits. Can. Inst. Min. Metall., pp. 233–242. Special Paper 24.

Graves, M.C., Zentilli, M., 1988. Geological Survey of Canada Open Files Report 1829, 110.

Gray, M.D., Hutchinson, R.W., 2001. New evidence for multiple periods of gold emplacement in the Porcupine mining district, Timmins area, Ontario, Canada. Economic Geology 96, 453–475.

Gray, P., Scott, S.K., 1990. Chemical Oscillations and Instabilities: Non Linear Chemical Kinetics. Oxford University Press, London.

Gromet, L.P., Dymek, R.F., Haskin, L.A., Korotev, R.L., 1984. The "North American shale composite": its compilation, major and trace element characteristic. Geochimica et Cosmochimica Acta 48, 2469–2482.

Gross, G.A., 1988. Gold Content and Geochemistry of Iron-Formation in Canada. Geological Survey of Canada. Paper 86-19, 54 p.

Groves, D.I., Foster, R.F., 1991. Archean lode gold deposits. In: Foster, R.P. (Ed.), Gold Metallogeny and Exploration. Blackie, London, pp. 63–103.

Groves, D.I., 1993. The crustal continuum model for late Archaean lode gold deposits of the Yilgarn Block, West Australia. Mineralium Deposita 28, 366–374.

Groves, D.I., Goldfarb, R.J., Gebre-Mariam, M., Hagemann, S.G., Robert, R., 1998. Orogenic gold deposits: a proposed classification in the context of their crustal distribution and relationship to other gold deposit types. Ore Geology Reviews 13, 7–27.

Groves, D.I., Goldfarb, R.J., Robert, F., Hart, C.J.R., 2003. Gold deposits in metamorphic belts: overview of current understanding, outstanding problems, future research, and economic significance. Economic Geology 98, 1–30.

Haeussler, C.J., Bradley, D.C., 1995. Link between ridge seduction and gold mineralization in southern Alaska. Geology 23, 995–998.

Hageman, S.G., Cassidy, K.F., 2000. Archean orogenic lode gold deposits. Reviews in Economic Geology 9–168.

Hagemann, S.G., Gebre-Mariam, M., Groves, D.I., 1994. Surface water influx in shallow-level Archean lode-gold deposits in western Australia. Geology 22, 1067–1070.

Hagemann, S.G., Mavrogenes, J.A., Moritz, R., 2006. The Au-Granite Connection Revisited. Goldschmidt Conference Abs. 2006. p. A222.

Hamilton, J.V., Hodgeson, C.J., 1986. Mineralization an d structure of the Kolar gold field, India. In: MacDonald, A.J. (Ed.), Gold 86, Toronto Proc, pp. 270–283.

Hannington, M.D., Scott, S.D., 1988. Mineralogy and geochemistry of a hydrothermal silica-sulfide-sulfate spire in the caldera of Axial Seamount, Juan de Fuca Ridge. Canadian Mineralogists 26, 603–625.

Hannington, M.D., Scott, S.D., 1989. Sulfidation equilibria as guides to gold mineralization in volcanogenic massive sulfides: evidence from sulfide mineralogy and the composition of sphalerite. Economic Geology 84, 1978–1995.

Hannington, et al., 2005. Sea-floor tectonics and submarine hydrothermal systems. Economic Geology 100, 111–141.

Harcouët-Menou, V., Guillou-Frottier, L., Bonneville, A., Adler, P., Mourzenko, V., 2009. Hydrothermal convection in and around mineralized fault zones: insights from two- and three-dimensional numerical modeling applied to the Ashanti belt, Ghana. Geofluids 9, 116–137.

Harris, A.C., White, N.C., McPhie, J., Bull, S.W., Line, M.A., Skrzecsynski, R., Mernagh, T.P., Tosdal, R.M., 2009. Early Archean hot springs above epithermal veins, North Pole, western Australia: new insights from fluid inclusion microanalysis. Economic Geology 104, 793–814.

Harris, J.R., Wilkinson, L., Grunsky, E., 2000. Effective use and interpretation of lithogeochemical data in regional mineral exploration programs: application of Geographic Information Systems (GIS) technology. Ore Geology Reviews 16, 107–143.

Harris, J.R., Wilkinson, L., Grunsky, E., Heather, K., Ayer, J., 1999. Techniques for analysis and visualization of lithogeochemical data with applications to the Swayze greenstone belt, Ontario. Journal of Geochemical Exploration 67, 301–334.

Haymon, R.M., Kastner, M., 1981. Hot spring deposits on the East Pacific Rise at 21°N: preliminary description of mineralogy and genesis. Earth and Planetary Science Letters 53, 363–381.

Haynes, S.J., 1983. Typomorphism of Turbidite-Hosted Auriferous Quartz Veins, Southern Guysborough County. Department Mines and Energy, Nova Scotia, pp. 183–224. Report 83-1.

Haynes, S.J., 1984. Polygenesis of Gold Deposits, Eastern Meguma Domain. Department Mines and Energy, Nova Scotia, pp. 215–223. Report 84-1.

Haynes, S.J., 1986. Geology and geochemistry of turbidite-hosted gold deposits, greenschist facies, eastern Nova Scotia. In: Keppie, J.D., et al. (Eds.), Aturbidite-Hosted Gold Deposits. Geological Association of Canada, pp. 161–177. Special Paper 32.

Haynes, S.J., February 1987. Classification of quartz veins in turbidite-hosted gold deposits, greenschist facies, eastern Nova Scotia. CIM Bulletin 80, 37–51.

Heather, K.B., Shore, G.T., van Breemen, O., 1996. Geological Investigations in the Swayze Greenstone Belt, Southern Superior Province, Ontario. NODA Summary Report 1995–1996. NRC and Ministry of Northern Ontario Development and Mines, pp. 61–72.

Heather, K.B., Shore, G.T., 1999. Geology, Swayze greenstone belt. Geological Survey of Canada Open File 3384, 2 sheets.

Heather, K.B., 2001. The Geological Evolution of the Archean Swayze Greenstone Belt, Superior Province, Canada (Ph.D. thesis). Keele University, Keele, England, 370 p.

Heiligmann, M., William-Jones, A.E., Clark, J.R., 2008. The role of sulphate-sulphide-oxide-silicate equilibria in the metamorphism of hydrothermal alteration at the Hemlo gold Deposit, Ontario. Economic Geology 103, 335–351.

Hein, J.R., Clague, D.A., Koski, R.A., Embley, R.W., Dunham, R.E., 2008. Metalliferous sediment and a silica-hematite deposit within the Blanco fracture zone, Northeast Pacific. Marine Georesources Geotechnology 26, 317–339.

Hekinian, R., Fouquet, Y., 1985. Volcanism and metallogenesis of axial and off-axial structures on the East Pacific Rise near 13°N. Economic Geology 80, 221–249.

Helgeson, H.C., 1969. Thermodynamics of hydrothermal systems at elevated temperatures and pressures. American Journal of Science l267, 729–804.

Helgeson, H.C., Delany, J.M., Nesbitt, H.W., et al., 1978. Summary and critique or the thermodynamic properties of rock forming minerals. American Journal of Science 278, 229.

Helt, K.M., Williams-Jones, A.E., Clark, J.R., Wing, B.A., Wares, R.P., 2014. Constraints on the genesis of Archean intrusion-related Canadian Malarctic gold deposit, Quebec, Canada. Economic Geology 109, 713–735.

Henderson, J.R., 1983. Analysis of Structure as a Factor Controlling Gold Mineralization in Nova Scotia. Geological Survey of Canada, pp. 13–21. Current Research Part B, Paper 83–1B.

Henderson, M.N., Henderson, J.R., 1986. Constraints on the origin of gold in the Meguma Zone, Ecum Secum area; Nova Scotia. Maritime Sediments and Atlantic Geology 22, 1–13.

Henderson, J.R., Henderson, M.N., Wright, T.O., 1990. Water-sill hypothesis for the origin of certain veins in the Meguma Group, Nova Scotia, Canada. Geology 18, 654–657.

Henderson, J.R., Wright, T.O., Henderson, M., 1989. Mechanics of formation of gold-bearing veins, Nova Scotia, Canada—Comment. Tectonophysics 166, 351–364.

Henderson, J.R., Wright, T.O., Henderson, M.N., 1996. A history of cleavage and folding: an example from the Goldenville Formation, Nova Scotia. Geological Society of America Bulletin 9, 1354–1366.

Henley, R.W., Norris, R.J., Patterson, C.J., 1976. Multstage ore genesis in the New Zealand geosyncline: a history of post-metamorphic lode emplacement. Mineralium Deposita 11, 180–196.

Herzig, P.M., Becker, K.P., Stoffers, P., Bäcker, H., Blum, N., 1988. Hydrothermal silica chimney fields in the Galapagos Spreading Center at 86°W. Earth and Planetary Science Letters 89, 261–272.

Hinse, G.J., Hogg, G.M., Robertson, D.S., 1986. On the origin of Archean vein-type gold deposits with reference to the Larder Lake 'Break' of Ontario and Quebec. Mineralium Deposita 21, 216–227.

Hodgson, C., 1990a. Uses and abuses of ore deposit models in mineral exploration. In: Sheahan, P.A., Cherry, M.E. (Eds.), Ore Deposit Models, Volume II, Geoscience Canada, pp. 1–11. Reprint Series 6.

Hodgson, C.J., 1990b. Uses (and abuses) of ore deposit models in mineral exploration. Geoscience Canada 17 (2), 79–89.

Hodgson, C.J., 1989. The Structure of shear-related, vein-type gold deposits: a review. Ore Geology Reviews 4, 231–273.

Hodgson, C.J., Troop, D.G., 1988. A new computer aided methodology for area selection in gold exploration-a case study from the Abitibi Greenstone Belt. Economic Geology 83, 952–977.

Hofstra, A.H., Cline, J.S., 2000. Characteristics and models for Carlin-type gold deposits (Chapter 5). In: Hagemann, S.G., Brown, P.E. (Eds.), Gold in 2000: Society of Economic Geologists, Reviews in Economic Geology, vol. 13, pp. 163–220.

Holland, H.D., Malinin, S.D., 1979. In: Barnes, H.L. (Ed.), The Solubility and Occurrence of Non-ore Minerals; in Geochemistry of Hydrothermal Ore Deposits, second ed. J. Wiley & Sons, pp. 461–508.

Horne, R.J., Culshaw, N., 2001. Flexural-slip folding in the Meguma group, Nova Scotia. Journal of Structural Geology 23, 1631–1652.

Hronsky, J.M.A., Groves, D.I., Loucks, R.R., Graham, B., 2012. A unified model for gold mineralisation in accretionary orogens and implications for regional-scale exploration targeting method. Mineralium Deposita 47, 339–358.

Huang, Z.L., Liu, C.Q., Yang, H.L., Xu, C., Han, R.S., Xiao, H.Y., Zhang, B., Li, W.B., 2002. The geochemistry of lamprophyres in the Laowangzhai gold deposits, Yunnan Province, China: implications for its characteristics of source region. Geochemical Journal 36, 91–112.

Hunt, T.S., 1868. The Gold Regions of Nova Scotia, Report of Dr. T. Sterry Hunt Addressed to Sir W.E. Logan, Director of the Geological Survey of Canada; Printed by Order of the House of Commons, Ottawa, 48 pages.

Hurst, M.E., 1935. Vein formation at Porcupine. Economic Geology 30, 103–127.

Hutchinson, R.W., 1965. Genesis of Canadian massive sulphides reconsidered by comparison to Cyprus deposits. CIM Bulletin 58, 972–986.

Hutchinson, R.W., 1987. Metallogeny of Precambrian gold deposits: space and time relationships. Economic Geology 82, 1993–2007.

Hutchinson, R.W., 1993. A multi-stage, multi-process genetic hypothesis for greenstone-hosted gold lodes. In: Haynes, S.J. (Ed.), Vein-Type Ore Deposits. Ore Geology Review 8, 347–381.

Hutchinson, R.W., Burlington, J.L., 1984. Some broad characteristics of greenstone belt gold lodes. In: Foster, R.P. (Ed.), Gold 1982. Geological Society of Zimbabwe, pp. 339–372. Special Publication No. 1.

Hutton, J., 1795. Theory of the Earth.

Irvine, T.N., Baragar, W.R.A., 1971. A guide to the chemical classification of the common volcanic rocks. Canadian Journal of Earth Sciences 8, 523–548.

Janecky, D.R., Seyfried Jr., W.E., 1984. Formation of massive sulfide deposits on oceanic ridge crests: incremental reaction models for mixing between hydrothermal solutions and seawater. Geochimica et Cosmochimica Acta 48, 2723–2738.

Jarrad, R.D., 2003. Subduction fluxes of water, carbon dioxide, chlorine, potassium. Geochemistry, Geophysics, Geosystems 4, 50 p.

Jébrak, M., LeQuentrec, M.F., Mareschal, J.C., Blais, D., 1991. A gravity survey across the Bourlamaque massif, southeastern Abitibi greenstone belt, Québec, Canada: the relationship between the geometry of tonalite plutons and associated gold mineralization. Precambrian Research 50 (3,4), 261–268.

Jemelieta, A., Spooner, E.T.C., 1989. An Evolutionary Trend of Na Enrichment in Archean Trondhjemite-tonalite-granodiorite (TTG) Intrusions Which Host Au- Quartz Vein Mineralization in the Renabie Area, Wawa, Ontario [abstract]. Geological Association of Canada - Mineralogical Association of Canada. Program with Abstracts, 14, 1989, A1 18.

Jensen, E.P., Barton, M.D., 2000. Gold deposits related to alkaline magmatism. Reviews in Economic Geology 13, 279–314.

Jensen, L.S., 1976. A new cation plot for classifying subalkalic volcanic rocks. Ontario Department of Mines Miscellaneous Paper 66, 22.

Johnson, J.P., McCulloch, M.T., 1995. Sources of mineralising fluids for the Olympic Dam deposit (South Australia): Sm-Nd isotopic constraint. Chemical Geology 12 (1), 177–199.

Jolliffe, A.W., 1966. Stratigraphy of the Steeprock Group. Steeprock Lake, Ontario, Geol. Ass. Can. Sp. paper No. 3, pp. 77–98.

Jones, J.B., Segnit, F.R., 1971. The nature of opal. 1. Nomenclature and consistent phases. Journal of the Geolological Society of Australia 18, 57–68.

Jowett, C., 1986. Review of ore deposit geology and its influence on mineral exploration by Richard P. Edwards and Keith Atkinson. In: Geoscience Canada, vol. 15. Chapman and Hall Ltd, Toronto. No. 4, p. 310.

Karvinen, W.O., September 1978. The Porcupine Camp – a model for gold exploration in the Archean. Canadian Mining Journal 49–53.

Karvinen, W.O., 1980. Geology and evolution of gold deposits, Timmins area, Ontario. In: Roberts, R.G. (Ed.), Genesis of Volcanic-Hosted Gold Deposit. Ontario Geological Survey, pp. 1–43. Open File Report 5293.

Karvinen, W.O., 1981. Geology and evolution of gold deposits, Timmins area, Ontario; P.29-46 in genesis of Archean, volcanic-hosted gold deposits. In: Symposium Held at the University of Waterloo, March 7, 1980. Ontario Geological Survey. Miscellaneous Paper 97, 175 p.

Karvinen, W.O., 1982. Geology and Evolution of Gold Deposits, vol. 24. CIM, Timmins, Ontario, pp. 101–112.

Katz, L.R., Kontak, D.J., Dubé, B., Rogers, J.R., McNicoll, V., Creaser, R.A., 2014. The Archean Côté Gold Intrusion-related Au (-Cu) Deposit, Ontario, Canada: a Large-Tonnage, Low-Grade Deposit Centred on a Magmatic-hydrothermal Breccia. Abstract. GAC-MAC Meeting. Fredericton 2014. Session SY5: Gold in 2014: A Holistic Approach. http://gac.esd.mun.ca/gac_2014/search_abs/sub_program.asp?sess=98&for%20m=10&abs_no=156.

Keays, R.R., Scott, R., 1976. Precious metals in ocean ridge basalts as source rocks for gold mineralization. Economic Geology 71, 705–720.

Keith, M., Haase, K.M., Schwarz-Schampera, U., Klemd, R., Petersen, S., Bach, W., 2014. Effects of temperature, sulfur, and oxygen fugacity on the composition of sphalerite from submarine hydrothermal vent. Geology. http://dx.doi.org/10.1130/G35655.1. First published online June 30, 2014.

Kempe, U., Belyatski, B., Krymski, R., Kremenetski, A., 2001. Sm–Nd and Sr isotope systematics of scheelite from the giant Au (–W) deposit Muruntau (Uzbekistan): implications for the age and sources of Au mineralization. Mineralium Deposita 36, 370–392.

Keppie, J.D., 1976. Structural model for the saddle reef and associated gold veins in the Meguma Group, Nova Scotia. CIM Bulletin 69 (774). http://dx.doi.org/10.1130/G35655.1103-116.

Keppie, D.J., Boyle, R.W., Haynes, S.J. (Eds.), 1986. Turbidite-Hosted Gold Deposits. Geological Association of Canada. Special Paper 32.

Kerrich, R., 1983. Geochemistry of Gold Deposits in the Abitibi Greenstone Belt, vol. 27. The Canadian Institute of Mining and Metallurgy, 75 p.

Kerrich, R., 1986. Fluid transport in lineaments. Philosophical Treansactions of the Royal Society of London, Series A 317, 219–251.

Kerrich, R., 1989. Geodynamic setting and hydraulic regimes: shear zone hosted mesothermal gold deposits. In: Bursnall, J.T. (Ed.), Mineralization and Shear Zones, vol. 6. Geological Association of Canada, pp. 89–128. Short Course Notes.

Kerrich, R., 1993. Perspectives on genetic models for lode gold deposits. Preferred Mineralium Deposita 28, 362–365.

Kerrich, R., Cassidy, K.F., 1994. Temporal relationships of lode gold mineralization to accretion, magmatism, metamorphism and deformation–Archean to present: a review. Ore Geology Reviews 9, 263–310.

Kerrich, R., Fryer, B.J., 1979. Archean precious metal hydrothermal systems, Dome Mine, Abitibi Greenstone Belt. IIREE and oxygen isotope relations. Canadian Journal of Earth Sciences 16, 440–458.

Kerrich, R., Fyfe, W.S., 1981. The gold carbonate association: source of CO_2 and CO_2 fixation reactions in Archaean lode gold deposits. Chemical Geology 33, 265–299.

Kerrich, R., Goldfarb, R., Groves, D., Garwin, S., 2000. The geodynamic of world-class gold deposits: characteristics, space-time distribution and origins. In: Hagemann, S.G., Brown, P.E. (Eds.), Gold in 2000: Society of Economic Geologists, Reviews in Economic Geology, vol. 13, pp. 501–555.

Kerrich, R., Hodder, R.W., 1982. Archean lode gold and base metal deposits: chemical evidence for metal fractionation into independent hydrothermal reservoirs. In: Hodder, R.W., Petruk, W. (Eds.), Geology of Canadian Gold Deposits, Special Volume 24. The Canadian Institute of Mining and Metallurgy, pp. 144–165.

Kerrich, C., Wyman, D., 1990. Geodynamic setting of mesothremal gold deposits: an association with accretionary tectonic regimes. Geology 18, 882–885.

Kerrich, R., Wyman, D., 1994. The mesothermal gold-lamprophyre association: significance for an accretionary geodymanic, supercontintal cycles and metallogenic processes. Mineralogy petrology 51, 147–172.

Kerswill, J.A., 1988. Lithogeochemical indicators of gold potential in the eastern Meguma Terrane of Nova Scotia. In: MacDonald, D.R., Brown, Y. (Eds.), Mines and Minerals Branch Rep. Activities 1988, Part A. Department of Mines and Energy, Nova Scotia, pp. 215–218. Rep. 88-3.

Kerswill, J.A., 1993. In: Kirkham, R.V., Sinclair, W.D., Thorpe, R.I., Duke, J.M. (Eds.), Models for Iron-Formation-Hosted Gold Deposits. Geological Association of Canada, pp. 171–179. Special Paper 40.

Kerswill, J.A., 1996. Iron formations strata-bound gold: in geology of Canada mineral deposit types. NDAG, Geology of Canada 8, 367–382.

Kessler, S.E., 2010. Hydrothermal Gold Resources for the Next Millennium. Economic Geologists. Special Publication 15, 5–18.

Kharaka, Y.K., Thordsen, J.J., Evans, W.C., Kennedy, B.M., 1999. Geochemical and hydromechanical interactions of fluid associated with the San Andreas Fault System. In: hanegerg, W.C., Mosley, P.S., Moore, J.C., Goodwin, B.L. (Eds.), Faults and Subsurface Fluid Flow in Shallow Crust, Geoph. Monogr., vol. 113, pp. 129–148.

Kinkel Jr., A.R., 1948. The Buffalo Ankerite mine. In: Structural Geology of Canadian Ore Deposits, a Symposium. CIMM, pp. 515–519.

Kita, I., Taguchi, S., Matsubaya, O., 1985. Oxygen isotope fractionation between amorphous silica and water at 34-93°C. Nature 314, 83–84.

Klominski, J., Partington, G.A., McNaughton, N.J., Ho, S.E., Groves, D.I., 1996. Radiothermal granites of the Cullen Batholith and associated mineralization (NT Australia). In: Czech Geol. Surv. Sp. Paper No. 5. Czech Geol. Surv., Prague, 44 p.

Knauth, P., Epstein, S., 1975. Hydrogen and oxygen isotope ratios in silica from the JOIDES Deep Sea Drilling Project. Earth and Planetary Science Letters 25, 1–10.

Knight, C.L., 1957. Ore genesis-the source bed concept. Economic Geology 52, 808–817.

Kociumbas, M.W., Power-Fardy, D., May 26, 2010. A Technical Review and Mineral Resource Estimate of the Hercules Property, Elmhirst and Rickaby Townships. WGM Ltd, Ontario, Canada. Prepared for Kodiak Exploration Ltd., as posted on. www.sedar.com.

Kolb, J., Kisters, A.F.M., Hoernes, S., Meyer, F.M., 2000. The origin of fluids and nature of fluid-rock interaction in mid-crustal auriferous mylonites of the Renco Mine, southern Zimbabwe. Mineralium Desposita 35, 109–125.

Kontak, D.J., Creaser, R.A., Hamilton, M.A., 2013. Geological and geochemical studies of the côté Lake Au(-Cu) Deposit area, Chester Township; Section 2, pp XX to YY in Ontario geological survey. In: Ayer, J.A., Kontak, D.J., Linnen, R.L., Lin, S. (Eds.), Miscellaneous Release Data 294: Results from the Shining Tree, Chester Township and Matachewan Gold Projects and the Northern Cobalt Embayment Polymetallic Vein Project.

Kontak, D.J., Horne, R.J., 2011. A multi-stage origin for the Meguma lode gold deposits, Nova Scotia, Canada: a possible global model for Slate-Belt hosted gold mineralization. In: Mihir, D., Goldfarb, R.J. (Eds.), Gold Metallogeny, India and Beyond. Alpha Geoscience International, Oxford, UK, ISBN 978-1-84265-646-4, pp. 58–82, 350 pages.

Kontak, D.J., Horne, R.J., Kyser, K., 2011. An oxygen isotope study of two contrasting orogenic veins systems in the Meguma Terrane, Nova Scotia, Canada, with implications for fluid sources and genetic models. Mineralium Deposita 46, 289–304. http://dx.doi.org/10.1007/s00126-011-0332-0.

Kontak, D.J., Horne, R.J., Smith, P., 2001. Meguma Gold Deposits, Nova Scotia: Overview of Past Work with Implications for Future Work. Gangue, G.A.C. Mineral Deposits Div., Issue 71, pp. 1–9.

Kontak, D.J., Kerrich, R., April 1997. An isotopic (C, O, Sr) study of vein gold deposits in the Meguma Terrane, Nova Scotia; implication for source reservoirs. Economic Geology 92 (2), 161–180.

Kontak, D.J., MacDonald, M.A., O'Reilly, G.A., Horne, A.J., Smith, P.K., 2004. Gold in the Meguma Terrane, southern Nova Scotia: is there a continuum between mesothermal lode gold and intrusion-related gold systems? (abstr). In: Geological Association of Canada, Annual Meeting Program with Abstracts. CD-ROM.

Kontak, D.J., Smith, P.K., 1993. A Metaturbidite-hosted lode gold deposit: the Beaver Dam deposit, Nova Scotia. 1. Vein paragenesis and Mineral chemistry. Canadian Mineralogist 31, 471–522.

Kontak, D.J., Smith, P.K., Kerrich, R., Williams, P.F., 1990. Integrated model for Meguma group lode gold deposits, Nova Scotia, Canada. Geology 18, 238–242.

Kontak, D.J., Smith, P.K., Reynolds, P.H., 1993. Geology and 40Ar/39Ar geochronology of the Beaver Dam gold deposit, Meguma terrane, Nova Scotia, Canada: evidence for a 370 Ma age of mineralization. Economic Geology 88, 139–170.

Koons, P.O., Craw, D., 1991. Evolution of fluid driving forces and composition within collisional orogens. Geophysical Research Letters 18 (5), 935–938.

Kretschmar, U.H., 1983. Turbidite-Hosted Gold Veins in Nova Scotia. Unpublished report for Jack Henderson, Geological Survey of Canada. 163 pages.

Kretschmar, U.H., 1984. Program with abstracts: Meguma turbidite-hosted gold deposits in Nova Scotia. In: Presented at Meeting, N.E. Section of Geol. Society of America. Providence, Rhode Island. March 16, 1984. http://geology.gsapubs.org/content/early/2012/10/18/G33301.1.abstract?pa%20petoc.

Kretschmar, U.H., 1999. Report on Clam Lake. Ontario Geological Survey (Assessment File. Pinewest Resources. Chester Twp).

Kretschmar, U.H., 2011. Syngenetic gold: lode vein geology and exploration implications. In: Deschênes, G., Dimitrakoupolos, R., Bouchard, J. (Eds.), World Gold 2011. Proceedings of the 50th Conference of Metallurgists. October 2–5, 2011. Canadian Institute of Mining and Metallurgy, Montreal, pp. 849–863.

Kretschmar, U.H., 2012. Program with abstracts: a syngenetic mode for lode gold vein formation and implications. In: Presented at Meeting, GAC and MAC, St. John's (2012), Session SS2: Gold Metallogeny of the North Atlantic Borderlands and Beyond.

Kretschmar, U., Scott, S.D., 1976. Phase relations involving arsenopyrite in the system Fe-As-S and their applications. The Canadian Mineralogist. 14, 364–386.

Ku, T.L., 1976. The uranium series methods of age determination. Annual Review of Earth and Planetary Sciences 4, 347–399.

Kuhn, T.S., 2012. The Structure of Scientific Revolutions. 50th Anniversary. Ian Hacking (intro.), fourth ed. University of Chicago Press, ISBN 978-0-226-45811-3, p. 264. LCCN 2011042476. See review in:http://en.wikipedia.org/wiki/The_Structure_of_Scientific_Revolutions.

Kuhns, R.J., Kennedy, P., Cooper, P., Brown, P., Mackie, B., Kusins, R., Friesen, R., 1986. Geology and mineralization associated with the Golden giant Deposit, Hemlo, Ontario, Canada. In: James MaDonald, A. (Ed.), Gold, 1986: An International Symposium on the Geology of Gold Deposits, pp. 327–354.

Ladeira, E.A., 1991. Genesis of gold in Quadrilatera Ferrifero: a remarkable case of permanency recycling and inheritance-a tribute to Dajcleia Guimeros, Pierre Routhier and Hans Ramberg. In: Ladeire, E.A. (Ed.), Proceedings of Brazil Gold 1991, a Symposium on the Geology of Gold. A.A. Balkama, Rotterdam, pp. 11–30.

Lafrance, B., DeWolfe, J., Stott, G., 2004. A structural reappraisal of the Beardmore-Geraldton belt at the southern boundary of the Wabigoon subprovince, Ontario, and implications for gold mineralization. Canadian Journal of Earth Science 41.

Laird, H.C., 1932. Geology of the Three Duck Lakes area. Ontario Department of Mines Annual Report 41 (Pt 3), 1–34.

Lambeck, A., Huston, D., Barovich, K., 2010. Typecasting prospective Au-bearing sedimentary lithologies using sedimentary geochemistry and Nd isotopes in poorly exposed Proterozoic basins of the Tanami region, Northern Australia. Mineralium Deposita 45 (2010), 497–515.

Langford, G.B., 1928. Geology of the Beardmore-Nezah Gold Area, Thunder Bay District, Ontario. Ontario Department of Mines Annual Report 37 (pt 4), 83–108.

Large, et al., 2007. Multistage sedimentary and metamorphic origin of pyrite and gold in the giant Sukhoi Log Deposit, Lena gold province, Russia. Economic Geology 102, 1233–1267.

Large, R.R., 1977. Chemical evolution and zonation of massive sulfide deposits in volcanic terrains. Economic Geology 72, 549–572.

Large, R.R., 2010. Stratiform Organic-rich V-As-Mo-Se-Ni-Ag-Zn-Au Horizons in the Popovich and Roberts Mountains Formations, Carlin District, Nevada. Abstract. Meeting of the Geological Society of Nevada May 14–17, 2010.

Large, R.R., Gregory, D.D., Steadman, J.A., Tomkins, A.C., Louejeva, E., Danushevsky, L.V., Halpin, J.A., Maslennikov, V., Sack, P.J., Mukerjee, I., Berry, R., Hickman, A., 2015. Gold in oceans through time. Earth and Planetary Science Letters 428, 139–150.

Lawrence, D.M., Treloar, P.J., Rankin, A.H., Harbridge, P., Holliday, J., 2013. The geology and mineralogy of the Loulo Mining District, Mali, West Africa: evidence for tow distinct styles of orogenic gold mineralization. Economic Geology 108, 199–227.

LeBas, M.J., Lemaitre, R.W., Streckeisen, A., Zanettin, B., 1986. A chemical classification of volcanic-rocks based on the total alkali silica diagram. Journal of Petrology 27 (3), 745–750.

Lemarchand, J., Tremblay, A., Ruffet, G., Gobeil, C., 2010. Caractérisation et datation 40 Ar/39 Ar des structures aurifères de la région de Val d'Or, Abitibi. Rapport Final 2010, Sous projet SC34, Divex (Diversification de l'Exploration Minerale au Quebèc). www.divex.ca.

Levitan, G., 2008. Gold Deposits of the CIS. Xlibris Corp. www.Xlibris.com, 352 p.

Lindgren, W., 1933. Mineral Deposits, fourth ed. McGraw Hill, New York. 930 p.

Lindsay, J.F., Brasier, M.D., McLoughlin, N., Green, O.R., Fogel, M., Steele, A., Mertzman, S.A., 2005. The problem of deep carbon: an Archean paradox. Precision Research 143, 1–22.

Linnen, R.L., Lin, S., Zhang, J., Martin, R.D., Naderi, N., Davis, D.W., Hamilton, M.A., Creaser, R.A., Berger, B.R., Banerjee, N.R., Wing, B.A., Wu, C., 2013. Synthesis of the structure, petrology, geochronology and geochemistry of the young Davidson gold deposit and surrounding area. Section 3, pp. XX to YY. In: Ayer, J.A., Kontak, D.J., Linnen, R.L., Lin, S. (Eds.), Ontario Geological Survey, Miscellaneous Release Data 294: Results from the Shining Tree, Chester Township and Matachewan Gold Projects and the Northern Cobalt Embayment Polymetallic Vein Project.

Linnen, R.L., Van Lichtervelde, M., Černý, P., 2012. Granitic Pegmatites as Sources of Strategic Metals Elements, vol. 8, pp. 275–280.

Lobato, L., Rodriguez, L.R., Vierna, F., 2001. Brazil's premier gold province: Pt. 11 geology and genesis of gold deposits in the Achean Rio des Vallros greenstone belt, Quadrilatero Ferrifero. Mineralium Depsoita 36 (3–4), 249–277.

Lode Gold Deposits. http://gsc.nrcan.gc.ca/mindep/synth_dep/index_e.php%20(Haakon%20Fossen:%20www.geosociety.wordpress.com/2013/11/28/let-the-.

Lonsdale, P., Batiza, R., Simkin, T., 1982. Metallogenesis at seamounts on the East Pacific Rise at 18 25 S. Marine Technology Society Journal 16, 54–60.

Lopez-Pacheco, A., 2011. Interview with 2010 CIM Distinguished Lecturer Jim Franklin. Future Mineral Resource Discoveries. New Knowledge Need for Discoveries. CIM News. Canadian Institute of Mining and Metallurgy, 113 p. December 2010–January 2011.

Lord, C.S., 1934. List of channel assays from Silver Fox vein Newspaper Article, Mar. 12, 1935: "Bear Lake Gold Mines drilling is complete".

Loucks, R.R., Mavrogenes, J.A., June 25, 1999. Gold solubility in supercritical hydrothermal brines measured in synthetic fluid inclusions. Reports Science 284 (5423), 2159–2163. http://dx.doi.org/10.1126/science.284.5423.2159.

Lovelock, J., 2009. The Vanishing Face of Gaia: a Final Warning. Allen Lane, Penguin Books, 131 p.

Luff, W.M., 1973. Structural geology of the Brunswick No. 12 Open Pit. CIM Bulletin 68 (758), 64–74.

MacDonald, A.J., Hodgeson, C.J., 1986. The case for the magmatic-hydrotherma origin of Archaen lode gold deposits (abs.). In: Chater, A.M. (Ed.), Gold '86, an International Symposium on the Geology of Gold Deposits-Poster Paper Abstracts. Geol. Assoc. Can., Toronto, pp. 98–99.

MacDonald, M.A., O'Reilly, G.A., 1989. Gold enrichment associated with postmagmatic processes in the south mountain Batholith, southwestern Nova Scotia. In: MacDonald, D.R. (Ed.), Report of Activities. Department of Natural Resources, Nova Scotia, pp. 13–15. Report 89-1.

Mackasey, W.O., Wallace, H., 1978. Geology of Elmhirst and Rickaby Townships. Ontario Geological Survey, District of Thunder Bay. GR168, Accompanied by Map 2373.

Malcolm, W., 1912. Gold Fields of Nova Scotia. Geological Survey of Canada, Memoir, 20-E, 331 p.

Marquis, P., Hubert, C., Rigg, D.M., 1990. An evaluation of genetic models for gold deposits of the Bousquet district, Quebec, based on their mineralogical, geochemical and structural characteristics. In: Rive, M., Verplast, P., Gagnon, Y., Lulin, J.-M., Riverin, G., Simard, A. (Eds.), The Northwestern Quebec Polymetallic Belt: A summary of 60 Years of Mining Exploration; Can. Inst. of Min. and Metal, vol. 24, pp. 343–399.

Martin, R., 2012. Syenite-Hosted Gold Mineralization and Hydrothermal Alteration at the Young-Davidson Deposit (Master thesis). University of Waterloo, Matachewan, Ontario, 172.

Matthai, S.K., Henley, R.W., Bacigalupo-Rose, S., Binns, R.A., Andrew, A.S., Carr, G.R., French, D.H., McAndrew, J., Kavanaugh, M.E., 1995. Intrusion related high temperature gold quartz veining in the Cosmopolitan Howley metasedimentary rock-hosted gold deposit, Northern Territory, Australia. Economic Geology 90, 1012–1245.

Mawer, C.K., 1986. The bedding-concordant gold-quartz veins of the Meguma group, Nova Scotia. In: Keppie, J.D., Boyle, R.W., Haynes, S.J. (Eds.), Turbidite-Hosted Gold Deposits. Geological Association of Canada, pp. 135–148. Special Paper 32.

Mawer, C.K., 1987. Mechanics of formation of gold-bearing quartz veins, Nova Scotia, Canada. In: Zwart, H.J., Martens, M., van der Molen, I., Passhier, C.W., Spiers, C., Visser, R.L.M. (Eds.), Tectonic and Structural Processes on a Macro-, Meso and Micro-scale, Tectonophysics, vol. 135, pp. 99–119.

Mawer, C.K., 1989. Comments on: Henderson et al. (1989): mechanics of formation of gold-bearing veins, Nova Scotia, Canada—Comment. Tectonophysics 166, 355–356.

McAllister, A.L., 1960. Massive sulphide deposition in New Brunswick. CIM Bulletin 53 (574), 88–98.

McBride, D.E., 1972. The Macex Deposit: Geological and Minor Element Study of a Porphyry Copper-molybdenum Deposit (M.Sc. thesis). Queen's Univ., 159 p.

McBride, D.E., 1976a. Tectonic setting of the Tetagouche Group, host to the New Brunswick polymetallic massive sulphide deposits. In: Metallogeny and Plate Tectonics; Geol. Assoc. Can. Sp. vol. 14, pp. 472–485.

McBride, D.E., 1976b. The Structure and Stratigraphy of the B-zone Heath Steele Mines (Ph.D thesis). Univ. of New Brunswick, Newcastle, New Brunswick, 227 p.

McBride, D.E., 1978. Geology of the Ecum Secum Area, Halifax and Guysborough Counties, Nova Scotia. Paper 78-1. Nova Scotia Department of Mines.

McBride, D.E., 1980a. Report on the 1980 Exploration of the Beardmore Property by Abitibi-Price Inc. Ont. Geol. Sur. Assessment File No. 2-3577, 10 p. and map.

McBride, D.E., 1980b. Report on the Drill Program of Abitibi-Price Inc, Castlewood Lake Area. Ont. Geol. Sur. Assessment File No 23.

McBride, D.E., 1981. Geological Report, Pan Empire Claim Group, Beardmore Area, NTS 42E, 12 and 13. Ont. Geol. Sur. Assessment Report, No. 2-4505, 9 p.

McBride, D.E., 1987. Gold mineralization and its position in the geological evolution of the Beardmore-Tashota area. CIM Bulletin 80 (100), 51–58.

McBride, D.E., 1991. Some observed relationships in deformed Mineral deposits and their importance. Geological Association of Canada, Abst., Ann. Mtg.

McBride, D.E., 1992. Geological Report on the Bannockburn Property. Ont. Geol. Surv. Assess Report.

McBride, D.E., 2002. Qualifying Report on the Chester Township Property for Northville Gold Corporation filed on. www.sedar.com. by Condor Gold Corp. July 2002, 39 p.

McBride, D.E., April 2009. When it comes to gold think outside the box. Canadian Mining Journal 8.

McBride, D.E., 2009a. Report on the Geological Mapping of Claim 4209871. Limerick Twp. Ont. Geol. Surv. Assess. Files, 7 p.

McBride, D.E., 2009b. NI 43-101 Report on the Jacobson Township Property of Cline Mining Corporation. Cline Mining Corp. December 10, 2009, 75 p. www.sedar.com.

McBride, D.E., 2010. Technical Report on the Chester Township Property for Trelawney Mining and Explorations Inc filed on. www.sedar.com. February 2010 48 p.

McBride, D.E., April 2014. NI 43-10-Technical Report on the Kaslo Property, Slocan Mining Camp. British Columbia for Agave Silver Corporation. www.sedar.com. Agave Silver Corp.

McCall, G.W., Nabelek, P.I., Bauer, R.L., Glascock, M.D., 1990. Petrogenesis of Archean lamprophyres in the southern Vermillion granitic complex, northeastern Minnesota with implications for the nature of their source. Contributions to Mineralogy and Petrology 104, 439–452.

McCartney, W.D., 1965. Metallogeny of post Cambrian geosynclines. Geological Survey of Canada Paper 65 (6), 33–41.

McCartney, W.D., Potter, R.R., 1962. Mineralization as related to structural deformation, igneous activity and sedimentation in folded geosynclines. Canadian Mining Journal 83 (4), 83–87.

McCuaig, T.C., Hronsky, J.M.A., 2000. The current status and future of the interface between the exploration Industry and economic geology. Chapter 16. Gold in 2000. In: Reviews in Economic Geology, vol. 13. Econ. Geol. Pub. Inc., pp. 553–559

McCuaig, T.C., Kerrich, R., 1998. P-T-t-deformation-fluid characteristics of lode gold deposits: evidence from alteration systematics. Ore Geology Reviews 12, 381–453.

McLennan, J.F., 1915. Quartz veins in lamprophyre intrusions. Engineering and Mining Journal 99, 11–13.

McNeil, A.M., Kerrich, R., 1986. Archean lamprophyre dikes and gold mineralization, Matheson, Ontario: the conjunction of LILE (large-ion lithophile element)-enriched mafic magmas, deep crustal structures and gold concentration. Canadian Journal of Earth Science 23 (3), 324–343.

Mercier-Langevin, P., Gibson, H.L., Hannington, M.D., Goutier, J., Monnecke, T., Dube, B., Houle, M.G., 2014. Preface: a special issue on Archean magmatism, volcanism, and ore deposits: Part 2. Volcanogenic massive sulfide deposits. Economic Geology 109, 1–9.

Meyer, C., 1981. Ore forming processes in geological history. Economic Geology 75th Anniversary Volume, 6–41.

Milkereit, B., 2007. In: . Proceedings of Exploration 07: Fifth Decennial International Conference on Mineral Exploration, Toronto, 2007.

Milne, V.G., 1967. Geology of Cirrus Lake – Bamoos lake area. Ontario Department of Mines Geological Report 43, p. 61.

Mitchell, R.H., 1994. Suggestions for revisions to the terminology of kimberlites and lamprophyres from a genetic viewpoint. In: Meyer, H.O.A., Leonardos, O.H. (Eds.), Kimberlites, Related Rocks and Mantle Xenoliths. Companhia de Pesquisa de Recursos Minerais, Rio de Janeiro, pp. 15–26.

Moiseenko, V.G., Neronskii, G.I., 1973. Confinement of gold mineralization to metamorphic complexes. In: Smirnov, A.M. (Ed.), Metamorf, Kompleksy Vostoka SSSR, Akad. Nauk, SSSR, Dal'nevost Nauch. Vladivostock, pp. 219–225.

Moore, D.W., 1977. Geology and Geochemistry of a Gold-Bearing Iron Formation and Associated Rocks, Back River, Northwest Territories (M.Sc. thesis). University of Toronto, Toronto, Canada (Unpublished) 186 p.

Moore, E.S., 1940. Genetic relations of gold deposits and igneous rocks in the Canadian shield. Economic Geology 48, 127–139.

Moorehouse, W.W., 1949. Geology of Osway Township. Ontario Department of Mines, Annual Report 58 (Pt 5), 27.

Morelli, R.M., Creaser, R.A., Selby, D., Kontak, D.J., Horne, R.J., 2005. Rhenium-osmium geochronology of arsenopyrite in Meguma group gold deposits, Meguma terrane, Nova Scotia, Canada: evidence for multiple-gold mineralizing events. Economic Geology 100, 1229–1242.

Morovek, P., Pertold, Z., Puncochar, M., Studincna, B., Zacharis, J., 1996. Gold Deposits of Bohemia. Czech Geological Survey, Prague, 96.

Morton, J.L., Sleep, N.H., 1985. A mid-ocean ridge thermal model: constraints on the volume of axial hydrothermal heat flux. Journal of Geophysical Research 90B, 11345–11353.

Mottl, M.J., 1983. Metabasalts, axial hot springs, and the structure of hydrothermal systems at mid-ocean ridges. Geological Society of America Bulletin 94, 161–180.

Mullen, E.D., 1983. $MnO/TiO_2/P_2O_5$: a minor element discriminant for basaltic rocks of oceanic environments and its implications for petrogenesis. Earth and Planetary Science Letters 62, 53–62.

Mullis, J., 1979. The system methane-water as a geological thermometer and barometer from the external part of the central Alps. Bulletin a'e Mineralogie 102, 526–536.

Mumin, A.H., Fleet, M.E., Chryssoulis, S.L., 1994. Gold mineralization in AS-rich mesothermal gold ores of the Bogosu-Prestea mining district of the Ashanti Gold Belt, Ghana, remobilization of "invisible" gold. Mineralium Deposita 29, 445–460.

Mumin, A.H., Scott, S.D., Somarin, A.K., Oran, K.S., 2007. Structural controls on massive sulphide deposition and hydrothermal alteration in the South Sturgeon Lake Caldera, Northwestern Ontario. Exploration and Mining Geology 1–2, 83–107.

Murphy, J.B., van Staal, C.R., Keppie, J.D., 1999. Middle to late Paleozoic Acadian orogeny in the northern Apalachians—a Laramide-style plume modified orogeny? Geology 27, 653–656.

Nekrasov, I.Y., 1996. Geochemistry, Mineralogy and Genesis of Gold Deposits. A.A. Balkema. ISBN: 90 5410 723 5.

Nesbitt, B.E., 1991. Phanerozoic gold deposits in tectonically active continental margins. In: Foster, R.P. (Ed.), Gold Metallogeny and Exploration. Blackie, London, pp. 63–103.

Nesbitt, B.E., Muroowchick, J.B., Muchlenbachs, K., 1986. Duel origins of lode gold deposits in the Canadian Cordillera. Geology 14, 506–509.

Newhouse, W.N., 1936. A zonal gold mineralization in Nova Scotia. Economic Geology 31, 805–831.

Nifontov, R.V., 1957. The Chemogenic Gold Content of Sedimentary Rocks in Some Mining Districts. In the collected papers entitled Methods of Investigating Raw Minerals. State Scientific and Technical Publishing House of Literature on Geology, Geodesy, and the Conservation of Mineral Resources.

Norris, R.J., Henley, R.W., 1976. Dewatering of a metamorphic pile. Geology 4, 333–336.

Olson, S.F., Diakit, Ç.K., Ott, L., Guindo, A., Ford, Ch R.B., Winer, N., Hanssen, E., Lay, N., Bradley, R., Pohl, D., 1992. Regional setting, structure and descriptive geology of the Middle Proterozoic Syama gold deposit, Mali, west Africa. Economic Geology 87, 310–331.

Orogenic Gold Deposits: Clive Willman and Richard Goldfarb. https://www.youtube.com/channel/UClcZXBMeHSw-4BwOfVQPK).

Pace, A., 2013. Gold Hosted in the Trondhjemitic, Tonalitic, and Granodioritic Phases of the Missanabie Lake Batholith; Ontario Geological Survey Resident Geologist Recommendations for Exploration, pp. 14–15.

Pace, A., McMillan, R., 2012. Sault Ste. Marie Resident Geologist's District 2012; in Report of Activities 2012, Resident Geologist Program, Timmins Regional Resident Geologist Report: Timmins and Sault Ste. Marie Districts. Ontario Geological Survey, pp. 40–41. Open File Report 6286.

Peccerillo, A., Taylor, S.R., 1976. Geochemistry of Eocene calc-alkaline volcanic rocks from the Kastamonu area, Northern Turkey. Contributions to Mineralogy and Petrology 58, 63–81.

Perring, C.S., MacNaughton, N.J., 1989. The relationship between Archean gold mineralization and spatially associated minor intrusions at the Kambalda and Norseman gold camps, Western Australia: lead isotope evidence. Mineralium Deposita 27, 10–22.

Perruoti, S., Lindsey, M.D., Jessel, M.W., Ailleres, L., Martin, R., Bourassa, Y., 2014. 3D modelling of the Ashanti belt, southwest Ghana: evidence for a Litho-stratigraphic control on gold occurrences within the Birimian Sefwi Group. Ore Geology Reviews 69, 252–264.

Petrov, B.V., Krendelev, F.P., Bobrov, V.A., Tsimbalist, V.G., 1972. Behavior of radioelements and gold during metamorphism of sedimentary rocks in the Patom Uplands. Geochemistry International 647–655.

Pettke, T., Diamond, L.W., Kramers, J.D., 2000. Mesothermal gold lodes in the North-western Alps: a review of genetic constraints from radiogenic isotopes. European Journal of Mineralogy 12, 213–230. www.geo.unibe.ch/diamond/.../Pettke,Diamond,Kramers_2000.pdf.

Phillips, G.N., Groves, D.I., 1983. The nature of gold-bearing fluids as deduced from gold deposits of Western Australia. Journal of the Geological Society of Australia 30, 25–39.

Phillips, G.N., Groves, D.I., Brown, I.J., 1987. Source requirements for the Golden Mile, Kalgoorlie: significance to the metamorphic replacement model for Archean gold deposits. Canadian Journal of Earth Sciences 24, 1643–1651.

Phillips, G.N., Hughes, M.J., 1998. Victorian gold deposits. Journal of Australian Geology and Geophysics 17 (4), 213–216.

Phillips, G.N., Powell, R., 1992. Gold Only Provinces and Their Common Features, vol. 43. James Cook Univ., Contributions of the Economic Geology Research Unit, 27 p.

Phillips, G.N., Powell, R., 2009. Formation of gold deposits: review and evaluatin of the continuum model. Earth Science Reviews 94, 1–21.

Pickthorn, W.J., Goldfarb, R.J., Leach, D.L., 1987. Dual origins of lode gold deposits in the Canadian Cordillera—discussion. Geology 15, 471–473.

Pirajno, F., 2009. Hydrothermal Processes and Mineral Systems, vol. XXXII. ISBN: 978-1-4020-8612-0 1250 P, 340 Illus, 70 Illus in Color. Hardcover. www.springer.com/978-1-4020-8612-0.

Pitcairn, J.K., Teagle, D.A.H., Kerrich, R., Craw, D., Brewer, T.S., 2005. The behavior of nitrogen and nitrogen isotopes during metamorphism and mineralization-evidence from the Otago and Alpine schists, New Zealand. Earth and Planetary News Letters 233, 229–246.

Planavsky, N.J., Reinhard, C.T., Wang, X., Thomson, D., McGoldrick, P., Rainbird, R.H., Johnson, T., Fischer, W.W., Lyons, T.W., 2014. Low Mid-Proterozoic atmospheric oxygen levels and the delayed rise of animals. Science 346 (6209), 635–638. http://dx.doi.org/10.1126/science.1258410.

Plank, T., Langmuir, C.H., 1998. The chemical composition of subducting sediment and its consequences for the crust and mantle. Chemical Geology 145, 325–394.

Poole, W.H., 1971. Graptolites, Copper and Potassium-Argon in Goldenville Formation. Geological Survey of Canada, Nova Scotia, 9 p. Paper 71-1A.

Poulsen, K.H., Hannigan, 1996. Volcanic associated massive sulphide gold. In: Eckstrand, R.O., Sinclair, W.D., Thorpe, R.I. (Eds.), Geology of Canadian Mineral Deposit Types, vol. 8. Geological Survey of canada, Geology of Canada, pp. 183–196.

Poulsen, K.H., Robert, F., Dubé, B., 2000. Geological Classification of Canadian Gold Deposits. Geological Survey of Canada. Bulletin 540.

Powell, R., Will, T.M., Phillips, G.N., 1991. Metamorphism in Archean greenstone belts: calculated fliud compositions and imptications for gold mineralization. Journal of Metamorphic Geology 9, 141–150.

Proudlove, C., Hutchinson, R.W., Rogers, D.S., 1989. Multphase mineralization concordant gold zones, Dome mines, south Porcupine Ontario, Canada. In: Keays, R.R., Groves, D.I. (Eds.), The Geology of Gold Deposits: the Perspective in 1988. Economic Geology 112–123. Monograph 6.

Pye, E.G., 1951. Geology of Errington township, Little Long Lac area. Ontario Department of Mines 50 (Pt 6), 140.

Pyke, D.R., 1982. Geology of the Timmins Area. Ontario Geological Survey. Report 219, 141 p.

Quiring, H., 1948. Geschichte des Goldes. F. Enke Verlag, Stuttgart, 318 p.

Rabeau, O., Legault, M., Cheilletz, A., Jebrak, M., Royer, J.J., Cheng, L.Z., 2011. Gold potential of a hidden Archean fault zone: the case of the Cadillac-Larder Lake Fault. Exploration and Mining Geology 19, 99–116.

Rabeau, O., Royer, J.-J., Jebrak, M., Cheilletz, A., 2013. Log-uniform distribution of gold deposits along major Archean fault zones. Mineralium Deposita 48, 817–824. http://dx.doi.org/10.1007/S00126-013-0470-7.

Ramdohr, P., 1969. The Ore Minerals and Their Intergrowths. Pergamon, New York.

Ramsay, J.G., 1974. The development of chevron folds. Bulletin Geological Society of America 85, 1741–1754.

Ramsay, J.G., 1980a. Shear zone geometry: a review. Journal of Structural Geology 2, 83–99.

Ramsay, J.G., 1980b. The crack and seal mechanism of rock deformation. Nature, London 284, 125–139.

Ramsay, J.G., Huber, M.I., 1983. The techniques of modern structural geology. In: Strain Analysis, vol. 1. Academic Press, 307 p.

Ramsay, J.G., Huber, M.I., 1987. The techniques of modern structural geology. In: Folds and Fractures, vol. 2. Academia Press, pp. 309–700.

Reid, D.F., Smea, B.W., Hood, W.C., 1986. Final Report on the Black Beaver Project, 1986 Exploration Program. NTS 52L/12SE. Southeastern Manitoba. Report 22–86. Manitoba Assessment File No. 71392.

Richkaby, H.C., 1934. Geology of the Swazye gold area. Ontario Department of Mines 58 (pt. 3), 1–36.

Ridler, R.H., 1970. Relationship of mineralization in volcanic stratigraphy in the Kirkland-Larder Lake area, Ontario. In: Geological Association of Canada, Proceedings, vol. 21, pp. 23–41.

Ridler, R.H., 1975. Regional Metallogeny and Volcanic Stratigraphy of the Superior Province. Geological Survey of Canada, pp. 353–358. Paper 75–1A.

Ridley, J.R., Diamond, L.W., 2000. Fluid chemistry of orogenic lode gold deposits and implications for genetic models. Reviews in Economic Geology 13, 141–162.

Rimsstidt, J.D., 1997. Quartz solubility at low temperatures. Geochimica et Cosmochimica Acta 61, 2553–2558.

Rimstidt, J.D., Barnes, H.L., 1980. The kinetics of silica-water reactions. Geochimica et Cosmochimica Acta 44, 1683–1699.

Robb, L., 2004. Introduction to Ore-forming Processes. Blackwell Science Ltd.

Robert, F., 1990. In: An Overview of Gold Deposits in the Easter Abitibi Subprovince. The Northwestern Quebec Polymetallic Belt: a Summary of 60 Years of Mining Exploration, vol. 43. Can. Inst. Min. Metall., pp. 93–105

Robert, F., Brommecker, R., Bourne, B.T., Dobak, P.J., McEwan, C.J., Rowe, R.R., Zhou, X., 2007. Models and exploration methods for major gold deposits, Paper 48, exploration in the new Millennium. In: Milkereit, B. (Ed.), Proceedings of the Fifth Decennial International Conference on Mineral Exploration, Ore Deposits and Exploration Technology, pp. 691–711.

Robert, F., Poulsen, K.H., 1997. World class Archean gold deposits in Canada: an overview. Australian Journal of Earth Sciences 44, 329–351.

Robert, F., Poulson, K.H., Cassidy, K.F., Hodgeson, C.J., 2005. Gold metallogeny of the Superior and Yilgarn cratons. In: Hedenquist, J.W., Thompson, J.F.H., Goldfarb, R.J., Richards, J.P. (Eds.), Econ. Geol. One Hundredth Ann. vol. 1905–2005, pp. 1001–1033.

Robert, F., Poulsen, K.H., Dube, B., 1994. Structural Analysis of Lode Gold Deposits in Deformed Terranes. Geological Survey of Canada. Open File Report 2850, 140 p.

Robie, R.A., Hemingway, J.C., Fisher, J.R., 1975. Thermodynamic properties of minerals and related substances at 298.15 K and 1 bar (10^5 Pascals) pressure and at higher temperatures. USGS Bulletin 1452, 456.

Rock, N.M.S., 1991. Lamprophyres. Blackie, Glasgow, UK, 284 pp.

Rock, N.M.S., Duller, R., Haszeldine, R.S., Groves, D.I., 1987. In: Ho, S.E., Groves, D.I. (Eds.), Lamprophyres as potential gold targets: some preliminary observation, speculations on recent advances in understanding Precambrian gold deposits, vol. II. Geology Department and University Extension, Univ. of Western Australia, pp. 271–286.

Rock, N.M.S., Groves, D.I., 1988a. Can lamprophyres solve the genetic controversy over mesothermal gold depsoits. Geology 16, 538–541.

Rock, N.M.S., Groves, D.I., 1988b. Do lamprophyres carry gold as well as diamonds? Nature 332, 253–255.

Rock, N.M.S., Groves, D.I., Perring, C.S., Golding, S.D., 1989. Lamprophyres and Prophyries: what does their association mean? In: Keay, R.R., Ramsey, W.R.H., Groves, D.I. (Eds.), Geology of Ore Deposits; the Perspective in 1988 Economic Geology 609–625. Monograph 6.

Rock, N.M.S., Hallberg, J.A., Groves, D.I., Mather, P.J., 1988. Archaean lamprphyres in the goldfields of the Yilgarn Block, western Australia. In: Ho, S.E., Groves, D.I. (Eds.), Advances in Understanding Precambrian Gold Deposits, vol. II. Geology Department and University Extension, The University of Western Australia, pp. 245–276. Publication Number 12.

Rona, P., Scott, S.D., 1993. A special issue on sea-floor hydrothermal mineralization: new perspectives. Preface. Economic Geology 88, 1933–1976.

Rona, P.A., 1984. Hydrothermal mineralization at seafloor spreading centres. Earth Science Reviews 20, 1–104.

Rona, P.A., 1988. Hydrothermal mineralization at oceanic ridges. Canadian Mineralogist 26, 431–465.

de Ronde, C.E.J., De Wit, M.J., 1994. Tectonic history of the Barbarton green stone belt, South Africa-490 million years of Archean crustal evolution. Tectonics 13, 983–1005.

de Ronde, C.E.J., Faure, K., Bray, C.J., Whitford, D.J., 2000. Round Hill shear zone hosted gold deposit, Macrae's Flat, Otego, New Zealand: -evidence of a magmatic ore fluid. Economic Geology 95, 1025–1048.

Roobol, M.J., Hackett, D., 1987. Paleovolcanic facies and exhalite geochemistry; guides for selecting exploration areas in volcano-sedimentary complexes. Economic Geology 82, 691–705.

Rosenbauer, R.J., Bischoff, J.L., 1983. Uptake and transport of heavy metal by heated seawater: a summary of the experimental results. In: Rona, P.A., Bostrom, K., Laubier, K., Smith, K.I. (Eds.), Hydrothermal Processes at Seafloor Spreading Centers, NATO Conference Series, vol. 4. Plenum Press, New York, NY, pp. 177–197.

Rowins, S.M., Groves, D.I., McNaughton, N.J., 1997. Neoproterozoic telfer-style Au (Cu) deposits. Journal of Australian Geology and Geophysics 17 (4), 217–223.

Rumble, D., 2014. Hydrothermal graphitic carbon. Elements 10 (6), 427–433.

Ryan, R.J., Smith, P.K., 1998. A review of mesothermal gold deposits of the Meguma Group, Nova Scotia, Canada. In: Ramsay, W.R.H., Bierlein, F.P., Arne, D.C. (Eds.), Mesothermal Gold Mineralization in Space and Time. Ore Geology Reviews 13, 153–183.

Saager, R., Meyer, M., Muff, R., 1982. Gold distribution in supracrustal rocks from Archean greenstone belts of southern Africa and from Palaeozoic ultramafic complexes of the European Alps, metallogenetic and geochemical implications. Economic Geology 77, 1–24.

Sage, P.R., Heather, B.K., 1991. The Structural, Stratigraphy and Mineral Deposits of the Wawa Area: Field Trip A6. Geological Association of Canada, pp. 81–97.

Sangster, A.L., 1990. Metallogeny of the Meguma terrane, Nova Scotia. In: Sangster, A.L. (Ed.), Mineral Deposit Studies in Nova Scotia, vol. 1. Geological Survey of Canada, pp. 115–162. Paper 90-8.

Sangster, A.L., Smith, P.K., 2007. Metallogenic summary of the Meguma gold deposits, Nova Scotia. In: Goodfellow, W.D. (Ed.), Mineral Deposits of Canada, a Synthesis of Major Deposit-Types, District Metallogeny, the Evolution of Geological Provinces, and Exploration Methods. Geological Association of Canada, Mineral Deposits Division, pp. 723–732. Special Publication No. 5.

Sawkins, F.J., Rye, D.M., 1974. Relationship of Homestake type gold deposits to iron-rich Precambrian sedimentary rocks. Transactions of the Institution of Mining and Metallurgy Section B 83, 56–59.

Schardt, C., Large, R.R., 2009. New insights into the genesis of volcanic-hosted massive sulfide deposits on the seafloor from numerical modeling studies. Ore Geology Reviews 35, 333–351.

Schwerdtner, W.M., Lu, S.J., Yang, J.F., 2010. Wall-rock structure at the present contact surfaces between repeatedly deformed thrust sheets, Grenville Orogen of central Ontario, Canada. Canadian Journal of Earth Science 47, 875–899.

Schwerdtner, W.M., Riller, U.P., Borowik, A., 2005. Structural testing of tectonic hypotheses by field based analysis of distributed tangential shear: examples from major high-strain zones in the Grenville Province and other parts of the southern Canadian Shield. Canadian Journal of Earth Science 42, 1927–1947.

Scott, S.A., 2012. NI43–101 Technical Report on the Little Bear Lake Gold Project for Canadian Imperial Ventures Corp., 42 p. www.sedar.com.

Scott, S.D., 1997. Submarine hydrothermal systems and deposits (Chapter 15). In: Barnes, H.L. (Ed.), Geochemistry of Hydrothermal Ore Deposits, third ed. J. Wiley and Sons, Inc.

Scott, S.D., 1983. Chemical behaviour of sphalerite and arsenopyrite in hydrothermal and metamorphic environments. Mineralogical Magazine 47, 427–435.

Scott, S.D., Chase, R.L., Hannington, M.D., Michael, P.J., McConachy, Shea, G.T., 1990. Sulphide deposits, tectonics and petrogenesis of Southern Explorer Ridge, northeast pacific Ocean. In: Malpas, J., Moores, E.M., Panayiotou, A., Xenphontos, C., Ophiolites (Eds.), Crustal Analogues, Proceedings of the Symposium "Troodos 1987" Nicosia: Geol. Suv. Dept., Min. Agric. and Nat. Res, Cyprus, pp. 719–733.

Sener, A.K., Young, C., Groves, D.I., Krapez, B., Fletcher, I.R., 2003. Major Orogenic Gold Episode Associated with Cordilleran-Style Tectonics Related to the Assembly of Paleoproertozic, Australia. Centre for Global Metallogeny, School of Geographical Sciences, Univ. of Western Australia, Crawley, WA, Australia.

Seward, T.M., 1973. Thio complexes of gold and the transport of gold in hydrothermal ore solutions. Geochimica et Cosmochima Acta 37, 379–399.

Shanmugam, G., 2006. Deep Water Processes and Facies Models: Implications for Sandstone Petroleum Reservoirs. Elsevier.

Shelton, K.L., So, C.-S., Chang, J.S., 1988. Gold-rich mesothermal vein deposits in the Republic of Korea-geochemical studies in the Jungwon gold area. Economic Geology 83, 1221–1237.

Shore, M., Fowler, A.D., 1996. Oscillatory zoning in minerals: a common phenomenon. Canadian Mineralogist 34, 1111–1126.

Sibson, R.H., Robert, F., Poulsen, K.H., 1988. High angle reverse faults, fluid-pressure cycling and mesothermal gold-quartz deposits. Geology 16, 551–555.

Sibson, R.H., Scott, J., 1998. Stress/fault controls on the containment and release of over pressured fluids: examples from gold quartz veins systems I Juneau, Alaska, Victoria, Australia, and Otago New Zealand. Ore Geology Reviews 13, 293–306.

Sillitoe, D., 2010. Porphyry copper systems. Economic Geology 105, 3–41.

Sillitoe, D., 2011. SEG Newsletter. Oct. 2010. No. 83.

Sillitoe, R.H., Hannington, M.D., Thompson, J.F.H., 1996. High sulphidation deposits in the volcanogenic massive sulphide environment. Economic Geology 91, 204–212.

Siragusa, G., 1981. Precambrian Geology of Chester and Yeo Twps and Parts of Neville and Potier Twps., Sudbury District, Ontario Geological Survey. Preliminary map P 2449, Geological Series, scale 1:15,840 or 1 inch to 3/4 mile. Geology 1980.

Siragusa, G., 1991. Precambrian Geology, Parts of Chester, Neville, Potier and Yeo Townships: Ontario Geological Survey. Open File Map 214; scale 1:15,840.

Siragusa, G., 1993. Mineral Prospects of the Swayze Greenstone Belt, Ontario Geological Survey. Open File Report 5844. 30 p.

Skilling, I.P., White, J.D.L., McPhie, J., May 15, 2002. Peperite: a review of magma–sediment mingling. Journal of Volcanology and Geothermal Research 114 (1–2), 1–17. http://dx.doi.org/10.1016/S0377-0273(01)00278-5.

Skirrow, R., Coleman, M.I., 1982. Origin of sulphur and geothermometry of hydrothermal sulphides from the Galapagos Rift, 86°W. Nature 299, 142–144.

Spiess, F.N., MacDonald, K.C., Atwater, T., Ballard, R.D., Carranza, A., Cordoba, D., Cox, C., et al., 1980. East Pacific Rise: hot springs and geophysical experiments. Science 207, 1421–1433.

Spooner, E.T.C., 1991. The magmatic model for the origin of Archean Au-quartz vein ore systems: an assessment of the evidence, Brazil Gold '91. In: Ladeira, E.A. (Ed.), The Economics, Geology, Geochemistry and Genesis of Gold Deposits. A.A. Balkema, pp. 313–318, 844 pages.

Spooner, E.T.C., 1993. Magmatic sulphide/volatile interaction as a mechanism for producing chalcophile element enriched, Archean Au-quartz, epithermal Au-Ag and Au skarn hydrothermal lore fluids. Ore Geology Review 7, 359–379.

Springer, J.S., 1985. Carbon in Archean rocks of the Abitibi region (Ontario-Quebec) and its relation to gold distribution. Canadian Journal of Earth Sciences 22, 1945–1951.

Spry, A., 1969. Metamorphic Textures. Pergamon Press, 350 p.

Stanton, R.L., 1959. Mineralogical features and possible mode of emplacement of the Brunswick mining and smelting ore bodies, Gloucester County, New Brunswick. CIM Bulletin 52, 631–642.

Stanton, R.L., 1972. Ore Petrology. McGraw-Hill, New York.

Stanton, R.L., 1988. The precursor principle and possible significance of stratiform ores and related chemical sediments in the elucidation of processes of regional metamorphic mineral formation. Philosophical Transaction of the Royal Society of London 528–646.

Steadman, J.A., Large, R.R., Meffre, S., Olin, P.H., Danyushevsky, L.V., Gregory, D.D., Belousov, I., Lounejeva, E., Ireland, T.R., Holden, P., 2015. Synsedimentary to early diagenetic gold in black shale-hosted pyrite nodules at the Golden Mile Deposit, Kalgoorlie, West Australia. Economic Geography 110, 1157–1191.

Stringer, P., 1975. Acadian slaty cleavage noncoplanar with fold axial surfaces in the Northern Appalachians. Canadian Journal of Earth Science 12 (6), 949–961.

Styrt, M.M., Brackmann, A.J., Holland, H.D., Clark, B.C., Pisutha-Arnond, V., Eldridge, C.S., Ohmoto, H., 1981. The mineralogy and isotope composition of sulfur in hydrothermal sulphide/sulfate deposits of the East Pacific Rise, 21°N latitude. Earth and Planetary Science Letters 53, 382–390.

Sun, S.S., McDonough, W.F., 1989. Chemical and isotopic systematics of oceanic basalts; implications for mantle composition and processes. In: Saunders, A.D., Norry, M.J. (Eds.), Magmatism in the Ocean BasinsGeological Society of London, London, 42. pp. 313–345.

Sweeney, M.A., Binda, P.L., Vaughan, D.J., 1991. Genesis of the ores of the Zambian Copper belt. Ore Geology Reviews 6, 51–76.

Swinden, S., McBride, D.E., Dube, B., 1990. Preliminary Geological and Mineralogical Notes on the Nugget Pond Gold Deposit, Baie Verte Peninsula, Newfoundland; Current Research. Newfoundland Dept. Mines and Energy. Rept 90–1, pp. 201–215.

Syngenetic Gold. www.syngeneticgold.com.

Taner, M.F., Trudel, P., 1989. In: . Bourlamaque batholith and its gold potential, Val d'Or, Quebec, vol. 82. Can. Inst. Min. Bull., pp. 33–42

Tappe, S., Foley, S.F., Jenner, G.A., Kjarsgaard, B.A., 2005. Integrating ultramafic lamprophyres into the IUGS classification of igneous rocks: rational and implications. Journal of Petrology 46, 1893–1900.

Taylor, F.C., Schiller, E.A., 1966. Metamorphism of the Meguma group of Nova Scotia. Canadian Journal of Earth Sciences 3, 959–974.

Taylor, S.R., McLennan, S.M., 1985. The Continental Crust: Its Composition and Evolution. Blackwell Scientific, Oxford, 312 p.

Taylor, R., 2009. Ore Textures, Recognition and Interpretation. Springer Verlag, ISBN 978-3-642-01783-4 (With CD-ROM) 288 p.

Taylor, W.R., Rock, N.M.S., Groves, D.I., Perring, C.S., Golding, S.D., 1994. Geochemistry of Archean shoshonitic lamprophyres from the Yilgarn Block, Western Australia: Au abundance and association with gold mineralisation. Applied Geochemistry 9, 197–222.

Tenthorey, E., Cox, S.I.F., 2003. Cohesive strengthening of fault zones during interseismic period; an experimental study. Journal of Geophysical Research 111. http://dx.doi.org/101029/2005JB004122.

Therriault, R.G., 2010. Report on work completed on the Little Bear lake Property, July–October, 2009, Manitoba Geol. Surv. Assessment report.

Thomas, H.V., Large, R.R., Bull, S.W., Maslenikov, V., Berry, R.F., Fraser, R., Froud, S., Moye, R., 2011. Pyrite and pyrrhotite textures and compositions in sediment, laminated quartz veins, and reefs at Bendigo gold mine, Australia: insights for ore genesis. Economic Geology 106, 1–31.

Thompson, A.B., Connolly, J.A.D., 1990. Metamorphic fluids and anomalous porosities in the lower crust. Tectonophysics 182, 47–55.

Thompson, J.E., 1941. Geology of McGarry and McVittie townships. Ontario Department of Mines 50 (7).

Thompson, J.F.H., Newberry, R.J., 2000. Gold deposits related to reduced granite intrusions. Economic Geology Review 13, 377–400.

Thurston, P., Ayer, J.A., Goutier, J., Hamilton, M.A., 2008. Deposition gaps in Abitibi belt greenstone sequences: a key to exploring for syngenetic mineralization. Economic Geology 103, 1097.

Tilling, R.I., Gottfried, D., Rowe, J., 1973. Gold abundances in igneous rocks: bearing on gold mineralization. Economic Geology 68, 168–186.

Tomilenko, et al., 2010a. Economic Geology 105 (2), 375–394.

Tomilenko, A., Gibsher, N., Dublyansky, Y., Dallai, L., 2010b. Geochemical and isotope properties of fluids from gold-bearing and barren quartz veins of the Sovetskoye gold deposit (Siberia, Russia). Economic Geology 105 (2), 375–394.

Tomkins, A., 2013. On the source of orogenic gold. Geology 41, 1225–1226.

Tomkins, A.G., Grundy, C., 2009. Upper temperature limits of orogenic gold deposit formation-constraints from granulite hosted Griffen's Find deposit, Yilgarn Craton. Economic Geology 104, 669–685.

Travis, G.A., Woodall, R., Bartrum, G.D., 1971. The geology of the Kalgoorlie goldfield. Geological Society of Australia 175–190. Special Publication 3.

Tremblay, A., 2001. Postmineralization faults in the Beaufor gold deposit, Abitibi greenstone belt, Canada: geometry, origin and tectonic implications for the Val-d'Or mining district. Economic Geology 96, 509–524.

Tulonen, P.V.G., King, B.R., 1997. 1997 OPAP Technical Summary Report, Wolfe Lake Exploration Project, Lots 8 and 9, Concession VI, Tudor Township, Southern Ontario Mining District, Claim 1194956. Ont Gov't. Ass Files, 7 p.

Tunnicliffe, V., Botros, M., de burgh, M.E., Dinet, A., Johnson, H.P., Juniper, S.K., McDuff, R.E., 1986. Hydrothermal vents of Explorer Ridge, northeast Pacific. Deep Sea Research 33A, 402–412.

Turner, J.S., Gustavson, L.B., 1978. The flow of hot Saline solution from vents in the seafloor- some implications for exhalative massive sulfide and other ore deposits. Economic Geology 73, 1082–1100.

Vallance, J., Boiron, M.-C., Catheineau, M., Fourcade, S., Varlet, M., Marignac, C., 2004. The granite hosted gold deposit of Moulin de Chend (Saint-Yrieix district, Central Massif, France)-petrographic, structural, fluid inclusion and oxygen isotope constraints. Mineralium Deposita 39, 265–281.

Valliant, R.I., Hutchinson, R.W., 1982. Stratigraphic distribution genesis of the gold deposits, Bousquet Region, Northwestern Quebec. CIMM 24, 27–41.

Valliant, R.I., Mongeau, C., Doucet, R., 1982. The Bousquet pyritic gold deposit, Bousquet Region, Quebec; descriptive geology and preliminary interpretations on genesis. In: Geology of Canadian Gold Deposits CIMM Sp. Vol. 24, pp. 41–49.

Vanko, D.A., Milby, B.J., Heinzquith, S.W., 1991. Massive sulfides with fluid-inclusion-bearing quartz from a young seamount on the East Pacific rise. Canadian Mineralogist 29, 453–460.

Vesselinov, I., Kerestedjian, T., 1995. Kinetic aspects of sector zoning in arsenopyrite: a case study. Mineralogy and Petrology 52, 85–106.

Viljoen, R.P., Saageer, R., Vijoen, M.J., 1969. Metallogenesis and ore control in the Steynsdorp goldfield, South Africa. Economic Geology 64, 778–797.

Viljoen, R.P., Saageer, R., Viljoen, M.J., 1970. Some thoughts on the origin and processes responsible for the concentration of gold in early Precambrian of southern Africa. Mineralium Deposita 5, 164–180.

Voicu, G., Bardoux, M., Jebrak, M., Crepeau, R., 1999a. Structural, mineralogical and geochemical studies of the Paleoproterozoic Omai gold deposit, Guyana. Economic Geology 94, 1277–1304.

Voicu, G., Bardoux, M., Jebrak, M., 1999b. Tellurides from the Paleoproterozoic Omai gold deposit, Guiana shield. Canadian Mineralogist 37, 559–573.

Von Damm, K.L., 1987. Chemistry of Submarine Hydrothermal Solutions at 21°N, East Pacific Rise, Guaymas Basin of California (Ph.D. thesis). WHOI/MIT, Woods Hole-Cambridge Mass.

Von Damm, K.L., Bishoff, J.L., 1987. Chemistry of hydrothermal solutions from the Southern Juan de Fuca Ridge. Journal of Geophysical Research 93B, 11334–11346.

Von Damm, K.L., Bischoff, J.L., Rosenbauer, R.J., 1991. Quartz solubility in hydrothermal seawater, an experimental study and equation describing quartz solubility for up to 0.5 m NaCl solution. American Journal of Science 291, 977–1007.

Von Damm, K.L., Edmund, J.M., Measures, C.I., Grant, B., 1985. Chemistry of submarine hydrothermal solutions at Gauymas basin, Gulf of California. Geochimica et Cosmochimica Acta 49, 2221–2237.

Vu, L., Darling, R., Beland, J., Popov, V., 1987. Structure of the Ferderber gold deposit, Belmoral Mines Ltd., Val d'Or, Québec. CIM Bulletin 80, 68–77.

Waldron, W.J.F., White, C.E., Barr, S.M., Simonetti, A., Heaman, L.M., 2009. Provenance of the Meguma terrane, Nova Scotia: rifted margin of early Palaeozoic Gondwana. Canadian Journal of Earth Sciences 46 (1), 1–8.

Wall, V., 1989. Fluids and Metamorphism (Ph.D. thesis). Monash University, Melbourne, Australia (Unpublished) 234 p.

Wall, V.J., Yentzen, R.W., Graupnert, T., 2004. Muruntau, Uzbebistan, Prelim. Rpt. on CERCRAMS Research Project.

Walshe, J.L., Halley, S.W., Hall, G.A., Kitto, P., 2003. Contrasting fluid systems, chemical gradients and controlson large tonnage high grade Au deposits, Eastern Goldfields Province, Yigarn craton, Western Australia. In: Eliopoulos, D.G., et al. (Eds.), Mineral Exploration and Sustainable Development. Mill Press, Rotterdam, pp. 827–830.

Walther, J.V., Helgeson, H.C., 1977. Calculation of the thermodynamic properties of aqueous silica and the solubility of quartz and its polymorphs at high pressures and temperatures. American Journal of Science 277, 1315–1351.

Webber, A.P., et al., 2013. Golden plumes: substantial gold enrichment of oceanic crust during ridge-plume interaction. Geology 41, 87–90.

Whalen, J.B., McNicoll, V., Longstaffe, F.J., 2004. Juvenile ca. 2.735-2.720 Ga high- and low-Al tonalitic plutons: implications for TTG and VMS petrogenesis, western Superior Province, Canada. Precambrian Research 132, 275–301. http://dx.doi.org/10.1016/j.precamres.2004.02.008.

Whalen, J.B., Percival, J.A., McNicoll, V.J., Longstaffe, F.J., 2002. A mainly crustal origin for tonalitic granitoid rocks, Superior Province, Canada: Implications for late Archean tectonomagmatic processes. Journal of Petrology 43 (8), 1551–1570.

White, C.E., Barr, S.M., 2004. Age and Petrochemistry of Mafic Sills in Rocks of the Northwestern Margin of the Meguma Terrain, Bear River-Yarmouth Area of Nova Scotia. In Mineral Resources Branch, Report of Activities 2003. Department of Natural Resources, Nova Scotia, pp. 97–117. Report 2004-1.

White, J.D.L., Skilling, I., McPhie, J., 2000. Peperite: a useful genetic term. Bulletin of Volcanology 62, 65–66.

Whitehead, R.E.S., 1973. Application of Rock Geochemistry to Problems of Mineral Exploration and Ore Genesis at Heath Steele Mines, New Brunswick (Ph.D. thesis). Unvi. of new Brunswick, Rrederiction, New Brunswick.

Whitehead, R.E.S., Govett, G.J.S., 1974. Exploration rock geochemistry—detection of trace element halos at Heath Steele Mines (N. B. Canada) by discriminant analysis. Journal of Geochemical Exploration 3, 371–386.

Wilkinson, S.H., Kesler, S.H., 2007. Tectonism and exhumation in convergent margin orogens; insights into ore deposits. Journal of Geology 115, 611–627.

William, H.R., Breaks, F.W., Schnieders, B.R., Smyk, M.C., Charlton, S.G., Lockwood, H.C., 1991. Field guide to the Manitouwadge area. In: Franklin, J.M., Schnieders, B.R., Koopman, E.R. (Eds.), Mineral deposits of the western Superior Province, Ontario (Field Trip 9), Geol. Sur. Can. Open File 2164, 141 p.

Williams, P.F., Hy, C., 1990. Origin and deformational metamorphic history of gold-bearing quartz veins on the eastern shore of Nova Scotia. In: Sangster, A.L. (Ed.), Mineral Deposit Studies in Nova Scotia, 1. Geological Survey of Canada, pp. 169–194. Paper 90-8.

Williams, D.A., Kerr, R.C., Lesher, C.M., 1989. Emplacement and erosion by Archean basalt at Kambalda: revisited. Journal of Geophysical Research 103 (131), 27533–27549.

Williams, H.W., McBirney, A.R., 1979. Volcanology. Freeman, Cooper and Company, 397 pages.

Williams, N., 2008. The Exploration Crisis-an Academic Perspective. Society of Economic Geologists Newsletter. October 2008, No. 75. 5 p.

Wilson, I.R., 1971. Progress Report on a Study on Geochemical Variations in Wall Rock. University of Leeds, Department of Earth Sciences, 15th Ann. Rept., Res. Inst. Afr. Geol., Ashanti Mine, Obuasi, Ghana, pp. 39–43.

Windley, B.F., 2015. Archean Eon: Encl. Britannica Online, 2 p.

de Wit, M.J., Hart, C., Armstrong, R.J., Roede, R.A., Green, C.E.J., Tredoux, R.E.W., Peberdy, E., Hart, R.A., 1992. Formation of an Archean continent. Nature 317, 553–562.

Witt, W.K., Hickman, A.h., Townsend, D., Preston, W.A., 1991. Mineral Potential of the Archean Plibara and Yilgarn Cratons, Western Australia. Journal of Australian Geology and Geophysics 17 (3), 201–221.

Wiwchar, W., 1957. Consolidated Discovery Yellowknife mine. In: Structural Geology of Canadian Ore Deposits, vol. II, pp. 201–209.

Wong, L., Davis, D.W., Krogh, T.E., Robert, F., 1991. U-Pb zircon and rutile chronology of Archean greenstone formation and gold mineralization in the Val d'Or region, Quebec. Earth and Planetary Science Letters 104, 325–336.

Wood, D.A., 1980. The application of a Th-Hf-Ta diagram to problems of tectonomagmatic classification and to establishing the nature of crustal contamination of basaltic lavas on the British Tertiary Volcanic Province. Earth and Planetary Science Letters 50, 11–30.

Wood, B.L., Large, R.R., 2007. Syngenetic gold in western Victoria: occurrence, age, dimensions. Australian Journal of Geology 54, 711–732.

Woodall, R., 1983. Success in mineral exploration: confidence in science and ore deposit models. Geoscience Canada 11 (3), 127–133.

Woolery, T.J., Sleep, N.H., 1976. Hydrothermal circulation and geochemical flux at mid-ocean ridges. Journal of Geology 84, 249–275.

Wooley, A.R., Bergman, S.C., Edgar, A.D., Le Bas, M.J., Mitchell, R.H., Rock, N.M.S., Scott Smith, B.A., 1996. Classification of lamprophyres, lamproites, kimberlites and the kalsilitic, melilitic, and leucitic rocks. Canadian Mineralogist 34, 175–196.

Wulff, K., Dziggel, A., Kolb, J., Vennemann, T., Bottcher, M.E., Meyer, M., 2009. Origin of mineralizing fluids of the sediment-hosted Navachab Gold Mine, Namibia: constraints from stable (O, H, C, S) isotopes. Economic Geology 105 (2), 285–302.

Wyborn, A.J., 1988. Neoproterozoic Telpher-style Au (Cu) deposits. Journal of Australian Geology and Geophysics 17 (4), 217–223.

Wyman, D.A., Kerrich, R., 1988. Lamprophyres a source of gold. Nature 332, 209–210.

Wyman, D., Kerrich, R., 1989. Archean shoshonitic lamprophyres associated with Superior Province gold deposits: distribution, tectonic setting, noble metal abundances, and significance for gold mineralization. In: Keays, R.R., Ramsay, W.R.H., Groves, D.I. (Eds.), The Geology of Gold Deposits: The Perspective in 1988. Economic Geology 651–667. Monograph 6.

Wyman, D., Kerrich, R., 1993. Archean shoshonitic lamprophyres of the Abitibi Subprovince, Canada: petrogenesis, age, and tectonic setting. Journal of Petrology 34 (Part 6), 1067–1109.

Wyman, D.A., Kerrich, R., Groves, D.I., 1999. Lode gold deposits and Archean mantle plume-island arc interaction, abitibi subprovince, Canada. Journal of Geology 107, 715–725.

Wyman, D., Kerrich, R., Sun, M., 1995. Noble metal abundances of late Archean (2.7 Ga) accretion-related shoshonitic lamprophyres, Superior Province, Canada. GCA 59, 47–57.

Yamagishi, H., Dimroth, E., 1985. A comparison of Miocene and Archean rhyolite hyaloclastites: evidence for a hot and fluid rhyolite lava. Journal of Volcanology and Geothermal Research 23, 337–355.

Yang, X.M., 2013. Bedrock Geology of the Cat Lake-Euclid Lake Area, Bird River Greenstone Belt, Southeastern Manitoba (Parts of NTS 52L11,12). Manitoba Mineral Resources, Manitoba Geological Survey. Preliminary Map PMAP-7, scale 1:10,000.

Yang, X.-M., Gilbert, H.P., 2014. Archean tonalite-trondhjemite-granodiorite (TTG) suite in the Bird River greenstone belt, southeastern Manitoba. Lithogeochemical characteristics, geodynamic evolution and potential for porphyry Cu-(Au) mineralization. In: (Abstract) Program of 2014 GAC-MAC Annual Meeting, Fredericton, NB.

Yardley, B.w D., Banks, D.A., Munz, I.A., 1992. Halogen compositions of fluid inclusions as tracers of crustal fluid behavior. In: International Symposium on Water-Rock Interaction, Inter. Assoc. of Geochem. and Cosmochemistry and Alberta Research Council, Proceedings, vol. 7, pp. 1137–1140.

Zhou, T., Phillips, G.N., Denn, S., Burke, S., 2003. Woodcutter's goldfield: gold in Archean granite, Kalgoorlie, Western Australia. Australian Journal of Earth Science 50, 553–569.

Zierenberg, R.A., Shanks III, W.C., Bishoff, J.L., 1984. Massive sulphide deposits at 21°N, East Pacific Rise: chemical composition, stable isotopes and phase equilibria. Geological Society of American Bulletin 95, 922–929.

Index

Note: Page numbers followed by "f" and "t" indicate figures and tables, respectively.

F

G

H

Kolibri 360 GmbH
Ferdinand-Jühlke-Straße 7
99095 Erfurt, Deutschland
produktsicherheit@kolibri360.de